Prentice Hall Advanced Reference Series

Engineering

ANTOGNETTI AND MILUTINOVIC, EDS. *Neural Networks: Concepts, Applications, and Implementations, Volume I*
DENNO *Power System Design and Applications for Alternative Energy Sources*
ESOGBUE, ED. *Dynamic Programming for Optimal Water Resources Systems Analysis*
FERRY, AKERS, AND GREENEICH *Ultra Large Scale Integrated Microelectronics*
GNADT AND LAWLER, EDS. *Automating Electric Utility Distribution Systems: The Athens Automation and Control Experiment*
HALL *Biosensors*
HAYKIN, ED. *Advances in Spectrum Analysis and Array Processing, Vol. I*
HAYKIN, ED. *Advances in Spectrum Analysis and Array Processing, Vol. II*
HAYKIN, ED. *Selected Topics in Signal Processing*
HENSEL *Inverse Theory and Applications for Engineers*
JOHNSON *Lectures on Adaptive Parameter Estimation*
MILUTINOVIC, ED. *Microprocessor Design for GaAs Technology*
QUACKENBUSH, BARNWELL III, AND CLEMENTS *Objective Measures of Speech Quality*
ROFFEL, VERMEER, AND CHIN *Simulation and Implementation of Self-Tuning Controllers*
SASTRY AND BODSON *Adaptive Control: Stability, Convergence, and Robustness*
SWAMINATHAN AND MACRANDER *Materials Aspects of GaAs and InP Based Structures*
TOMBS *Biotechnology in the Food Industry*

BIOSENSORS

Elizabeth A. H. Hall

Prentice Hall
Englewood Cliffs, New Jersey 07632

Library of Congress Cataloging-in-Publication Data

HALL, ELIZABETH A. H.
Biosensors / Elizabeth A. H. Hall.
p. cm. — (Prentice Hall advanced reference series)
Includes bibliographical references and index.
ISBN 0-13-084526-4
1. Biosensors. I. Title.
R857.B54H35 1991
610'.28—cc20 90-47814
CIP

Cover design: *Karen A. Stephens*
Manufacturing buyer: *Kelly Behr*

First published by Open University Press
Milton Keynes, England

This edition for sale in North America only.

Prentice Hall Advanced Reference Series

North American edition published by Prentice Hall
A Division of Simon & Schuster
Englewood Cliffs, New Jersey 07632

Printed in the United States of America

10 9 8 7 6 5 4 3 2 1

ISBN 0-13-084526-4

Prentice-Hall International (UK) Limited, *London*
Prentice-Hall of Australia Pty. Limited, *Sydney*
Prentice-Hall Canada Inc., *Toronto*
Prentice-Hall Hispanoamericana, S.A., *Mexico*
Prentice-Hall of India Private Limited, *New Delhi*
Prentice-Hall of Japan, Inc., *Tokyo*
Simon & Schuster Asia Pte. Ltd., *Singapore*
Editora Prentice-Hall do Brasil, Ltda., *Rio de Janeiro*

Contents

Introduction

In the beginning was the word, and the word was *biosensors*. Analysts, academics and men of commerce and industry all questioned the meaning of this word and asked what it would bring. Market researchers produced market projections and media coverage of *diagnostics* became focused on sensors; but the question remained, 'Would biosensors revolutionize methods of analysis; could sensors sense anything, anywhere, anytime.'

This text attempts to provide as clear, thorough and up-to-date a treatment of the areas of knowledge fundamental to the understanding of biosensors as possible. To this end it involves a depth of coverage suitable essentially for a graduate science student, with no postgraduate specialization. The actual writing of the book was completed in 1988 and the reader will therefore find references no later than that date. In such a rapidly advancing field it has to be acknowledged that a publication such as this will, to some extent, become 'out of date' as soon as it is published.

The development of biosensors is concerned with the production of diagnostic reagentless-analysis probes, which can be used by non-specialist operators, either continuously on-line or discretely as a 'throw-away' device. It has involved a multidisciplinary research effort and any textbook on the subject will reflect this collaboration. However, it is inevitable that in such an expanding field, many aspects have been omitted or treated in insufficient depth. This cannot be avoided when each concept concerned with the realization of a biosensor would itself be worthy of an entire volume.

This is therefore, in no way intended to be a complete work, but more an introduction to biosensors. In providing a *basic* theoretical and practical approach and addressing a few selected examples in some depth, the reader can gain an initial indication of some of the possibilities and limitations of the field. For further information on any individual topic, however, the reader is directed to specialist advanced texts and the research literature.

PART 1

Sensor Techniques: The Concepts and Analytical Principles

Chapter 1

Biosensors in Context

The Diagnostic Market

Diagnostics as a whole represent a very large, well established and continually expanding market. Particularly in the current climate of 'prevention rather than cure' the need for detection at increasingly lower limits in more diverse applications is being added to a continuing requirement for monitoring and control in more traditional areas. Estimates of the market size and future projections are notoriously difficult and inaccurate. However one of the biggest diagnostic markets is undoubtably that of *clinical testing* and within this field alone future estimates reveal an upward trend. The figures associated with this trend, however, vary from survey to survey. A typical study of the European market (Fig. 1.1) suggests a clinical testing products market in excess of 4000 million US$ in the 1990s (Biomedical Business International). Comparison of this figure with an estimated UK market potential for biosensors of 4 million US$ by the year 2000 shows their comparatively small, but nevertheless significant single contribution to diagnostics as a whole. The UK market alone is of course a relatively small one anyway, and it is perhaps more realistic to compare the US consumption, where the current biosensor market is already reported to be 12 million US$, and future prospects vary from 100 to 10 000 million US$ by the turn of the century. This compares with a world market in 1985 of 1.5 billion US$ with an estimated growth rate of 9.5%, achieving a world market of 2 billion in 1990 and then expanding upwards and outwards.

Technical Insights Inc. identifies nearly 50% of the market as belonging to the medical arena, with veterinary and agricultural applications amounting to a figure of half the size (Table 1.1). Although the actual figures may differ between surveys, the relative proportions for different applications seem to be in agreement

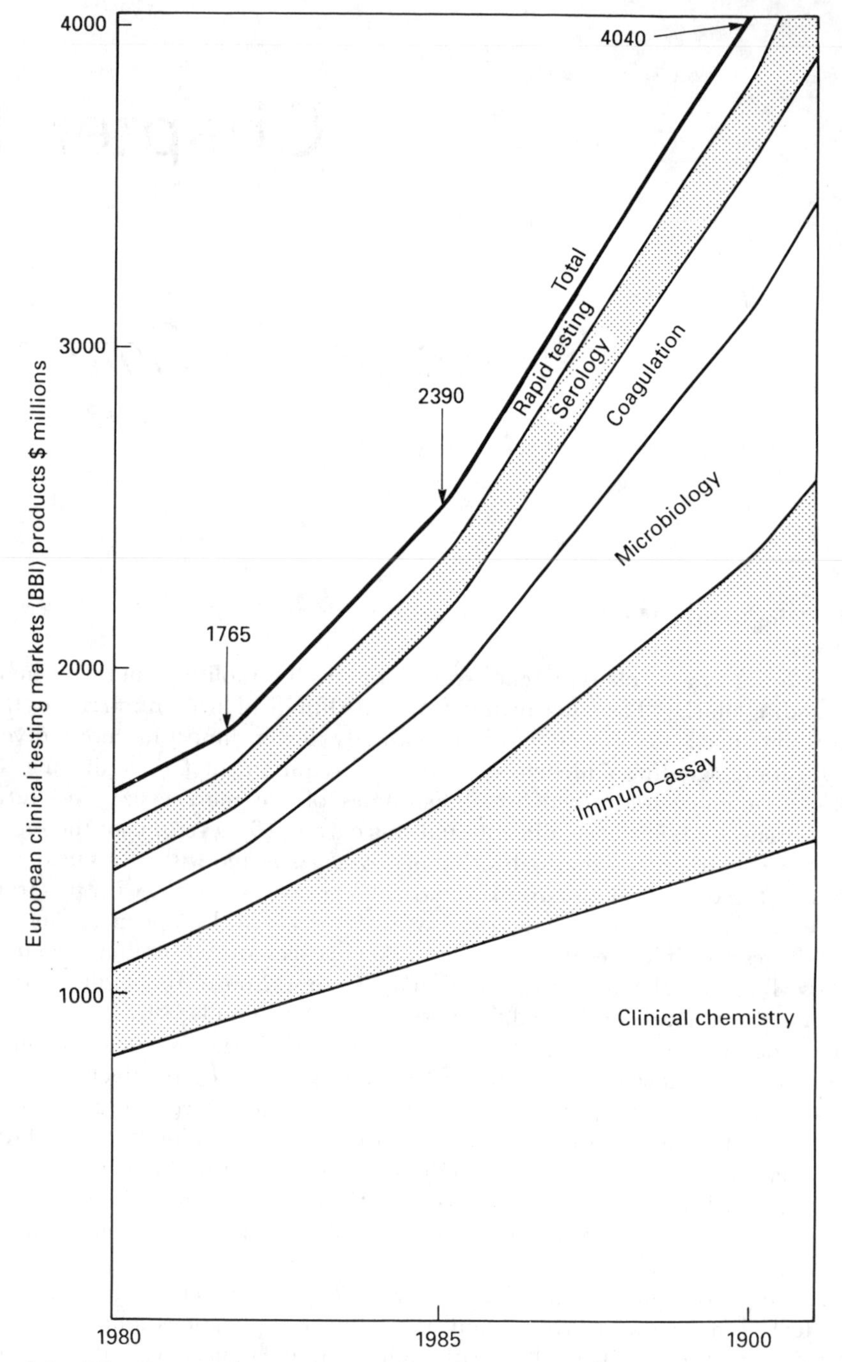

Fig. 1.1 The European market for clinical testing products. (Source: Biomedical Business International.)

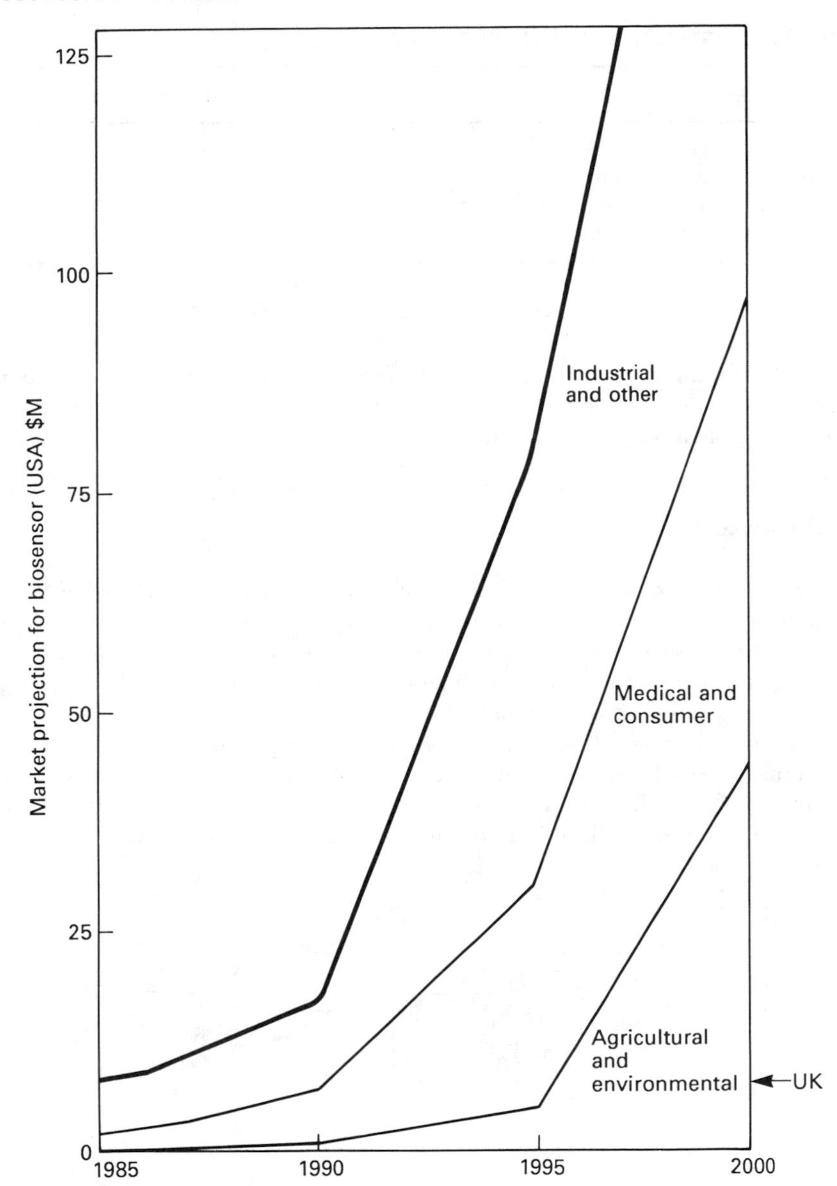

Fig. 1.2 USA market projection for biosensors.
(Source: Business Communications.)

for the 1990s. Figure 1.2 (Business Communications Co.) plots the US market projection into the next century and identifies a changing emphasis in applications with the agricultural and environmental component becoming increasingly significant.

Table 1.1 Estimated markets in the 1990s

Market	*Value* ($ *million*)
Medical and surgical	220
Veterinary and agricultural	105
Environmental and monitoring and safety	67
Industrial process monitoring	59

Source: Technical Insights Inc.

With this financial background establishing a market potential, it is perhaps pertinent to define the biosensor, and in view of this definition, to identify the diagnostic applications most relevant to their use.

The Biosensor

The unique feature of a biosensor is that the device incorporates a *biological sensing element* in close proximity or integrated with the signal transducer (Fig. 1.3), to give a reagentless sensing system specific for the target analyte. The utilization of the biological element capitalizes on the unique specificity of biological molecules for target species; the transduced signal in a biosensor is thus a secondary one, due to the reaction between *biorecognition molecule* and target analyte, rather than a direct signal due to the analyte itself. The use of this indirect means of assay means that chemically similar solution species can be identified by their biospecific reaction with an immobilized biomolecule such as an enzyme, antibody, nucleic acid, etc.

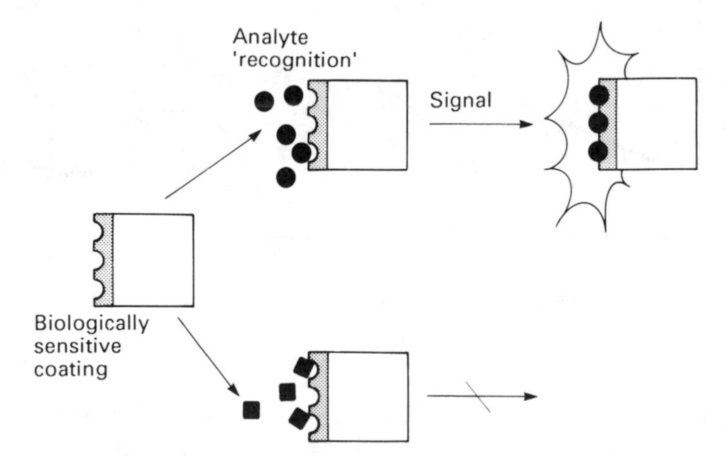

Fig. 1.3 The biosensor: surface-modified transducer which is reactive towards a specific chosen analyte.

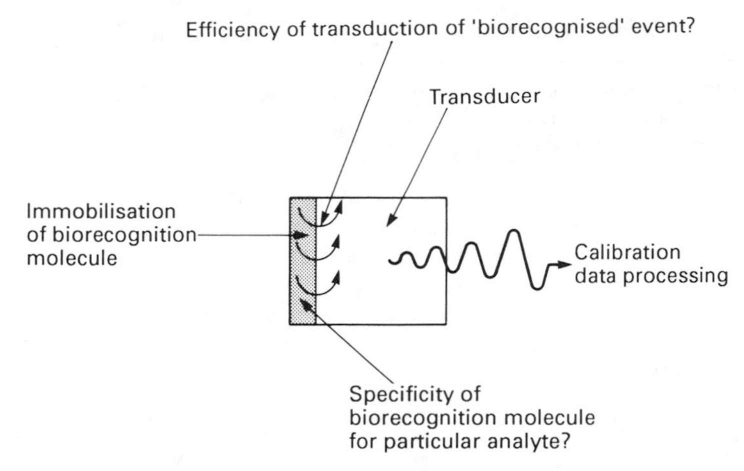

Fig. 1.4 The biosensor components: the problems associated with their union.

Immobilized in the vicinity of a transducer, only minute quantities of biological recognition molecule are needed, but the reliability of the sensor in the final instance, requires that this element show (Fig. 1.4):

- a high degree of specificity
- a good stability to operating conditions (e.g. temperature, pH, ionic strength, etc.)
- retention of biological activity in the immobilized state
- no undesirable sample contamination

In general, however, the characteristics required of a biosensor are very much those demanded of sensors as a whole (Table 1.2), where it should perhaps be noted that after successful research and development of a device, the ultimate adoption of a particular sensor, even when supplying a hitherto unfeasible but nevertheless desirable assay, lies in the acceptability by the user.

Applications

HEALTH CARE

The major initial impetus for the advancement of sensor technology came from health care developments, where it is now generally accepted that measurements of blood gases, ions and metabolites are sometimes essential and will anyway allow a better appreciation of local metabolic state. In intensive care units for example, patients frequently show rapid variations in biochemical levels that require an appropriate action be taken urgently. Even in less severe patient handling, more successful treatment can be achieved by obtaining 'instant' assays. In fact, at present the list of the most commonly required 'instant' analyses is not vast

Table 1.2 Fifteen characteristics required in a commercial sensor

- Relevance of output signal to measurement environment
- Accuracy and repeatability
- Sensitivity and resolution
- Dynamic range
- Speed of response
- Insensitivity to temperature (or temperature compensation)
- Insensitive to electrical and other environmental interference
- Amenable to testing and calibration
- Reliability and self-checking capability
- Physical robustness
- Service requirements
- Capital cost
- Running costs and life
- Acceptability by user
- Product safety—sample host system must not be contaminated by sensor

(Fig. 1.5), although this is probably strongly influenced by a current feasibility factor. In practice, they have sometimes been at least partially realized by the establishment of *en suite* analytical laboratories, where discrete samples are analysed, frequently using the more traditional analytical techniques.

Indeed, the aid to diagnostic and prognostic medicine has proved invaluable, but there is an ever increasing demand for inexpensive and reliable sensors aimed at further target analytes, thus allowing not only routine monitoring in the central or satellite laboratory, but also analysis with greater patient contact, in the hospital ward, casualty department, doctor's surgery, or by the patients themselves, in the monitoring and control of some treatable condition, such as diabetes. In fact it is probably true to say that the major biosensor market may be found where an immediate assay is required, without substantial *en suite* laboratory facilities. However if economic considerations of laboratory maintenance are coupled with the more direct analytical costs, then low-cost biosensor devices can be desirable in the whole spectrum of analytical applications from hospital to home.

Continuous *in vitro* realtime monitoring is also an aim, particularly in the treatment of chronic conditions, where direct feedback with a controlled drug-release system, giving constant administration in a closed-loop, could give better patient management. It can therefore be anticipated that with greater miniaturization and signal optimization, the 'on-line' application field will expand.

ARTIFICIAL PANCREAS

The 'classic' and most widely explored example of closed-loop drug control is probably to be found in the development of an artificial pancreas. Diabetic patients have a relative or absolute lack of insulin, a polypeptide hormone produced by the beta-cells of the pancreas, which is essential to the metabolism of

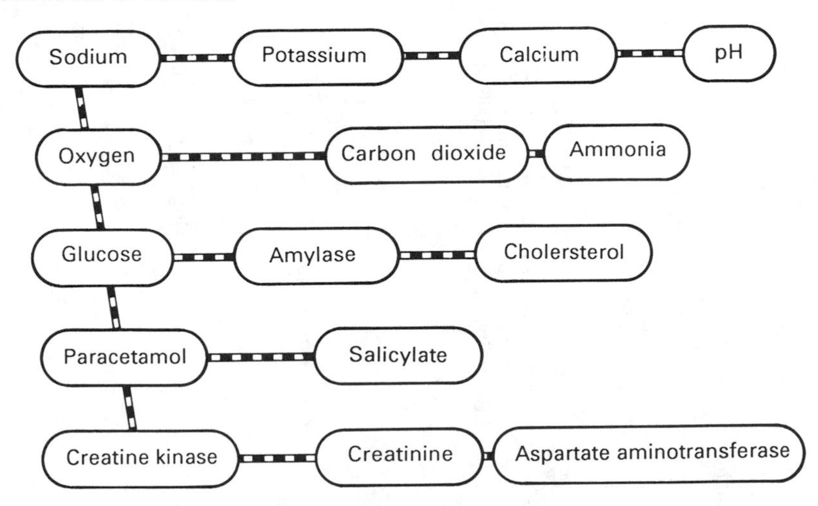

Fig. 1.5 Current commonly required instant assays in patient diagnosis.

a number of carbon sources. This deficiency causes various metabolic abnormalities, including higher than normal blood glucose levels, and where patients have suffered a complete destruction of the insulin-secreting islets of Langerhans, then insulin must be supplied. This has usually been achieved by subcutaneous injection, but fine control is difficult and hyperglycaemia cannot be totally avoided, or even hypoglycaemia sometimes induced, causing impaired consciousness and the serious long-term complications to tissue associated with this intermittent low glucose condition.

Better methods for the treatment of insulin-dependent diabetes have been sought and infusion systems for continuous insulin delivery have been developed. However, regardless of the method of insulin therapy, its induction must be made in response to information on the current blood glucose levels in the patient. Three schemes are possible (Fig. 1.6), the first two dependent on discrete manual glucose measurement and the third a 'closed-loop' system, where insulin delivery is controlled by the output of a glucose sensor which is integrated with the insulin infuser. In the former case, glucose has been estimated on 'finger-prick' blood samples with a colorimetric test strip or more recently with an amperometric 'pen'-size biosensor device by the patient themselves. Obviously these diagnostic kits must be easily portable, very simple to use and require the minimum of expert interpretation. However, even with the ability to monitor current glucose levels, intensive conventional insulin therapy requires multiple daily injections and is unable to anticipate future states between each application, where diet and exercise may require modification of the insulin dose. Dimitriadis and Gerich (1983) showed, for example, that administration of glucose by subcutaneous injection, 60 min before a meal provides the best glucose/insulin management.

The introduction of a closed-loop system, where integrated glucose measurements provide feedback control on a pre-programmed insulin administration,

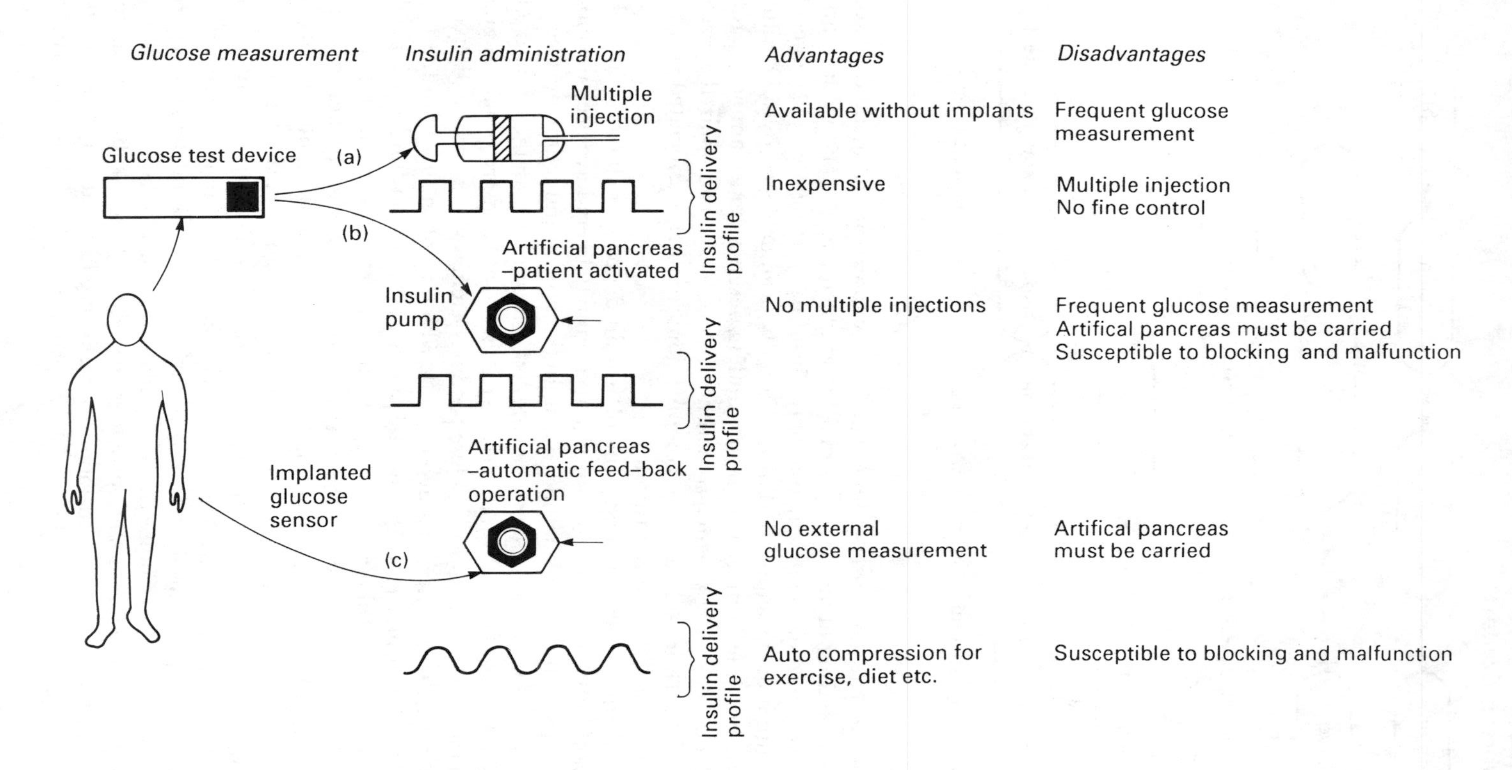

Fig. 1.6 Schemes for insulin therapy.

Table 1.3 Seven requirements for an implantable glucose sensor

- Linear in 0–20 mM range with 1 mM resolution
- Specific for glucose; not affected by changes in metabolite concentrations and ambient conditions
- Biocompatible
- Small—causes minimal tissue damage during insertion and there is better patient acceptability for a small device
- External calibration and $< 10\%$ drift in 24 h
- Response time < 10 min
- Prolonged lifetime—at least several days, preferably weeks in use

based on habitual requirement (Calabrese *et al.*, 1982) would therefore relieve the patient of frequent assay requirements and perhaps more desirably frequent injection.

Ultimately, the closed-loop system becomes an artificial pancreas, where the glycaemic control is achieved through an implantable glucose sensor. Obviously, the requirements for this sensor are very different to those for the one-off discrete measurement kits; these are summarized in Table 1.3, where the prolonged life-time and biocompatibility represent major differences from *ex vivo* throw-away devices.

MODES OF OPERATION

It follows that distinct modes of operation can be identified, taking the determination from the central analytical laboratory facility: off-line devices at various degrees of proximity to the sample host, giving discrete measurements, and on-line devices giving either discrete or continuous real-time measurements in both closed-loop and open-loop systems.

PROCESS CONTROL

These distinctions are not only applicable to health care (Fig. 1.7b) but are relevant to the ever increasing arena of biosensor applications. Real-time monitoring of carbon sources, dissolved gases, products etc., in fermentation processes (Fig. 1.7a) could lead to optimization of the procedure giving increased yields at decreased materials cost. Whilst real-time monitoring with feedback control involving automated systems does exist, currently only a few common variables are measured on-line (e.g. pH, temperature, CO_2, O_2), which are often only indirectly concerned with the process under control.

Three zones of influence can be identified in fermenter control:

Zone 3: Off-line distant—central laboratory coarse control with significant time lapse
Zone 2: Off-line local—fine control with short time lapse
Zone 1: On-line—real-time monitoring and control

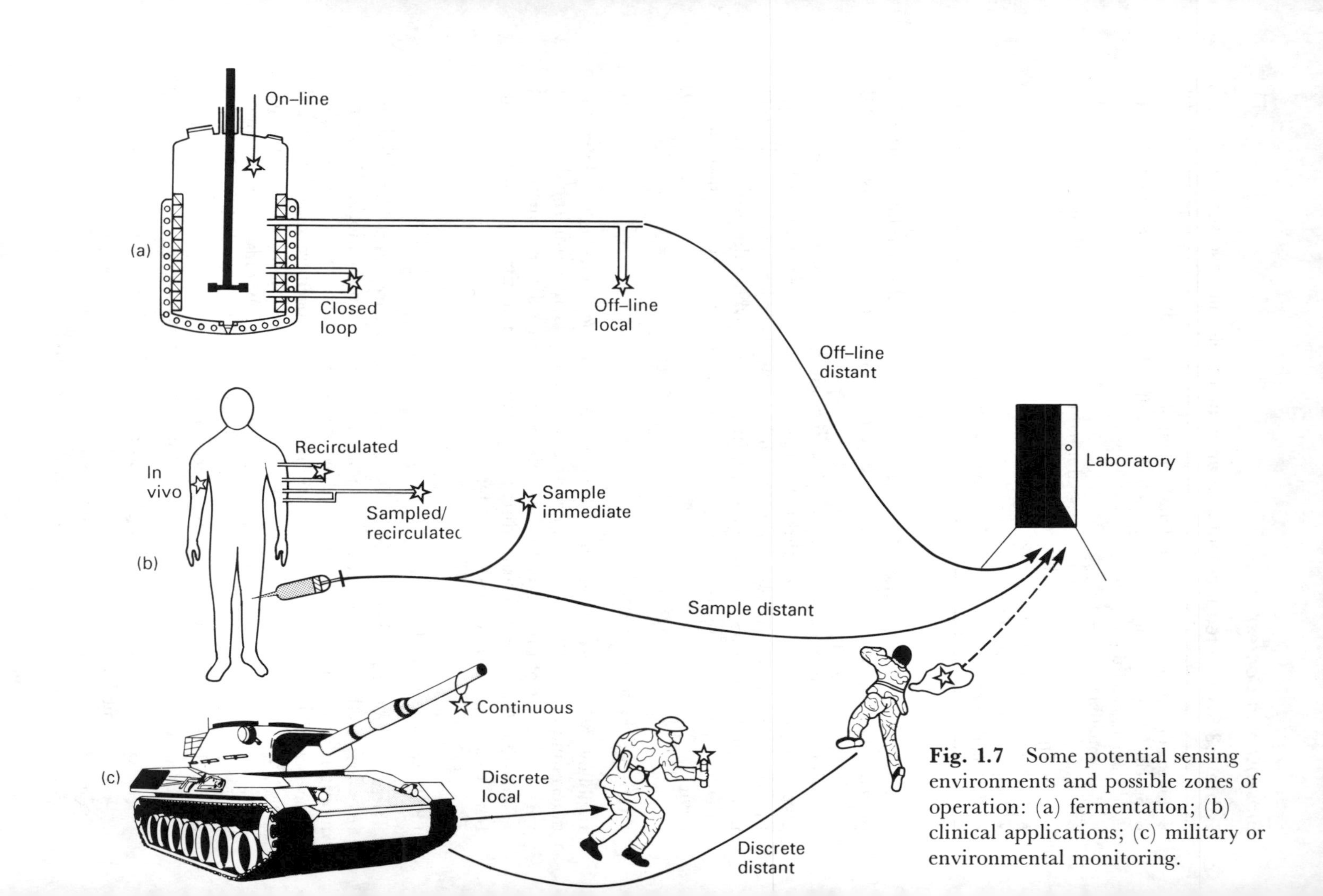

Fig. 1.7 Some potential sensing environments and possible zones of operation: (a) fermentation; (b) clinical applications; (c) military or environmental monitoring.

The further population of Zone 1 would of course be particularly desirable, ultimately allowing the process to follow an ideal pre-programmed fermentation profile to give maximum output. However, many problems are concerned with on-line measurements including *in situ* sterilization, sensor life-time, sensor fouling, etc. Some of the problems can be overcome if the sensor is situated so that the sample is run to waste, but this causes a volume loss, which can be particularly critical with small volume fermentations.

Although Zone 1 may be the ultimate aim, considerable advantage can be gained in moving from Zone 3 to Zone 2 giving a rapid analysis and thus enabling finer control of the fermentation. The demands of the sensor are perhaps not as stringent in this zone as in Zone 1, but nevertheless their use here is more analogous to the patient under self-control therapy, and their development must reflect both the sample environment and the potential user.

The benefits which are achievable with process-control technology are considerable:

- Improved product quality; reduction in rejection rate following manufacture
- Increased product yield; process tuned in real time to maintain optimum conditions throughout and not just for limited periods
- Increased tolerance in quality variation of some raw materials. These variations can be compensated in the process-control management
- Reduced reliance on human 'seventh sense' to control process
- Improved plant performance—processing rate and line speed automated, so no unnecessary dead-time allocated to plant
- Optimized energy efficiency

In fact the use of biosensors in industrial process control in general could facilitate plant automation, cut analysis costs and improve quality control of the product.

MILITARY APPLICATIONS

The requirement for rapid analysis can also be anticipated in military applications. The US army, for example, have looked at dipstick tests based on monoclonal antibodies. While these dipsticks are stable and highly specific (Q-fever, nerve agents, yellow rain fungus, soman, etc.) they are frequently two-step analyses taking up to 20 min to run. Such a time lapse is not always suited to battlefield diagnostics; the resulting consequences are suggested in Fig. 1.7(c).

A particularly promising approach to this unknown hazard detection seems to be via acetylcholine receptor systems. It has been calculated that with this biorecognition system, a matrix of 13–20 proteins are required to give 95% certainty of all toxin detection.

ENVIRONMENTAL MONITORING

Another assay situation which may involve a considerable degree of the unknown is that of environmental monitoring. The primary measurement media here will

Table 1.4 Summary of potential applications for biosensors

- Clinical diagnosis and biomedicine
- Farm, garden and veterinary analysis
- Process control:
  - fermentation control and analysis
  - food and drink production and analysis
- Microbiology: bacterial and viral analysis
- Pharmaceutical and drug analysis
- Industrial effluent control
- Pollution control and monitoring
- Mining, industrial and toxic gases
- Military applications

be water or air, but the variety of target analytes is vast. At sites of potential continual pollution, such as in factory effluent, it would be desirable to achieve on-line real-time monitoring and alarm, targeted at specific analytes, but in many cases random or discrete monitoring of both target species or general hazardous compounds is all that is required. The list ranges from biological oxygen demand (BOD) which provides a good indication of the general degree of pollution, atmospheric acidity and river water pH, to more specific targets such as detergent preparations, herbicides and fertilizers (organophosphates, nitrates, etc.). The survey of market potential has identified the increasing significance of this area and this is now substantiated by a developing interest from industry in the technology required to enable environmental monitoring.

TUNING SENSOR TO APPLICATION

The potential for biosensor technology is enormous and is likely to revolutionize analysis and control of biological systems with a wide sphere of influence (Table 1.4). It is possible therefore to identify very different analytical requirements and biosensor developments must be viewed under this constraint. It is often tempting to expect a single sensor targeted at a particular analyte, to be equally applicable to on-line closed-loop operation in a fermenter and pin-prick blood samples. In practice, however, the parallel development of several types of sensor, frequently employing very different measurement parameters is a more realistic approach, and one that is reflected in subsequent chapters, where a single analyte is the subject of several different genera of biosensor devices.

SENSITIVITY REQUIREMENTS

The range and type of analytes are also varied and cannot be considered under a single umbrella. The particular application imposes a final concentration range requirement, but initially the concentration level that must be achieved can be estimated by the type of analyte of interest. Metabolites, for example, are commonly found at a level $>10^{-6}$ mol/litre, whereas hormones may be in the

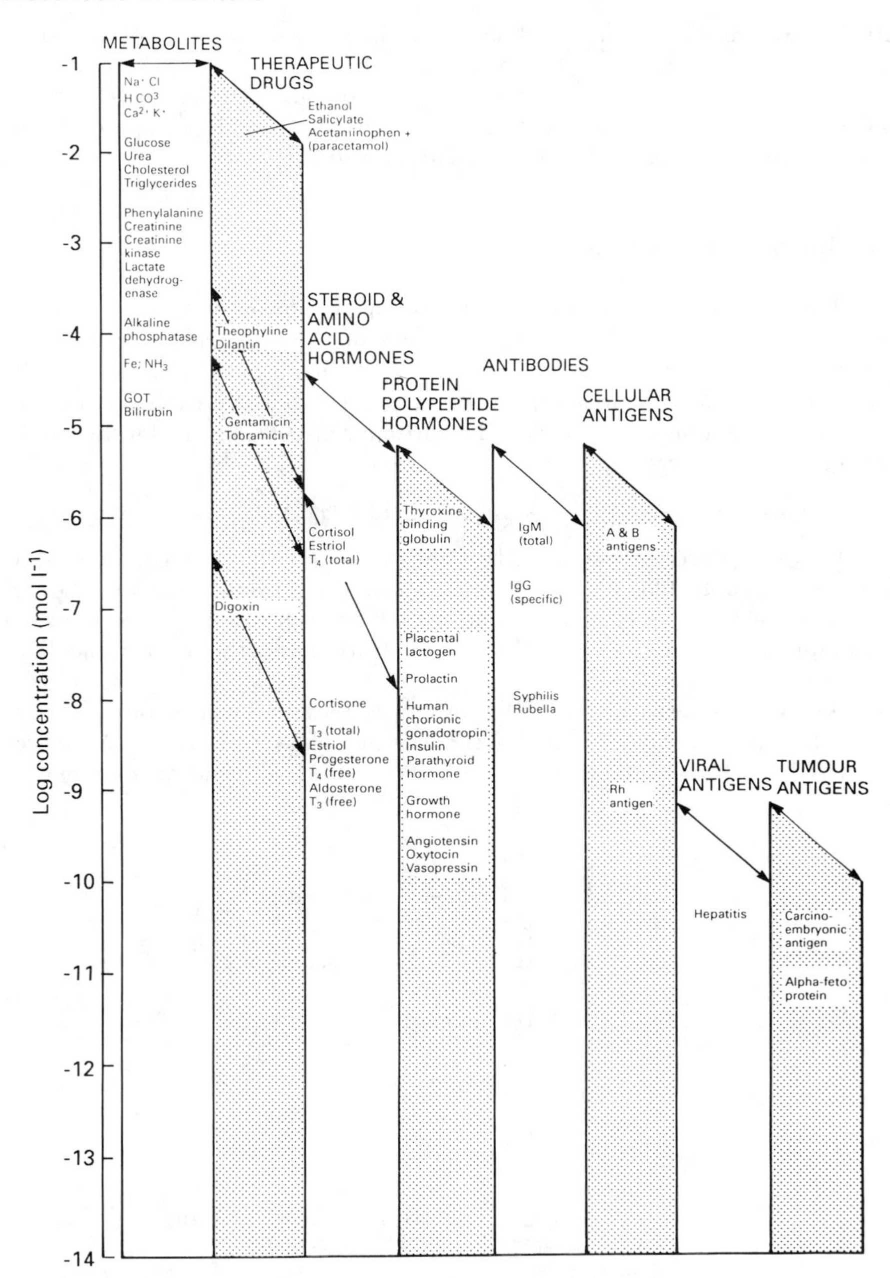

Fig. 1.8 Detection ranges required for some clinically significant analytes.

10^{-10}–10^{-5} mol/litre range and levels as low as 10^{-20} mol/litre would be desirable, or at any rate 10^{-12} mol/litre for viruses. This vast range of concentrations is summarized in Fig. 1.8. It is obvious from this figure that, just based on detection limits, very different approaches would be required for an antigen sensor to those measuring ion concentration.

The Birth of the Biosensor

The biosensor was first described by Clark and Lyons in 1962, when the term *enzyme-electrode* was adopted. In this first enzyme electrode, an oxido-reductase enzyme was held next to a platinum electrode in a membrane sandwich (Fig. 1.9). The platinum anode polarized at +0.6 V responded to the peroxide produced by the enzyme reaction with substrate. The primary target substrate for this system was glucose:

$$\text{glucose} + O_2 \xrightarrow{\textit{glucose oxidase}} \text{gluconic acid} + H_2O_2$$

and led to the development of the first glucose analyser for the measurement of glucose in whole blood. This Yellow Springs Instrument (Model 23 YSI) appeared on the market in 1974, and as will be seen in Chapter 8, the same technique as employed here has been applied to many other oxygen mediated oxido-reductase enzyme systems.

A key development in the YSI sensor was the employment of membrane technology in order to eliminate interference by other electro-active substances. Polarized at +0.6 V, the major interferent to the peroxide measurement is

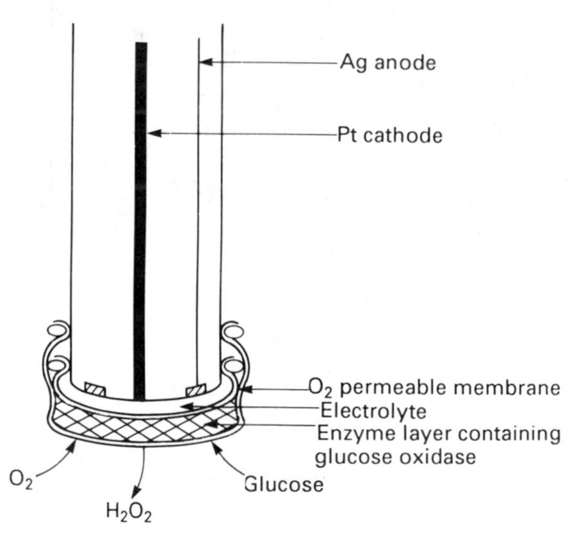

Fig. 1.9 The Clark enzyme electrode. Modification of the Clark oxygen electrode to give an enzyme electrode.

ascorbic acid. Various combinations of membrane–enzyme sandwich have been developed, all satisfying the following criteria:

- the membrane between electrode and enzyme layer should allow the passage of H_2O_2, but prevent the passage of ascorbate or other interferents
- the membrane between enzyme layer and sample should allow substrate/analyte to enter the enzyme layer

This was accomplished in the YSI, for example, with an enzyme layer sandwiched between a cellulose acetate membrane and a Nucleopore polycarbonate membrane (Grooms *et al.*, 1980).

The Growth of the Biosensor

It is probably true to say that that majority of traditional bioanalyses are based on a photometric method, where the biorecognition reaction is linked to a colorimetric, fluorescent or luminescent indicator molecule (Fig. 1.10a). On the other hand, chemical sensors have been more traditionally confined to electrochemical devices in the form of the potentiometric pH electrode (Fig. 1.10b) (Chapter 3) and the amperometric Clark-type oxygen electrode (Fig. 1.10c) (Chapter 5).

We have already highlighted acceptability by user as an important parameter in the characteristics of a biosensor. Adaption and exploitation of these three routes (*photometric*, *potentiometric* and *amperometric*), where user acceptability is already established, has therefore been an obvious approach to the development of reagentless biosensor devices with a high specificity and selectivity.

In general, however, it is clear that the achievement of a successful biosensor must in the first instance be subdivided into two requirements and we may consider a biosensor as a multicomponent system requiring the solution and the union of these individual subunits:

(1) Characterization of the bioassay principle
(2) Characterization of the base sensor (e.g. pH electrode, O_2 electrode, etc.) which can link with the bioassay

In view of this compartmentalized approach and the importance of the individual components, Part I of this book is devoted to the treatment of the units and Part II is concerned with their union.

Emerging and Advancing Multidisciplinary Technologies

Successful biosensors are of course not restricted to the three routes described above. In principle any variable which is concerned with the biorecognition reaction could be used to generate the transduced signal.

The biosensor field in fact is developing at the interface between existing and emerging technologies, combining physical and biological disciplines with state-

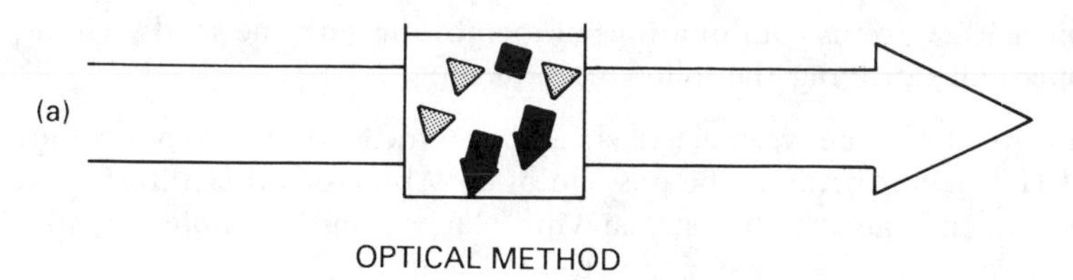

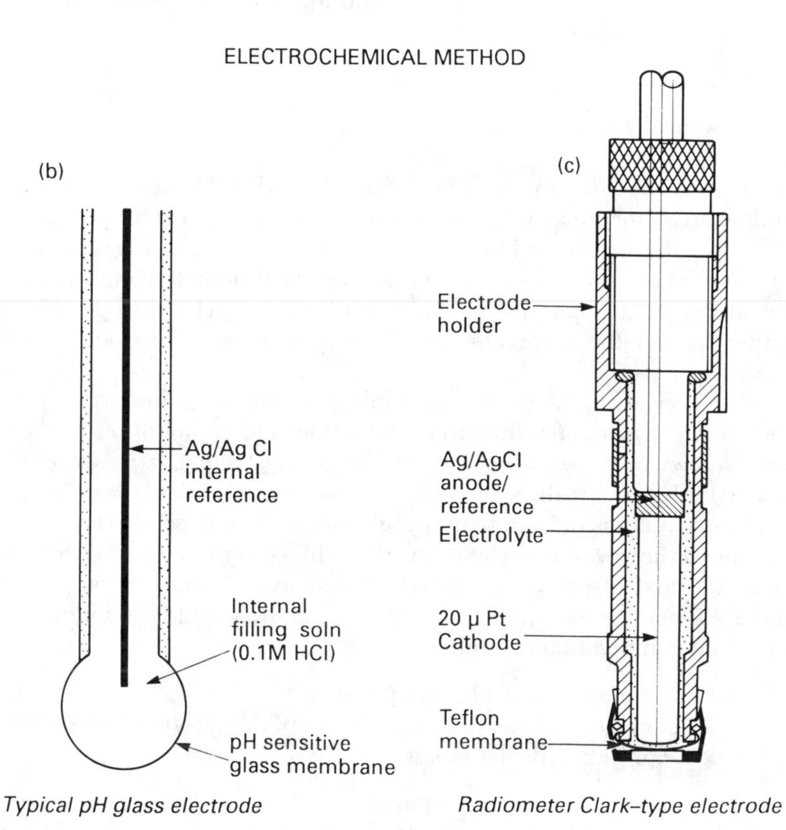

Fig. 1.10 The growth of the biosensor from established assay methods. (a) Photometric assay using optical labels; (b) potentiometric assay using pH electrode; (c) amperometric assay with Clark O_2 electrode.

of-the-art electronics. Recent advances in silicon technology, polymer fabrication, optics and processing and communication techniques have revealed new materials and methods suitable for exploitation as biosensors. We can expect therefore that the measurement parameters concerned with biosensor devices will become ever more diverse and employ techniques which have hitherto not been applied in reagentless diagnostics.

The production of piezoelectric materials and surface acoustic wave devices offers a surface which is susceptible to changes in mass at the interface (Chapter 11), while plasmon excitation of a thin metal film or other plasmon material is particularly sensitive to the dielectric constant of the layer immediately next to that film (Chapter 10). Utilization of such techniques, which are more familiar to the physics laboratory than applied to a biological system, exemplifies the multidisciplinary research effort that is required, involving all branches of science, in order to fully exploit these techniques for diagnostics.

Less recently developed technological interfaces can also be exploited. Monitoring solution conductance was originally applied as a method of determining reaction rates. The technique involves the measurement of changes in conductance due to the migration of ions. Many enzyme-linked reactions result in a change in total ion concentration and this would imply that they are suitable for conductimetric biosensors. This theme will be pursued in Chapter 11.

Even thermodynamic data can be useful: chemical reactions are accompanied by the absorption (endothermic) or evolution (exothermic) of heat. Measurements of ΔH, the enthalpy of reaction at different temperatures allows one to calculate ΔS (entropy) and ΔG (Gibbs free energy) for a reaction and therefore collect basic thermodynamic data.

The hydrolysis of ATP for example, follows the route:

$$ATP^{4-} + H_2O \rightarrow ADP^{3-} + HPO_4^{2-} + H^+; \ \Delta H_{298} = -22.2\ \text{kJ (pH 7)}$$

or the immunoreaction between anti-HSA and its antigen HSA yields -30.5 kJ/mol (Benzinger and Kitzinger, 1960; Sturtevant and Lyons, 1969; Menkins *et al.*, 1969). For this latter reaction, measured by heat burst calorimetry, the total increase in temperature for 1 μmol of antibody was of the order of 10^{-5} K, but many enzyme-catalysed reactions have $\Delta H > 1000$ kJ/mol, and produce more easily measurable changes in temperature.

For a true biosensor device the biorecognition compound must be immobilized on a temperature-sensing element capable of detecting very small temperature changes. The major initiative in this area has come from the Mosbach group at the University of Lund (Mosbach and Danielsson, 1981; Danielsson *et al.*, 1981, 1988). Initially they immobilized glucose oxidase or penicillinase in a small column, so that temperature changes in the column effluent were monitored by thermistors to give an *enzyme thermistor* sensitive to glucose and penicillin, respectively. They have also applied the technique to other substrates and to immunoassay using an enzyme-labelled antigen.

The application of the enzyme thermistor to the monitoring of glucose in a complex fermentation medium under computer control has also been tested by Wehnert *et al.* (1987). Off-line, for a cultivation medium of *Cephalosporium acremonium*, the results showed good correlation between glucose estimation made by the YSI and the thermistor, and the on-line detection closely reflected these measurements for the first 60 min, but were then inhibited by alterations in the enzyme column and sampling system, due to protein precipitation and general clogging. These are problems typical of Zone 1 operation.

Muramatsu *et al.* (1987) have employed the same principle in the modification of an integrated circuit thermal sensor to be a glucose biosensor. Glucose sensitivity was achieved by casting a cellulose triacetate/1,8-diamino-4-aminomethyloctane membrane over a thermal sensor, and covalently attaching the enzyme, glucose oxidase, via a glutaraldehyde reaction. The output voltage of the modified sensor was measured with respect to an unmodified sensor, and was thus related to the enthalpy change associated with the enzyme reaction. Fourier transform treatment of the signal to remove high frequency noise, identified a detection limit of 2 mM, but the authors anticipate that amplification techniques would greatly improve this device.

The arena of expertise required for biosensor development can therefore be considered to be sustained by collaboration from many areas of academia and industry alike (Fig. 1.11). The resulting output of this collaboration is likely in many cases to be a slow process, but is probably the only realistic route to successful future advances.

Immunoassay and DNA Probes

An important stream of analytical developments, which has been widely applied, is the immunoassay techniques and the DNA probes. The major problem with these biorecognition assays is the identification of a physicochemical parameter that can be monitored and related to binding of antibody and antigen or hybridization of strands of DNA, respectively. These two types of assay can both be characterized by an increase in molecular mass and volume associated with the binding event, and although current biosensor research is investigating the transduction of this parameter, the event has usually been followed with a photometric, radioactive or even enzyme marker.

There are many major current reasons for replacing the radioisotopic labels with non-radioactive ones, but the direct use of photometric indicators have rarely provided the same degree of sensitivity, so that enzymes have to date frequently been proven to be the most promising form of labelling. The principle of these label-linked assays is similar for both immunoassays and DNA probes (see Chapter 2).

Both these techniques are heterogeneous assays—so that they are already developed along the lines of the biosensor concept. It is the integration of the transducer with the assay which is required in order to make the step into the reagentless biosensor device.

The Biosensor Family

Scheller *et al.* (1985, 1987) have classified biorecognition coupled sensors into three generations according to the degree of integration of the separate components, i.e. the method of attachment of the biorecognition molecule to the base indicator (transducer) element. In the first generation, the biorecognition molecule is retained in the vicinity of the base sensor behind a dialysis membrane,

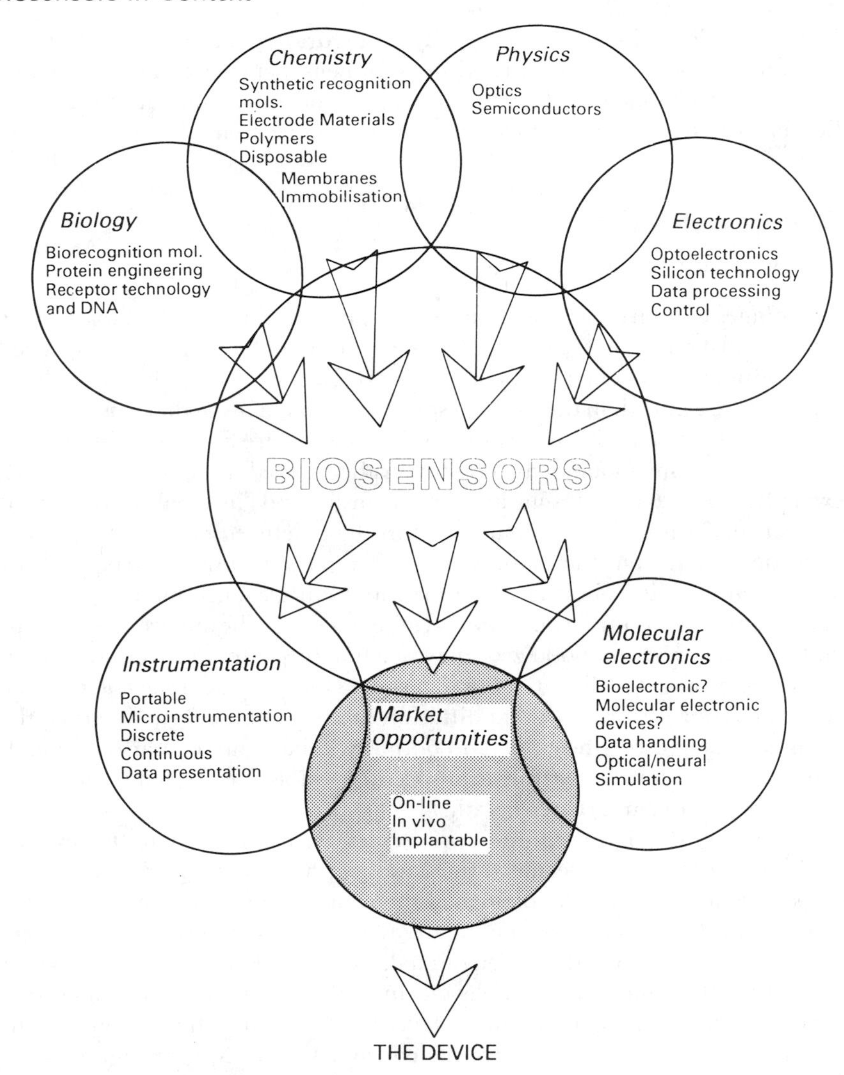

Fig. 1.11 The integration of the multidisciplinary expertise and the potential areas of participation involved in biosensor developments.

while in subsequent generations immobilization is achieved via cross-linking reagents or bifunctional reagents at a suitably modified transducer interface or by incorporation into a polymer matrix at the transduction surface. In the second generation, the individual components remain essentially distinct (e.g. control electronics—electrode—biomolecule), while in the third generation the biomolecule becomes an integral part of the base sensing element (e.g. ChemFET, see Chapter 4).

While these definitions were probably inaugurated for enzyme electrode systems, similar classifications appropriate to biosensors in general can be made, and the family tree followed. It is in the second and third generations of these families that the major development effort can now be seen.

The Bio↔Sensor Union

DESIGN OF THE UNION

The advantages of an irreversible immobilization procedure for the biomolecule are numerous. In particular, since the biorecognition molecule remains attached to the transducer, it may be possible to re-use it. In some cases the immobilized molecule is more stable than the solution species, but it is a myth that this is always the case. Anyway the kinetics of the immobilized species are likely to be influenced by the microenvironment and may be considerably altered from solution kinetics. For example, both 'external' solution mass transfer and 'internal' immobilized layer mass transfer must be accommodated in the kinetic equations, the form of which is dependent on the nature of the immobilization matrix and its dimensional profile (i.e. obeying one-dimensional diffusion, tubular, spherical, etc.). For example, slow mass transfer invariably increases the apparent Michaelis constant, K_{Mapp}, for an immobilized enzyme, but this often has the analytical benefit of increasing the linear dynamic range for enzyme-linked assays. If, however, mass transfer is so slow that diffusion limits the rate of reaction then the enzyme is not being used efficiently, and output will be reduced. Various models have been proposed for the treatment of specific immobilized systems, and these will be discussed later in greater detail.

Integration of the biorecognition molecules with the sensor element by immobilization of the biomolecule at the sensor surface is not only dependent for its success on the nature of the biological molecule or the characteristics of the transducer interface, but must be made with due regard for the efficient transduction of the analyte-dependent signal. Where pH is to be monitored for example, then the immobilized molecule must be resistant to the micro-pH environment created in the immobilized layer by the bio-linked reaction. On the other hand, if a redox enzyme is employed, and the transduced signal is the electron transfer involved in the enzyme–substrate reaction

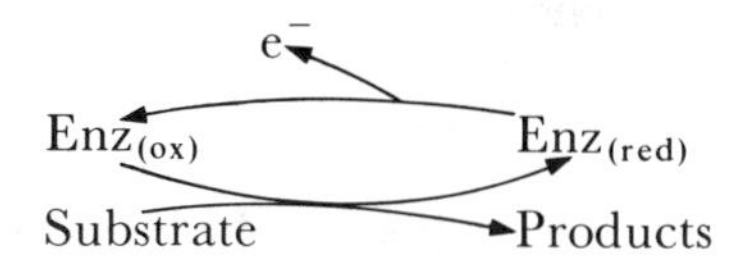

then efficient electron transfer must be possible between enzyme and transducer. This cannot usually be achieved simply by intimate contact between an unmodified electrode surface and the redox protein, but has been accomplished by the use of mediator sandwiches (electrode | mediator | enzyme) to shuttle the electrons between enzyme and electrode. The mediator is a low-molecular-weight component, with a redox couple close to that of the enzyme prosthetic group (see Chapter 8). The need for its presence must be accommodated in the immobilization procedure.

IMMOBILIZATION TECHNIQUES

General immobilization techniques can be illustrated by consideration under five main classes:

(1) Retention by an inert membrane
(2) Physical adsorption at a solid interface
(3) Crosslinking with bifunctional agents
(4) Entrapment in polymer matrices
(5) Covalent binding to a functionalized support

The first class represents the first-generation biosensors as already defined and the major effort is now directed towards the other methods.

Covalent Bonding to a Functionalized Support

Covalent attachment of the biorecognition molecule to the sensor surface is generally the most irreversible of the techniques. In a protein, bonding must be effected through the use of the nucleophilic functional amino acid groups in the biomolecule, that are not concerned with the function and activity of the 'biorecognition' site itself. Assistance can sometimes be gained in directing the bonding away from the active site by performing the immobilization in the presence of the biorecognition substrate (thus keeping the active site otherwise 'occupied' during the immobilization).

The techniques employed in order to achieve this attachment are, on the whole, classical synthetic coupling methods (Fig. 1.12). The mechanisms of reaction are diverse and so usually allow a reaction to be selected for a particular biological system which will bind the biomolecule without loss of activity. However, no single immobilization coupling reaction has emerged which is universally employable and the 'best' result is only found by experimentation. Even for similar biological systems, a single procedure cannot always be employed.

Immobilization of single-stranded DNA has been developed for probe hybridization analyses mostly on supports of cellulose papers or nylon filters. The synthetic procedures are however largely analogous to those employed for protein immobilization (Fig. 1.13). Where nylon is the support, for example, the amino groups on the membrane can be employed directly to interact with thymine residues of the nucleic acid.

The limitations for general applicability of these covalent binding techniques, together with the multi-step nature of many of the immobilization reactions, do not always favour their adoption from a commercial viewpoint, and the relative irreversibility of the technique must be compared and contrasted critically with the merits and weaknesses of other methods.

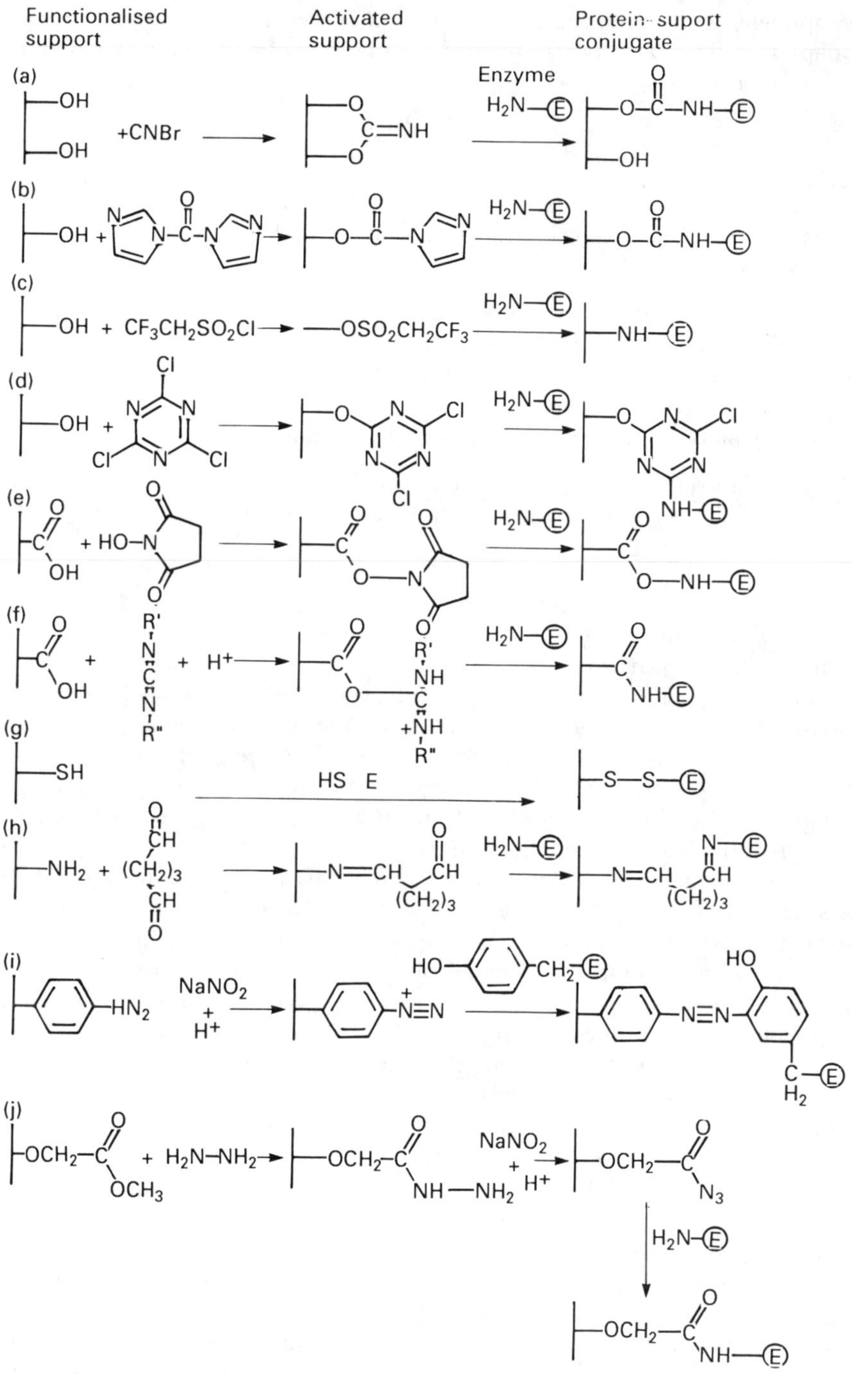

Functionalised support | Activated support | Protein-suport conjugate

(a) (b) (c) (d) (e) (f) (g) (h) (i) (j)

Fig. 1.12 Summary of some commonly employed coupling methods for immobilization of proteins to functionalized supports.

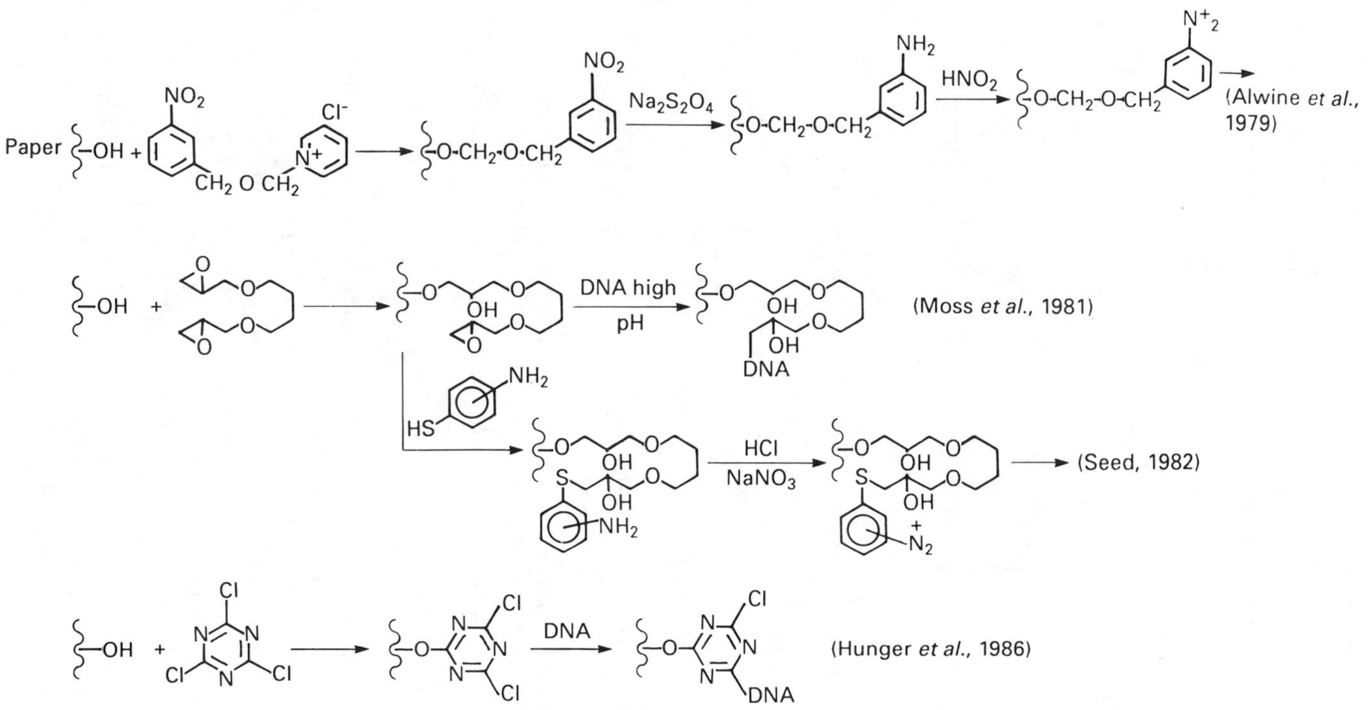

Paper
Na$_2$S$_2$O$_4$
HNO$_2$
(Alwine *et al.*, 1979)
DNA high pH
(Moss *et al.*, 1981)
HCl
NaNO$_3$
(Seed, 1982)
DNA
(Hunger *et al.*, 1986)

Fig. 1.13 Some immobilization methods developed for DNA.

Table 1.5 Some common bifunctional reagents for cross-linking proteins

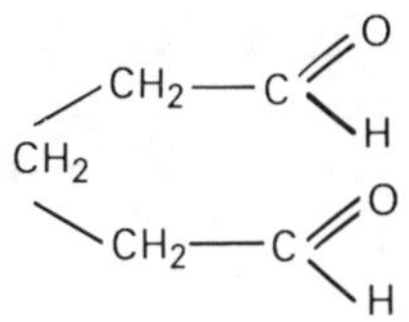

Gluteraldehyde	*Hexamethylene diisocyanate*
Toluene-2-isocyante 4-isothiocyanate	1, 5-Difluoro-2, 4-dinitrobenzene
Bisdiazobenzidine-2, 2'-disulphonic acid	N-ethyl-5-phenylisoxazolium 3'-sulphonate

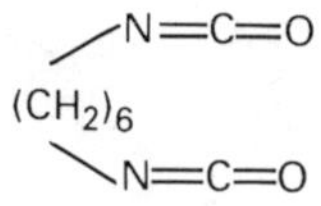

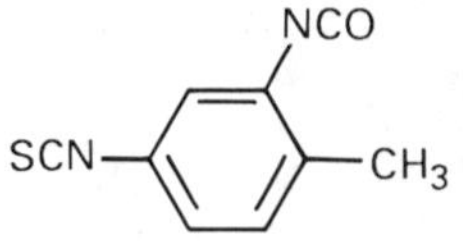

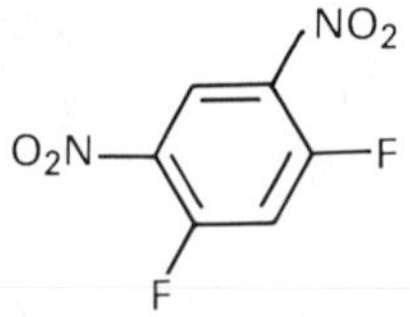

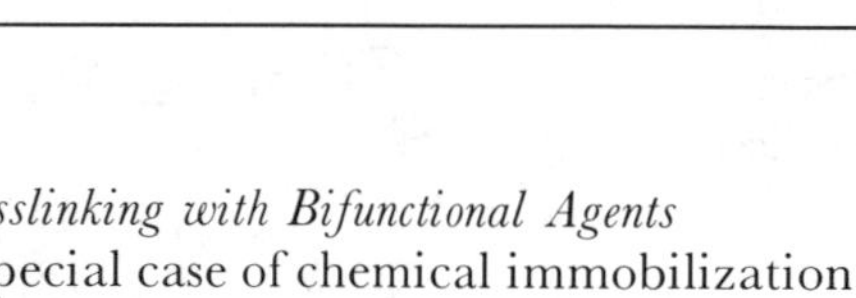

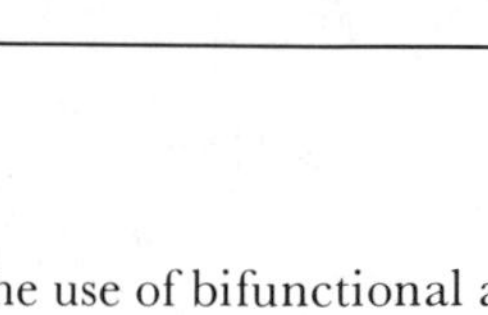

Crosslinking with Bifunctional Agents

A special case of chemical immobilization requires the use of bifunctional agents (Table 1.5), crosslinking protein to protein, and in some instances also to the support. Glutaraldehyde in particular is a crosslinking reagent which is widely employed. This reagent will react with the lysine amino groups in the enzyme, but overloading of the surface with the biological component must be avoided, since although activity increases with loading initially, restricted access and other inhibition effects at high loadings decrease the net activity. In practice, suitable levels of activity are achieved by the co-immobilization of an inert protein. The immobilization layer will have physical characteristics (e.g. thickness, porosity, etc.), dependent on the exact reaction conditions, but will in any case be considerably thicker than a monolayer.

Physical Adsorption at a Solid Interface

The great advantage of physical adsorption is that no reagents are required for the immobilization. Only weak interactions are involved between the support and the biomolecule due to Van der Waals forces, dipole–dipole interactions, hydrogen bonding, or the formation of electron transition complexes. Unfortunately, except in the last case, the reversible nature of the binding equilibrium is highlighted by its susceptibility to changes in the ambient conditions (e.g. pH, ionic strength, temperature, etc.). When a protein adsorbs on a solid surface for example, it will be influenced by hydrophobic, hydrophilic, ionic and polar interactions between the surface and the protein. It is also apparent that the protein could interact with the surface in a variety of different ways, depending on the orientation with which it approaches the surface.

In an unperturbed solution, this process is under mass transport control, as is the reverse process of desorption. Assuming that every molecule that collides with the surface is adsorbed, then a concentration gradient rapidly develops at the surface, due to the depletion of the layer immediately next to the interface. The rate of adsorption therefore becomes proportional to the rate of diffusion (Macritchie, 1978):

$$\frac{\mathrm{d}n}{\mathrm{d}t} = C_0\left(\frac{D}{\pi t}\right)^{1/2}$$

where n = the number of molecules, C_0 = bulk concentration and D = diffusion coefficient. This model does not accommodate the situation where the layer immediately next to interface is only partially saturated and the rate of adsorption falls below the rate of diffusion. In fact when the immobilized species is no longer present in the surrounding solution (as would be the case for the sensor), the predominant process will be desorption, and the biorecognition surface will decay as the immobilized species is lost to the bulk solution.

However the ease of immobilization by this method and its general applicability for many surfaces, makes it particularly attractive as a pilot method, or for devices not requiring long-term stability. It is even feasible in fact that such a method might be developed to give a certain degree of long-term stability. Nitrocellulose, for example, is a classical support for DNA immobilization (Denhardt, 1966; Southern, 1975). Single-stranded DNA is assumed to be attached to the support by non-covalent interactions, and yet the probed DNA sequence can be thermally eluted and the immobilized DNA re-used up to six times (Meinkoth and Wahl, 1984).

Entrapment in Polymer Matrices

Entrapment of the biomolecule in a three-dimensional polymer matrix covers a variety of different polymerization methods and polymer characteristics. The polymer may be an inert support, or may itself perform some function essential to the transduction of the analyte-dependent signal. In the former case polymers such as polystyrene, PVC, polyacrylamide, etc. are deposited on the sensor surface together with the recognition molecule, whereas in the latter case these support matrices may be modified to include a chemically active functionality (e.g. redox group), or formed with some residual chemical function.

Conducting polymers and in particular electrochemically deposited polymers have become of interest as support matrices, especially for electrochemical sensors, where they can be grown *in situ* under easily controlled electrochemical conditions (see Chapter 8). The electron transport properties of these polymers may in some cases be exploitable in the signal transduction process for the biosensor device.

Future Directions

If a comparison is made with naturally occurring 'biosensors' then it is only possible to conclude that the evolution of the man-made biosensor has only just begun. Present biosensor devices are very crude compared with the complexity of say, the nose, or the translation of a light stimulus by the eye. The 'recognition' molecules employed in these sensors are not necessarily highly specific but transduction of the signal via integrated circuits devised of biomolecules shows a degree of sophistication that at present can only be dreamed of. Specificity appears to be induced through data processing and pattern recognition analysis via a continual learning process. Probably, in this latter field of data processing techniques we can expect to see significant progress, applicable to biosensors in the near future.

The apparent desire for a continual increase in the density of electronic components to give ever smaller 'packages' has been limited, not by the microlithographic technique employed for their fabrication, but by the minimum size possible for a transistor, without incurring loss of electrons. The inherent property of many biological molecules to be able to synthesize complex self-organizing molecules with apparently just the required electronic properties, might suggest that the solution to this problem is to be found in the replacement of silicon with biomolecular components. This supposition has led to the proposition of many *molecular electronic* systems.

In the same way that conducting and semiconducting polymers, synthesized and investigated for microelectronic applications, have been exploited in sensor developments, so may any future developments in molecular electronics or insights into biological signal transduction make the migration into biosensor technology.

References

Alwine, J.C., Kemp, D.J., Parker, B.A., Reiser, J., Renart, J., Stark, G.R. and Wahl, G.M. (1979). *Methods Enzymol.* **68**, p. 220.

Benzinger, T.H. and Kitzinger, L. (1960). *Methods Biochem. Anal.* **8**, p. 309.

Biomedical Business International, Alternate Site Clinical Testing, Report No. 7088.

Business Communications Co., Biosensors and Chemical Sensors, Report No. C-053. Stamford, USA.

Calabrese, G., Bueti, A., Zega, G., Giombolini, A., Bellomo, G., Antonella, M.A., Massi-Benedetti, M. and Brunetti, P. (1982). *Horm. Metabol. Res.* **14**, p. 505.
Clark Jr, L.C. and Lyons, C. (1962). *Ann. N.Y. Acad. Sci.* **102**, p. 29.
Danielsson, B., Mattiasson, B. and Mosbach, K. (1981). *Appl. Biochem. Bioeng.* **3**, p. 97.
Danielsson, B., Mosbach, K., Winquist, F., and Lundström, I. (1988). *Sensors and Actuators* **13**, p. 139.
Denhardt, D.T. (1966). *Biochem. Biophys. Res. Commun.* **23**(5), p. 641.
Dimitriadis, G.D. and Gerich, J.E. (1983). *Diabetes Care* **6**, p. 374.
Grooms, T.A., Clark, L.C. and Weiner, B.J. (1980). *Enzyme Engineering*, vol. 5, Eds Weetal, H.H. and Royer, G.R., p. 217. Plenum Press, New York.
Hunger, H.D., Coutelle, C., Behrendt, G., Flackmeier, C., Rosenthal, A., Speer, A., Breter, H., Szargan, R., Franke, P., Stahl, J., Cuong, N.V. and Barchend, G. (1986). *Anal. Biochem.* **156**(2), p. 286,
Macritchie, F. (1978). *Adv. Protein Chem.* **32**, p. 283.
Meinkoth, J. and Wahl, G. (1984). *Anal. Biochem.* **138**(2), p. 267.
Menkins, R.M., Watt, G.D. and Sturtevant, J.M. (1969). *Biochemistry* **8**, p. 1874.
Mosbach, K. and Danielsson, B. (1981). *Anal. Chem.* **53**, p. 1.
Moss, L.G., More, J.P. and Chan, L. (1981). *J. Biol. Chem.* **256**(24), p. 12 655.
Muramatsu, H., Dicks, J.M. and Karube, I. (1987). *Anal. Chim. Acta* **197**, p. 347.
Scheller, F.W. and Schubert, F. (1987). *Bioengineering* **3**(1), p. 30.
Scheller, F.W., Schubert, F., Renneberg, R., Muller, H.-G., Janchem, M. and Weise, H. (1985). *Biosensors* **1**(1), p. 135.
Seed, B. (1982). *Nucl. Acids Res.* **10**(6), p. 1799.
Southern, E.M. (1975). *J. Molec. Biol.* **98**, p. 503.
Sturtevant, J.M. and Lyons, P.A. (1969). *J. Chem. Thermodyn.* **1**, p. 201.
Wehnert, G., Sauerbrei, A., Bayer, Th., Scheper, Th., Schüger, K. and Herold, Th. (1987). *Anal. Chim. Acta* **200**, 73.

Chapter 2

The Biomolecule Reviewed

The Ingredients (Stryer, 1981; Alberts *et al.*, 1983)

Most of the chemical components of living organisms are organic compounds of carbon, many also containing nitrogen. Although each living species contains its own combination of these biomolecules, unique only to itself, the diversity can be reduced to a few building blocks of common structure. It is possible to organize these building blocks on a hierarchical ladder, according to molecular weight (Fig. 2.1).

The bottom of the ladder is occupied by the low-molecular-weight gases (oxygen, nitrogen and carbon dioxide) and by water. These molecules together with the monoatomic ions, in particular Na^+, K^+, Mg^{2+}, Ca^{2+}, Cl^- and the elements P, S, Mn, Fe, Co, Cu and Zn are most common in further involvement as progress is made up each rung of the molecular-weight ladder. The monitoring of these ions is frequently used to follow the metabolic state of a patient under care and is commonly achieved using ion selective electrode sensors, where the recognition surface of the sensor is provided by a membrane, containing a molecule with a selectivity for the target ion (Chapter 3). The oxygen (Chapter 5) and carbon dioxide (Chapter 3) electrodes form two of the well-established base sensors to which analyte specific reactions have been linked via a biorecognition macromolecule.

In fact, some of the larger biomolecules shown in Fig. 2.1 will feature frequently in subsequent chapters, due to their *molecular-recognition capabilities*. By reference to the schematic biosensor (see Fig. 1.3), it can be seen that it is these recognition characteristics that will be exploited and it is important therefore, to consider these biorecognition compounds and their potential targets.

This chapter is a glossary of terms and a brief background of some typical

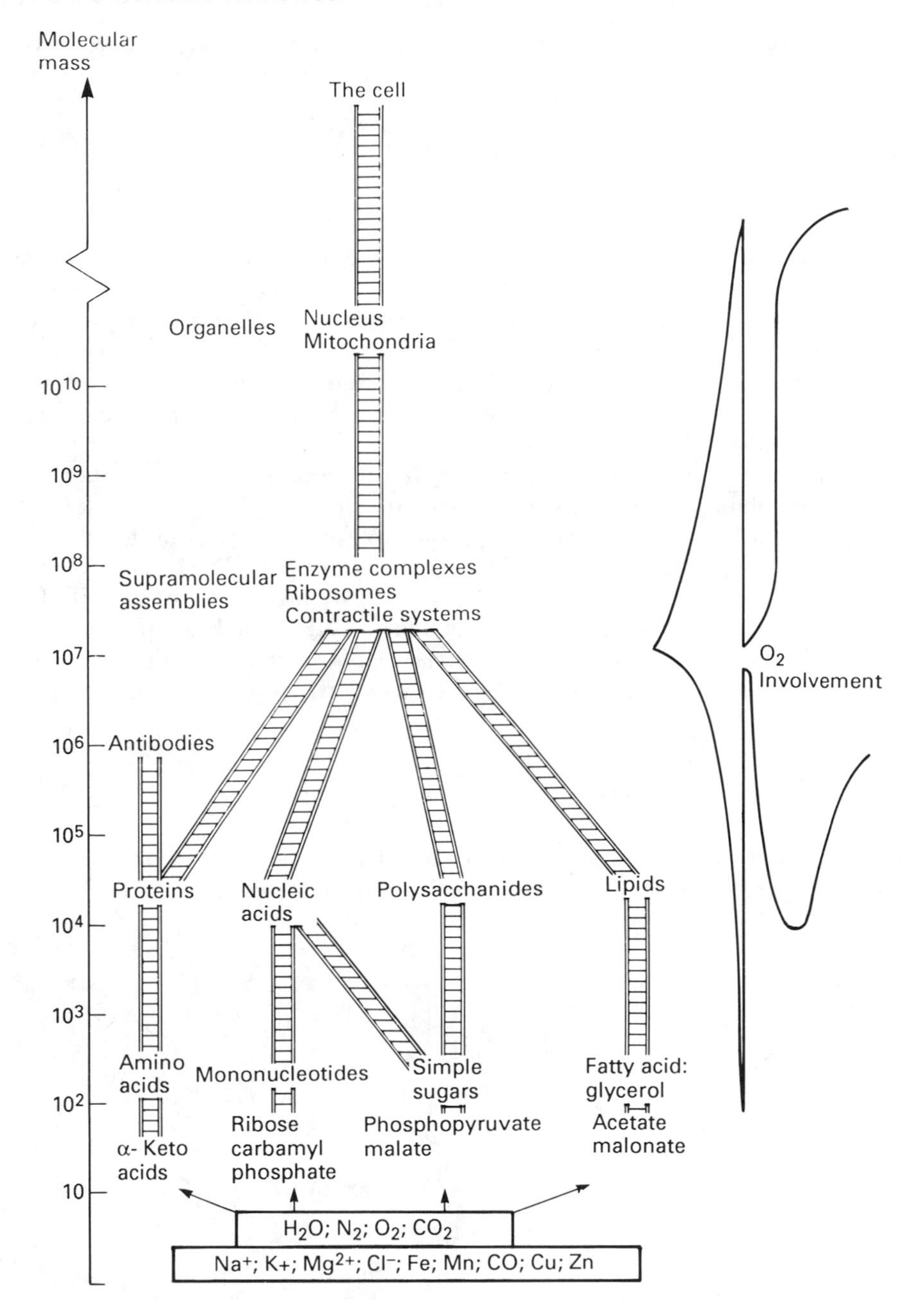

Fig. 2.1 The biomolecular ladder.

physico-chemical features of biomolecules that might be exploited in order to provide an immobilized recognition surface, capable of transducing an analyte-dependent signal in a reagentless biosensor.

Proteins

Proteins are polypeptides whose structure and perhaps most importantly whose function is dependent on not only the amino acid sequence, that is the primary structure, but also on the conformation (the secondary, tertiary and quaternary structures). The peptide bond (Fig. 2.2) and the disulphide bridges impart certain restrictions on the structure, and the peptide strands are further organized by interactions between residue side chains, which may be inter- or intra-molecular; they are summarized in Fig. 2.3.

The net result of all the interactions between the amino acids is that there is spontaneous folding of the protein to give a unique structure (Pauling *et al.*, 1951). All the amino acids have at least two groups capable of existing in ionic form. The α-carboxyl group, –COOH, can lose H^+ to become –COO–. The reaction is pH dependent, and is characterized by a pK_a typically in the range 2 to 3. Similarly the α-amino group, $-NH_2$, can be protonated to give NH_3^+ and has a pK_a value of about 10. It follows, therefore, that between about pH 4 and 9 the amino acid will exist as a dipolar ion; that is in a *zwitterionic form* with little net charge:

$$H_2N-\underset{R}{\overset{|}{C}H}-COOH \rightleftharpoons H_3N^+-\underset{\text{(zwitterion)}}{\overset{R}{CH}}-COO^-$$

At the isoelectric point, p*I*, there will be no net charge, and the molecule will not move in an electric field. Where R contains no ionizable groups,

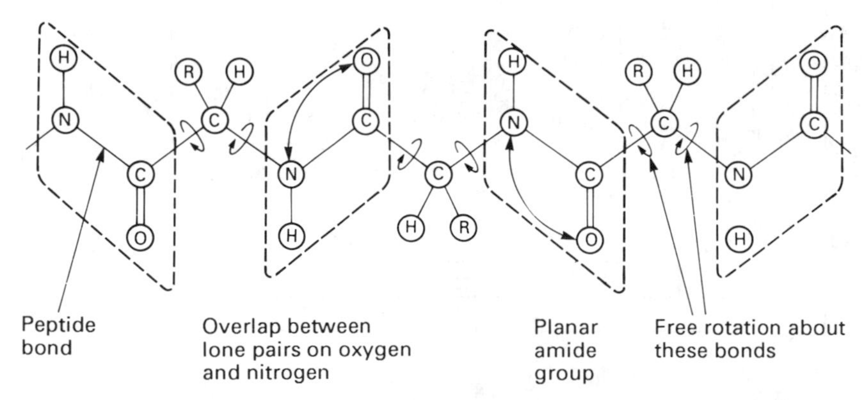

Fig. 2.2 Limited rotation in the peptide chain. The peptide bond shows some double-bond character due to overlap of lone pairs on nitrogen and oxygen.

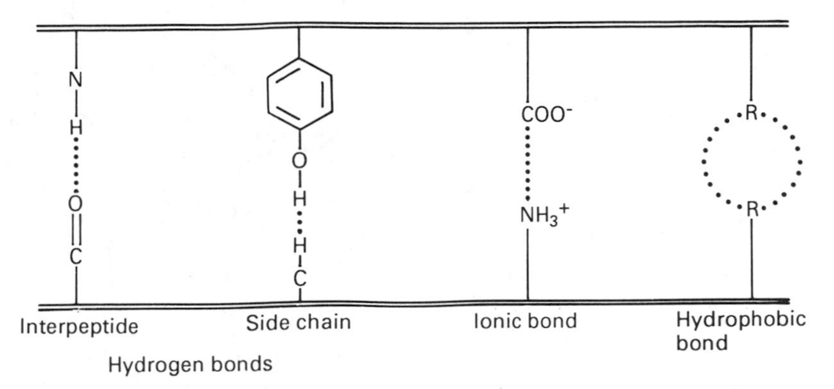

Fig. 2.3 Non-covalent inter- and intra-molecular bonds in peptide strands.

$$pI = \frac{(pK_a^{NH_2} + pK_a^{COOH})}{2}$$

The movement of the amino acids under the influence of an electric field allows their separation and identification; it can be used as a powerful assay technique.

The dual polarity feature accounts for many of the properties of amino acids, e.g. the large dipole moments, the high solubility in water and low solubility in organic solvents.

By analogy, it would be expected that each peptide strand would also contain at least two ionizable groups, but since the α-carboxyl or α-amino groups are now involved in peptide-bond formation, they are not available for ionization. The zwitterionic behaviour is therefore more restricted to the terminal amino group and the terminal carboxyl group. These groups are considerably further away from one another than they would be in free amino acids and so the electrostatic interactions between them are diminished and their pK_a values are lower than in the α-amino acid. It follows that the groups in proteins that are principally involved in acid–base equilibria are the side-chain R groups. Ordinarily there may be 50–60 titratable groups per 100 000 molecular weight of protein, and it is understandable, therefore, that the titration curves for proteins are complex and difficult to interpret.

The biorecognition properties of protein molecules will depend almost entirely on the amino acids of the exposed surfaces.

Weak non-covalent interactions can occur between the residues on the exposed surfaces of the protein and other non-protein molecules (Fig. 2.4). If a sufficient number of these weak bonds are formed simultaneously with the incoming molecule, then the molecule can bind tightly to the protein. Obviously for this to occur the molecule must fit precisely into the *binding site* on the protein surface. This feature is analogous to the recognition surface of the model biosensor (see Fig. 1.3).

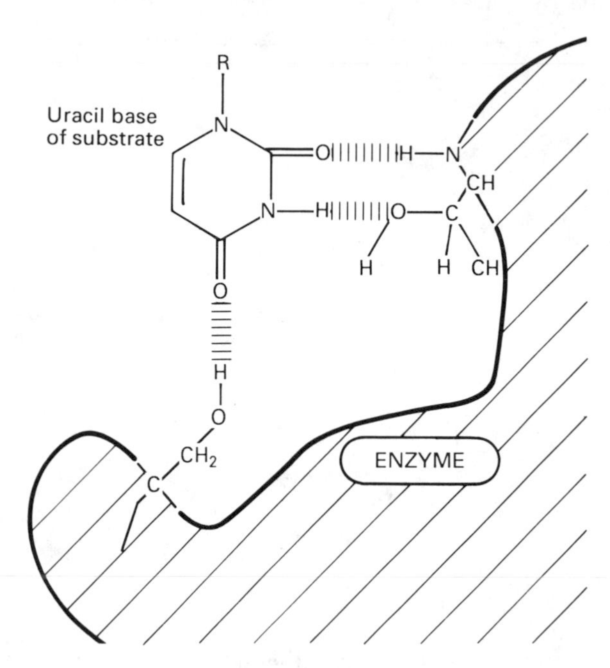

Fig. 2.4 Example of the interaction between enzyme surface and incoming substrate.

Enzyme Complexes (Boyer, 1975)

One of the most important functions of proteins is to act as catalysts or enzymes for chemical reactions (Wolfenden, 1972). These enzymes, which have proved particularly useful in the development of biosensors, are able to stabilize the transition state between a substrate and its products by interactions at the binding site as described above. Substrate specificity by the enzyme is provided by the surface interactions and this inherent characteristic can be exploited in the development of enzyme-based biosensors. The non-covalent binding of the enzyme–substrate transition state lowers the activation energy for the reaction and thus catalyses the reaction. Sometimes the surface cavity does not act as a catalytic site until it is modified by a second incoming molecule. These participants known as the *coenzymes* are non-peptide molecules capable of completing the binding site for the transition state.

The enzyme itself may also contain a non-amino acid component, which is concerned with the chemical function associated with that enzyme; this is the *prosthetic group*. The enzyme may therefore he divided into protein (*apoenzyme*) and non-protein (coenzyme or prosthetic) units.

Enzymes are classified according to their function, i.e. the nature of the chemical reaction that they catalyse. The classes will be relevant to different types of biosensor, according to their function, and may be divided into six major groups:

(i) *Oxido-reductases*—electron transfer for oxidation or reduction of groups such as:

- $>CH-OH$
- $>C=O$
- $-CH=CH-$
- $>CH-NH_2$
- $>CH-NH-$

and with the coenzymes, e.g. NADH and NADPH

(ii) *Transferases*—transfer functional groups such as:

aldehydes and ketones
glycosyls
acyls
phosphates
sulphur containing groups

(iii) *Hydrolases*—hydrolyses:

esters
anhydrides
peptides bonds
other C–N bonds
glycosides

(iv) *Lyases*—add to double bonds:

$>C=C<$
$>C=O$
$>C=N$

(v) *Isomerases*—isomerizes optical isomers, etc.

(vi) *Ligases*—ATP-linked bond formation:

C–O
C–S
C–N
C–C

Enzyme Kinetics

Since the transition state involves complexation with the protein, then there will be a maximum in the concentration of substrate that can be processed at once. When the enzyme is saturated the reaction rate is dependent only on the *turnover number* and typically is of the order of 1000 substrate molecules per second. Enzymes are normally characterized by the concentration of substrate required to achieve half this maximum rate. This is derived from consideration of the enzyme kinetics:

$$E + S \rightleftharpoons ES \rightarrow E + P$$

where E is enzyme, S is substrate and P is product, and E+P→ES is slow enough to neglect, so that at steady state,

$$\frac{d[ES]}{dt} = k_1[E][S] - k_{-1}[ES] - k_2[ES] = 0$$

If the reaction is described by the rate constant K_M, the *Michaelis constant*, where

$$K_M = \frac{(k_{-1} + k_2)}{k_1}$$

and the enzyme concentration is described as total, $[E_o]$, rather than unbound enzyme concentration,

$$[E_o] = [E] + [ES]$$

then it follows that

$$[ES] = \frac{[E_o][S]}{K_M + [S]}$$

and the rate of the enzyme-catalysed reaction would be given by the *Michaelis–Menten equation*:

$$v = \frac{-d[S]}{dt} = k_2[ES] = \frac{k_2[E_o][S]}{K_M + [S]}$$

When $[S] \gg K_M$ a maximum is reached in the rate, V_{max}, so that from the above equation $V_{max} = k_2[E_o]$; and when $[S] = K_M$ then $v = V_{max}/2$ (Fig. 2.5).

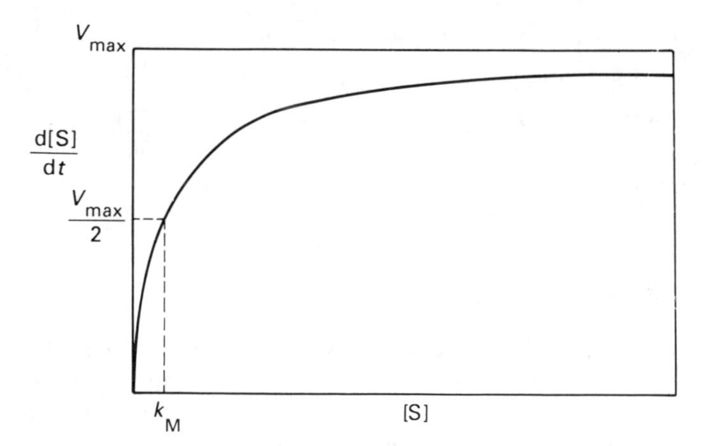

Fig. 2.5 Plot of the Michaelis–Menten relationship. $[S] = K_M$ when the rate $-d[S]/dt$, is half the maximum V_{max}.

Biosensors which employ enzymes as the recognition molecule require that enzyme in an immobilized state, and so the enzyme kinetics must also account for the diffusional limitations of the immobilization matrix. These modifications will be developed for specific models in the relevant sections.

The kinetic features themselves can also be exploited in the development of an assay technique for a biosensor. Various kinds of inhibition have been characterized, caused by a variety of agents, including drugs, nerve gases, toxins, insecticides, etc.

The inhibition may be *competitive* or *non-competitive*; in the latter case several mechanisms can be envisaged where the inhibitor alters the activity of the active site without actually blocking it. Competitive inhibition by the agent I can be considered as:

$$E+S \rightleftharpoons ES;\ (K_M):$$
$$E+I \rightleftharpoons EI;\ (K_I)$$
$$ES \rightarrow E+P$$

so that,

$$[E]=[E_o]-[ES]-[EI]$$

and the reaction rate becomes:

$$v=\frac{V_{max}[S]}{[S]+K_M\{1+([I]/K_I)\}}$$

The reciprocal plot of this relationship (*Lineweaver–Burke plot*),

$$\frac{1}{v}=\frac{K_M}{V_{max}}\left[1+\frac{[I]}{K_I}\right]\frac{1}{[S]}+\frac{1}{V_{max}}$$

for $1/v$ versus $1/[S]$ shows the same intercept as in the absence of inhibition, but a slope which is a linear function of inhibitor concentration, and may be used as a diagnostic test.

Models for non-competitive inhibition reveal characteristic changes in the kinetics which may also be related to inhibitor concentration.

Chapter 1 mentioned the use of the acetylcholine receptor system in the detection of nerve gases and toxins, suggesting that a matrix of 13–20 proteins would give a 95% probability of detecting the presence of all toxic substances. Acetylcholinesterase is one of the most active enzymes known. It has been intensively studied, not only because of its physiological importance, but also due to its inhibition by compounds such as insecticides and nerve gases.

These kinds of enzyme inhibition-based diagnoses are likely to be particularly relevant to environmental and pollution monitoring and control, where the exact nature of the target analyte is not known ahead of its detection.

Enzymes can therefore be identified as a naturally occurring class of molecular complexes whose function it is to recognize and manipulate smaller molecular-weight compounds with a high specificity. This is analogous to the surface of the modified transducer in Fig. 1.3, which is required in order to construct the biosensor. It is possible to imagine therefore, that if some measurable physico-chemical event could be identified that was involved in the reaction between enzyme and substrate, then the biospecificity of these polypeptide macromolecules could be exploited in the development of a biosensor. The parameter chosen is likely to be associated with the particular function of the enzyme. Oxido-reductases for example, would be associated with electron transfer.

It also follows that modelling of just the surface cavity of the binding site could produce similar biospecificity (M. Thompson, unpublished work), although without the prosthetic group function. Attempts by synthetic chemists to produce such selective surfaces may also be relevant to biosensor developments, particularly if they can be linked to different 'prosthetic group' transducers.

The Proteins of the Immune System (Kabat, 1976)

Antibodies represent one of the major classes of protein; they constitute about 20% of the total plasma protein and are collectively called *immunoglobulins* (Ig). The simplest antibodies are usually described as Y-shaped molecules with two

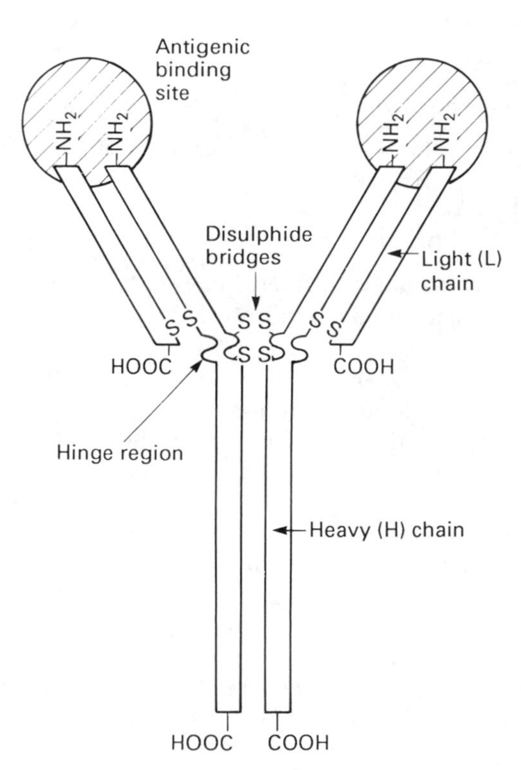

Fig. 2.6 Scheme for a typical antibody composed of two heavy chains and two light chains.

identical binding sites for *antigen* (Fig. 2.6) (Nisonoff *et al.*, 1975). An antigen can be almost any macromolecule that is capable of inducing an immune response. Smaller molecules, *haptens*, can also bind specifically to the antibody without inducing an immune response, but these can be coupled with a protein in order to make them antigenic.

The antibody has a basic structural unit consisting of four polypeptide chains (Edelman, 1970; Porter, 1973) (Fig. 2.6), two light (L) chains and two heavy (H) chains that are about twice the length of the L chains. The peptide chains are held together as described above with covalent disulphide bridges and non-covalent interactions between amino acid residues of adjacent chains.

The antibody can be considered as a molecular-recognition complex which binds reversibly with a specific antigen. The kinetics of this binding between antibody (Ab) and antigen (Ag) is described by the affinity or association constant K_a:

$$\mathrm{Ab} + \mathrm{Ag} \rightleftharpoons \mathrm{AgAb}$$

$$K_a = \frac{[\mathrm{AgAb}]}{[\mathrm{Ag}][\mathrm{Ab}]}$$

Values of K_a range from about 10^4 to 10^{12} litres/mole. Immunoglobulins with K_a values $< 10^4$ for a particular antigen would be ineffective as antibodies against the antigen.

Unlike the enzyme proteins, the antibodies do not act as catalysts, stabilizing the transition state between substrate and products. Their purpose is rather to bind foreign substances—antigens—so as to remove them from the system. If the antigen contains multiple antigenic sites then chains of antibody–antigen can form and these aggregates precipitate readily (Fig. 2.7). Maximum precipitation is usually achieved at molar equivalence of antibody and antigen since excess of either will disturb the formation of these precipitatable complexes.

Antibodies in Assays

As with the enzymes, these biomolecules display the sort of degree of specificity that is a prerequisite for a biosensor. However, since the antibodies are concerned with removal of foreign substances often by precipitation as large aggregates, rather than the catalytic conversion of substrate to products, then the physico-chemical event that accompanies their reaction is likely to be concerned, for example, with changes in particle size, and it will be a parameter related to this change that must be transduced in the biosensor, if a direct assay is to be made. Alternatively, labels can be employed as indicated in Chapter 1. Labelled immunoassay can be subdivided into two categories: heterogeneous and homogeneous assay. The former, following more closely to the concept of biosensor-based assays, is of more interest here. Enzyme-linked immunosorbent assay (ELISA) has employed several enzymes which can be conjugated easily to antigens and

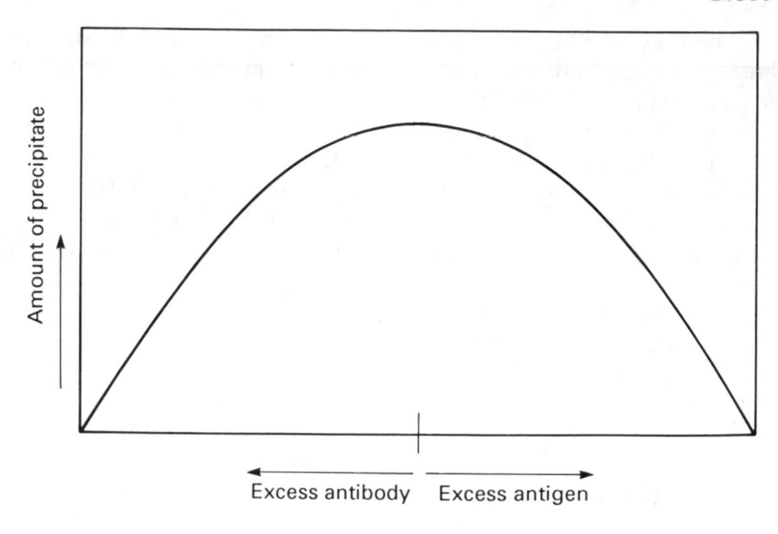

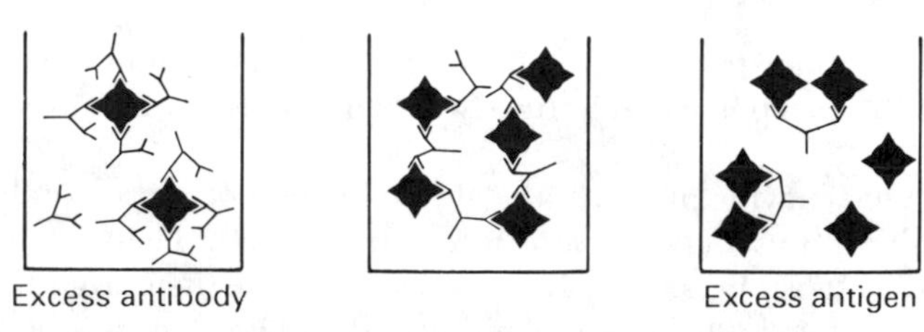

Fig. 2.7 Precipitation of antibody–antigen complexes. Maximum complexation with equimolar amounts of antigen and antibody.

antibodies and show simple kinetics. The most frequently employed enzymes have been alkaline phosphatase, β-D-galactosidase and horseradish peroxidase.

These assays are always indirect, but can be devised in many ways, for example:

Primary Binding Immunoassay

The sample antiserum is incubated with immobilized antigen (Fig. 2.8a), and the antigen-bound antibody then detected by incubation with labelled anti-immunoglobulin, then estimated via the bound label.

Conversely, similar protocol is suitable for the determination of antigen concentration, by incubation with immobilized antibody (Fig. 2.8b) and detection via a second labelled antibody.

Competitive Binding Immunoassay

A fixed amount of labelled antigen competes with sample antigen for a limited amount of antibody. Various constructions can be devised along this principle or

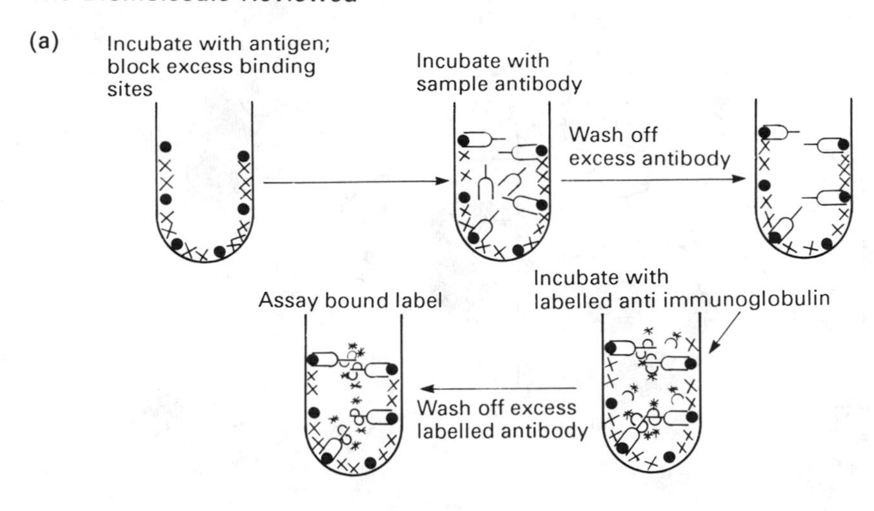

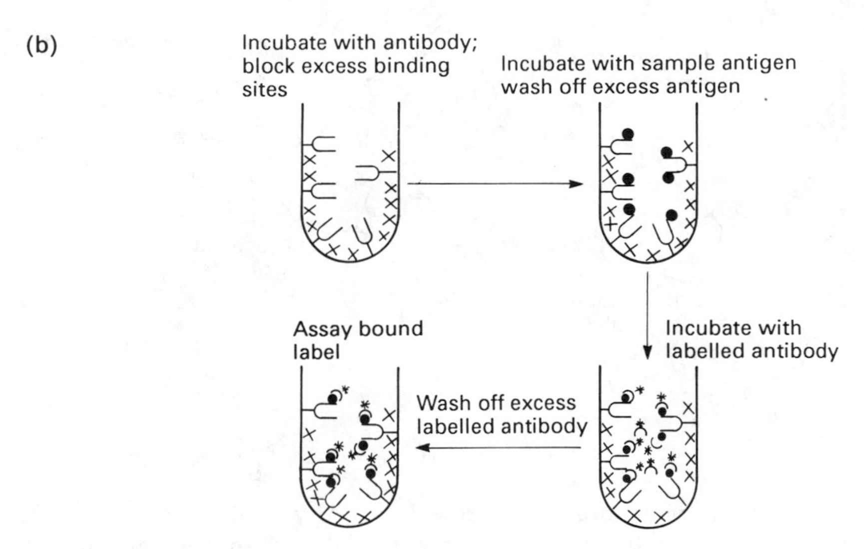

Fig. 2.8 Primary binding immunoassay. (a) Scheme for estimation of sample antiserum; (b) similar procedure for determination of antigen.

with a competition involving labelled antibody. Some of the possibilities are shown in Fig. 2.9.

Clearly, in conjunction with an enzyme sensor, immunoassay could also be designed around a single probe, so that a possible approach for its development is as an extension of an enzyme biosensor.

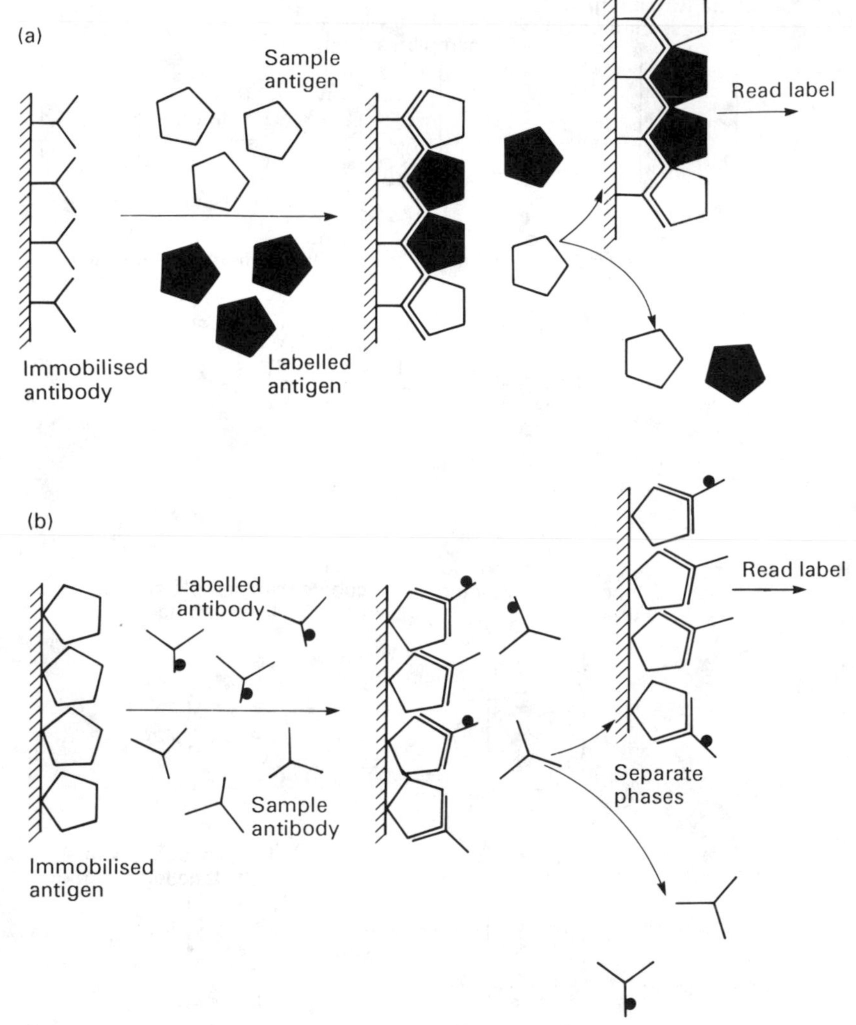

Fig. 2.9 Competitive binding immunoassay. (a) Antigen determined by competition for an immobilized antibody with a labelled antigen; (b) similar scheme for sample antibody using immobilized antigen.

MONOCLONAL ANTIBODIES IN IMMUNOASSAY

During the past two decades there has been an extraordinary increase in the use of diagnostic immunoassay using these techniques. This is attributable to the remarkable specificity and sensitivity of the antibody molecule. The antibody may be *polyclonal* (PCA) or *monoclonal* (Mabs). Immunization of an animal stimulates many different B cells to produce antibodies, so that a PCA will contain a mixture of antibodies, some of which will react with epitopes, some of which are specific for the particular antigen and some of which are common for many antigens. Mabs in contrast have been identified during screening as specific for a unique epitope and will be associated with only one immune response. This characteristic can have both advantages and disadvantages. It can, for example, be too specific so that only one strain of a particular bacteria is detected, failing to detect any other related but equally relevant strains.

A significant advantage in the use of Mabs however, is the flexibility and control obtainable in designing assay configurations. Two-site immunoassay, for example, requires antibodies specific for two distinct epitopes on the antigen, one of which will be labelled.

Non-peptide Biomolecules from Amino Acid Precursors

The amino acids are the precursors of various important biomolecules, e.g. vitamins, hormones, coenzymes, alkaloids, pigments, etc. Table 2.1 gives examples of the extent of the variation of function that these biomolecules achieve.

Particularly worthy of note are the porphyrins, derived from the amino acid glycine. This is the precursor to the prosthetic group of the *cytochromes*, a group of electron-transfer proteins, containing an iron porphyrin complex as coenzyme. This class of enzymes is principally involved with the transfer of electrons to molecular oxygen and other electron acceptors.

In heme, the four ligand groups of the porphyrin form a square planar complex with the six-coordination iron atom, and in the cytochromes the fifth and sixth ligands to iron are usually provided by proteins (Fig. 2.10). As can be seen, in cytochrome C one of these proteins is also linked covalently to the pyrrole ring via the addition of cysteine-SH across the two vinyl side chains.

In haemoglobin, the fifth coordination position is occupied by an imidazole group from a histidine residue, and the sixth may be unoccupied (deoxyhaemoglobin) or occupied with oxygen or other ligands such as carbon monoxide, cyanide, etc. Unlike the cytochromes, these complexes act as oxygen transporters without change in oxidation state.

Nucleic Acids and Derivatives (Watson and Crick, 1953)

The nucleotide/nucleoside family contains two main branches: those based on the sugar ribose and those based on 2-deoxyribose. Both branches of mononucleotides are strong acids due to a phosphoric acid group which has pK_a values between 1 and 6.

Mono-, di- and tri-phosphates all occur. In particular the derivatives of the base adenosine (AMP, ADP and ATP) are important as phosphate transfer agents. The tri- and di-phosphates also perform the function of a coenzyme-like covalently bound carrier for specific building blocks. For example, uridine

Table 2.1 Precursor functions of some amino acids

Arginine
Spermine
Spermidine
Putrescine

Aspartic acid
Pyrimidines

Glutamic acid
Glutathione

Glycine
Purines
Glutathione
Creatine
Phosphocreatine
Tetrapyrroles

Histidine
Histamine
Ergothionine

Lysine
Cadaverine
Anabasine
Coniine

Ornithine
Hyoscyamine

Serine
Sphingosine

Tyrosine
Epinephrine
Norepinephrine
Melanin
Thyroxine
Mescaline
Tyramine
Morphine
Codeine
Papaverine

Tryptophan
Nicotinic acid
Serotonin
Kynurenic acid
Indole
Skatole
Indoleacetic acid
Ommochrome

Valine
Pantothenic acid
Penicillin

diphosphate glucose (Fig. 2.11), in which the glucose is bound through the β-phosphate group, is the glucose donor for glycogen synthesis.

Polynucleotides are formed through a phosphodiester linkage, bridging between the sugars of two mononucleotides, and giving the deoxyribonucleic acids (DNA) and the ribonucleic acids (RNA). These comparatively simple polymers, involving sequences of nucleotides derived from just four bases—adenine (A), cytosine (C), guanine (G) and thymine (T)(DNA) or uracil (U)(RNA)—are the carriers of biological information. Through non-covalent interactions, termed base pairing, they can 'recognize' other nucleotides. The interactions are highly specific; that is, a purine base interacts with a pyrimidine base, to achieve the maximum number of hydrogen bonds (Fig. 2.12), i.e. between:

(i) adenine–thymine
(ii) cytosine–guanine

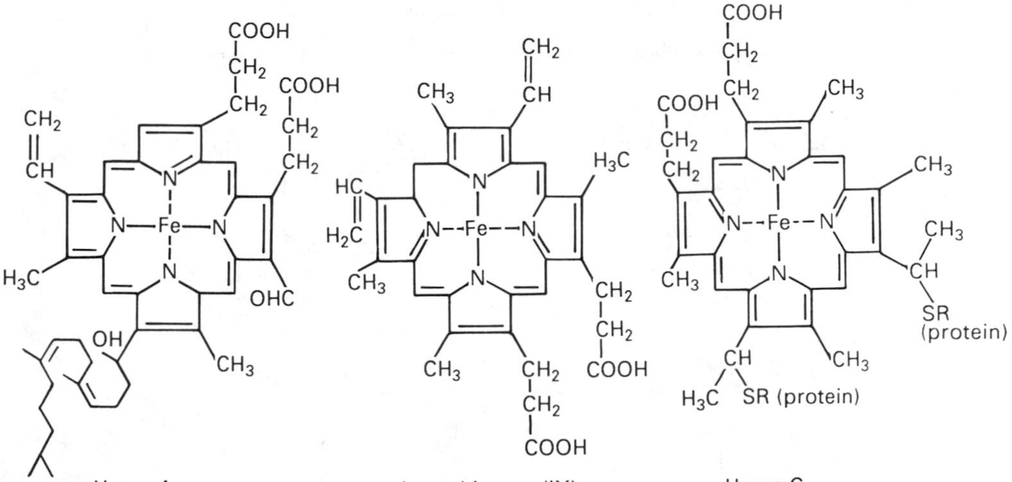

Heme A
prosthetic group
of cytochrome A
proteins

(proto) heme (IX)
prosthetic group
of cytochrome B
proteins

Heme C
prosthetic group
of cytochrome C
proteins

Fig. 2.10 Prosthetic groups of haemoproteins.

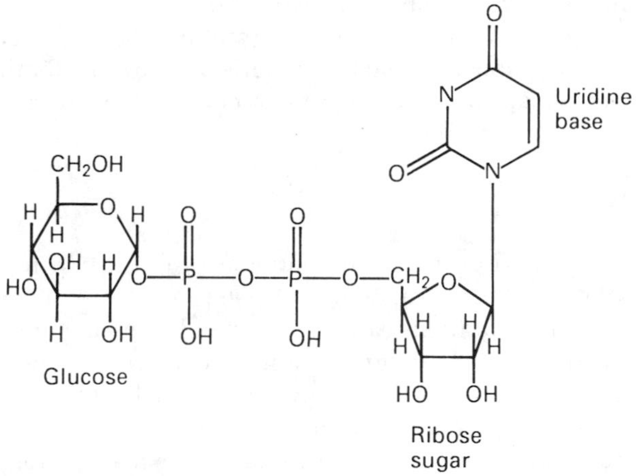

Uridine base

Glucose

Ribose sugar

Fig. 2.11 Uridine diphosphate glucose acts as a coenzyme; it donates the glucose residue for glycogen synthesis.

In the chromosomes DNA, containing the genetic information, exists as a double-stranded helix. The base-pairing restrictions mean that one strand will be the complement of the second strand, and as such on separation each strand can act as a template for the replication of the other strand (Fig. 2.13).

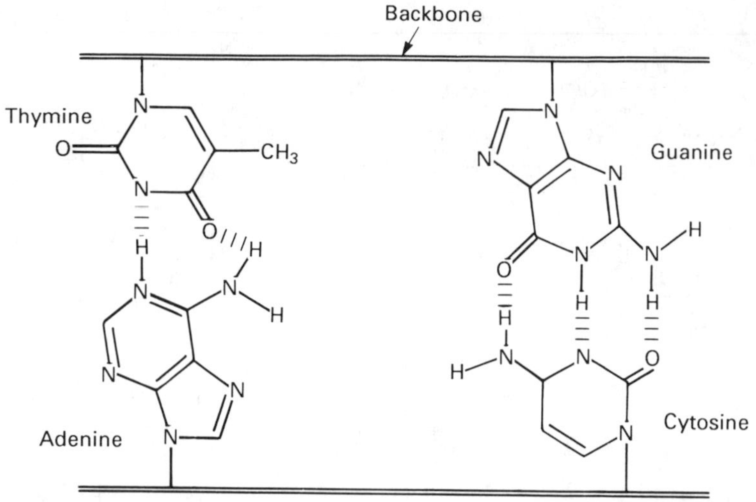

Fig. 2.12 Base pairing in DNA via hydrogen-bond formation. Thymine pairs with adenine, guanine with cytosine.

Once more, the analogy with the surface-modified transducer of the biosensor is suggested. Application of a single-stranded complementary polynucleotide template to any target analyte that can be described genetically should give an unrivalled specificity for that analyte. However, the identification of a physico-chemical event that follows this target recognition process presents rather a different transduction problem to that encountered above *in situ* for the cell, and is more analogous to the antibody–antigen interaction described above.

DNA in Assays

Human genetic diseases present an international problem. Up to one-third of children admitted to paediatric wards in the Western world are suffering from an inherited disease. If the mutations causing a particular disease are known, then the presence of that defect can be diagnosed using the relevant oligonucleotide as a DNA probe. Both 'long' and 'short' DNA probes are used. In the former case the DNA probe fragment is produced by insertion of the target sequence into a suitable vector, for example plasmid, and cloning. The desired fragment is thus replicated, supplying the probe which is then labelled with a reporter group such as a radioactive label, photometric group, enzyme, etc. In these probes the actual DNA sequence of the probe does not need to be known, but to produce the 'short' probes by chemical synthesis, the sequence itself must be identified.

The list of gene-specific probes increases constantly (Table 2.2) and the market for genetic disease testing is estimated to have reached US$ 150 million by 1995, while cancer-related probes are expected to have levelled off at US$ 400 million by the turn of the century (Frost and Sullivan, 1986).

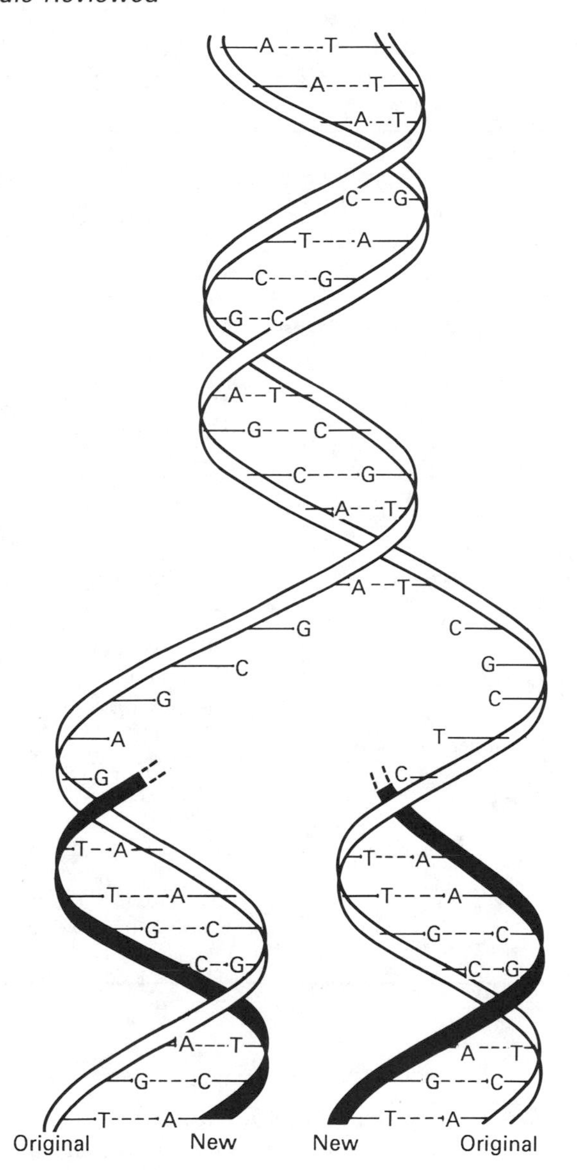

Fig. 2.13 Replication of DNA. Each strand acts as a template for the replication of a new complementary strand.

The general principle of DNA probe assay is similar to the immunoassay already described (Fig. 2.14). Indeed, even the applications of DNA probes and monoclonal antibody immunoassay frequently overlap, thus establishing a 'competition' between the two possible approaches.

For probes of infectious disease, it is assumed that all strains can still contain a

Table 2.2 Gene-specific probes of clinical relevance

Probe	Disease
Enzymes	
Glucose-6-phosphate dehydrogenase	Favism
Ornithine transcarbamylase	OTC deficiency
Hypoxanthine–guanine phosphoribosyl transferase	Lesch–Nyhan syndrome
Phenylalanine hydroxylase	Phenylketonuria
3-Hydroxy-3-methylglutaryl-CoA-reductase	
Hormones	
Insulin	Diabetes mellitus
Growth hormone	Dwarfism
Gonadotrophin	
Prolactin	
Cell-surface receptors	
Acetylcholine receptor	
Low-density lipoprotein receptor	Role in ischaemic heart disease
Blood and immune-system proteins	
Globins	Haemoglobinopathies
Clotting factors VIII, IX	Haemophilia A and B
Antithrombin III	Increased risk of thromoembolism
Histocompatibility antigens	Various immunodeficiency disorders
Complement components	
Immunoglobins	
T-cell receptor	
Other genes	
Collagen	Many collagen disorders
α_1-Antitrypsin	Emphysema, liver disease
Oncogenes	Cancer

common DNA sequence region, and thus be identified by a single probe. Mabs on the other hand can fail to detect other strains of a particular disease.

One obvious application for DNA probes is in testing for virus infections. Viruses appear to be almost uniquely DNA or RNA contained within an outer coat or capsid of protein. Recognized by the cell as a foreign body, they will induce an antigenic reaction causing antibody generation so they can also be detected in an immunoassay. Indeed, the present tests that are generally employed, detect the expression rather than the presence of the genetic information and this introduces a time lapse into the detection, which could be eliminated by the direct probe assay.

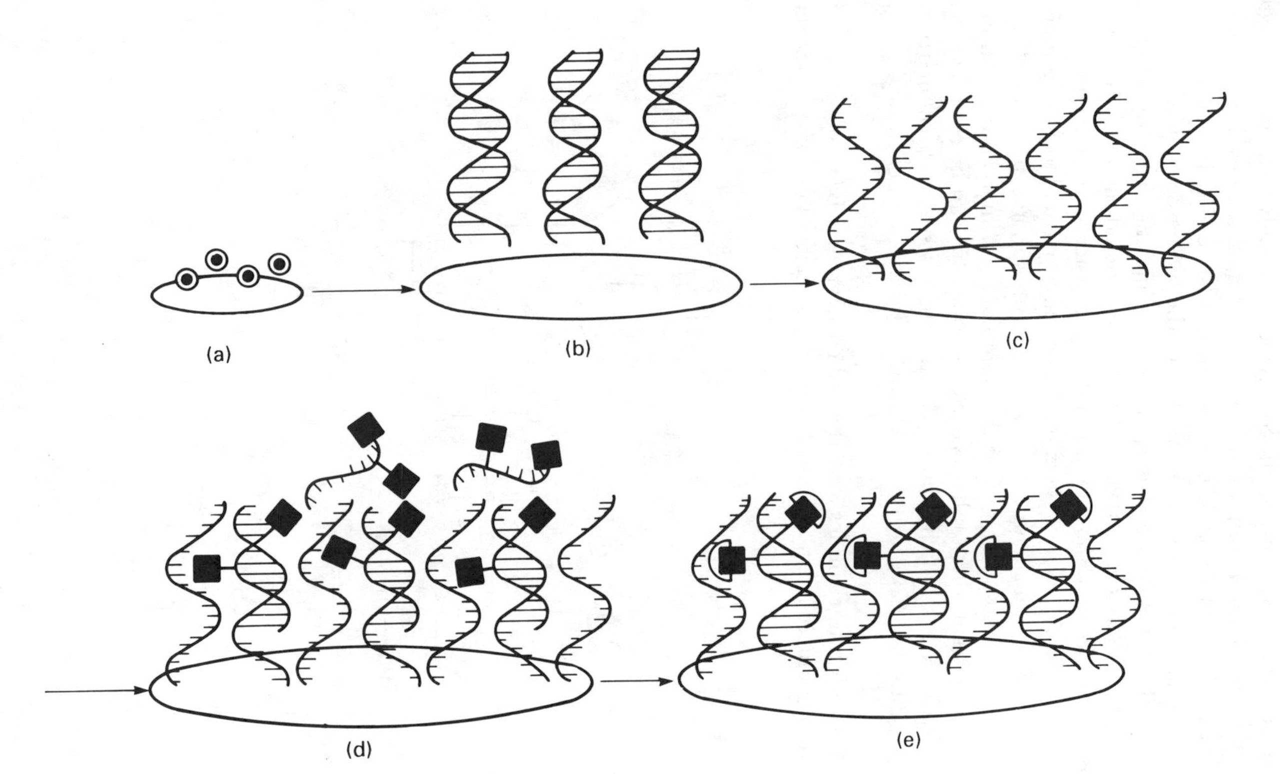

Fig. 2.14 DNA probe assay. (a) Deposit sample organism on immobilization matrix; (b) release DNA; (c) immobilize DNA to matrix and separate strands; (d) add labelled DNA probes and hybridize; (e) read label.

Coenzyme Nucleotides

A number of very different non-protein coenzyme chemically reactive molecules occur. These are particularly important in biosensor applications, because very often they act as a 'transduction' centre for the enzyme, and it is their physico-chemical behaviour which will be at the core of the biosensor device.

ATP has already been mentioned as a particularly important cofactor, essential to a large number of individual enzymes. Nucleotides, including some containing nitrogenous bases other than those found in the nucleic acids, are encountered in many of the coenzymes. Three such examples are particularly prominent. *Coenzyme A* (Fig. 2.15) is an adenine ribonucleotide, phosphorylated in the 3′ and 5′ positions, the β-5′ phosphate giving the pantothenyl-β-mercaptoethylamine ester which terminates in the sulphydryl –SH group. This can be esterified to a thiol ester, $-S-CO-CH_3$, which is more reactive than its oxygen counterpart, $-O-CO-CH_3$, allowing CoA/acetyl-CoA to function as an efficient acyl group carrier.

The nicotinamide dinucleotides were the first coenzymes to be recognized (Fig. 2.16), and are encountered in more than 200 enzyme systems. Two exist, *NAD* and

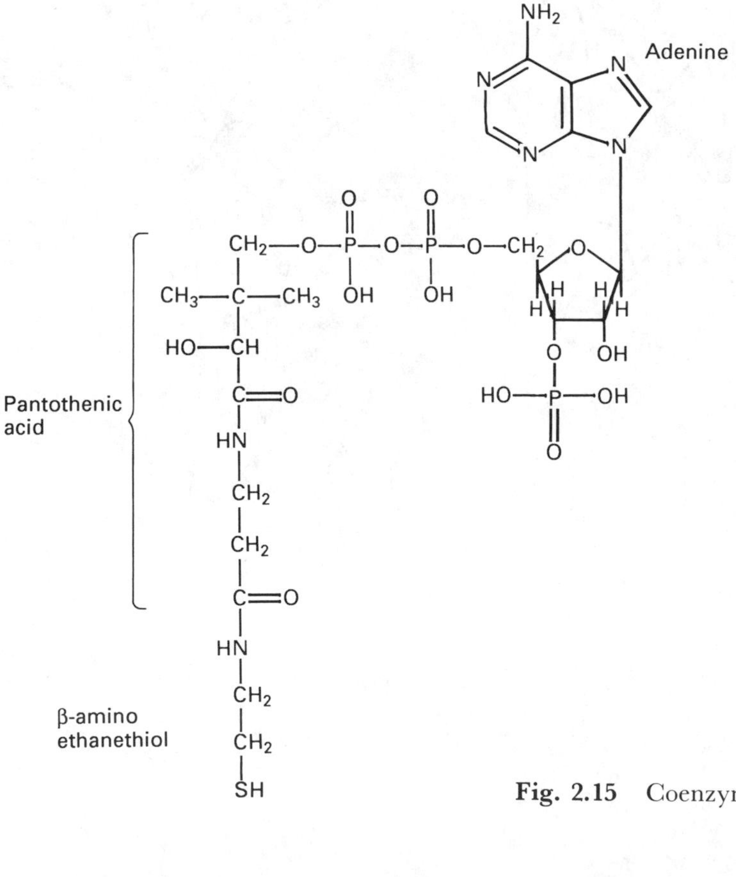

Fig. 2.15 Coenzyme A.

NADP where NADP is the 2′-phosphoric acid ester of NAD (nicotinamide adenine dinucleotide). In both cases the coenzymes function as redox systems, where the redox centre is positioned in the nicotinamide group. Formally NAD- and NADP-dependent reactions involve the transfer of hydride ion (H^-) and the equilibrium:

$$NAD^+ + H^- \rightleftharpoons NADH$$

and generally these nucleotides are reversibly dissociable from their apoenzymes, with dissociation constants between 10^{-4} and 10^{-7} M. In general the reduced form is more tightly bound by a factor of about 10^2 than the oxidized form.

In the coenzymes *flavin mononucleotide* (FMN) and *flavin adenine dinucleotide* (FAD) (see Fig. 8.10) the redox centre is an isoalloxazine derivative. The number of enzymes, known as flavoproteins, that utilize either FMN or FAD is large. Unlike CoA and NAD these cofactors form a more integral or permanent part of the enzyme complex itself. Rather than acting as a freely dissociating 'external' carrier therefore, these nucleotides can be considered as acting internally, as prosthetic groups.

The previous sections have identified molecular recognition complexes and discussed the role of surface interactions in achieving specificity. The enzymes, for example, are capable of binding the transition state between substrate and products and thus catalysing the reaction. The nature of this reaction can be identified by the coenzyme, or the prosthetic group, and so the coenzyme could be viewed as the 'engine' of the enzyme. In considering the analogy with the biosensor model, an enzyme protein gives a specificity for a chosen substrate, while transduction of the action of the coenzyme component provides a signal that monitors the substrate.

On the other hand, the binding event between antibody and antigen or single-stranded DNA and its complement is an equilibrium process, whose direct transduction must be concerned with changes associated with the complex formation, e.g. particle size, refractive index, etc.

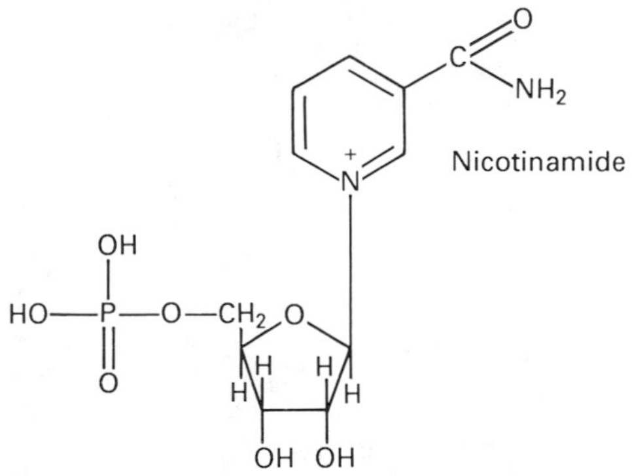

Fig. 2.16 Nicotinamide mononucleotide (NMN).

These then are some examples of the basic biorecognition units that will be involved in the bio–sensor union. Their successful immobilization at a transducer interface with retention of activity, and the efficient transduction of the analyte-dependent variable is a prerequisite to the development of a biosensor device.

References

Alberts, B., Bray, D., Lewis, J., Raff, M., Roberts, K. and Watson, J.D. (1983). *Molecular Biology of the Cell.* Garland, New York.
Boyer, P.D. (Ed.) (1975). *The Enzymes,* Vols I–XI. Academic Press, New York.
Edelman, G.M. (1970). *Sci. Am.* **223**(2), p. 34.
Frost and Sullivan Report No. 1479 (1986). 'DNA Probes in Medicine', No. 1479.
Kabat, E.A. (1976). *Structural Concepts in Immunology and Immunochemistry.* Holt, Rinehart and Winston, New York.
Nisonoff, A., Hopper, J.E. and Spring, S.B. (1975). *The Antibody Molecule.* Academic Press, New York.
Pauling, L. and Corey, R.B. (1951). *Proc. Natl. Acad. Sci. USA* **37**, p. 729.
Pauling, L., Corey, R.B. and Branson, H.R. (1951). *Proc. Natl. Acad. Sci. USA* **37**, p. 205.
Porter, R.R. (1973). *Science* **180**, p. 713.
Stryer, L. (1981). *Biochemistry.* Freeman, San Francisco.
Watson, J.D. and Crick, F.H.C. (1953). *Nature* **171**, p. 737.
Wolfenden, R. (1972). *Acc. Chem. Res.* **5**, p. 10.

Chapter 3

Ion-selective Potentiometric Measurement

Measurement of H^+

According to Brønsted and Lowry, acids may be defined as proton donors and bases as proton acceptors, such that:

$$\underset{\text{(acid)}}{HA} \rightleftharpoons \underset{\text{(proton)}}{H^+} + \underset{\text{(base)}}{A^-}$$

The dissociation of weak acids and bases are equilibrium processes, so that the equilibrium law may be applied to them:

$$\frac{[H^+][A^-]}{[HA]} = \text{dissociation constant, } K$$

For water,

$$H_2O \rightleftharpoons H^+ + OH^-$$

and

$$\frac{[H^+][OH^-]}{[H_2O]} = K$$

Since only a small fraction of water is dissociated, $[H_2O]$ may be regarded as being constant under all conditions and

$$[H^+][OH^-] = \text{constant} \times K = K_w$$

Conductivity measurements have shown that K_w, the ionic product of water, is

1×10^{-4} mol^2 · litre^{-2}. The acidity or alkalinity of a solution can be measured by its hydrogen ion concentration, but it is more convenient to refer to pH instead,

$$\mathrm{pH} = -\log[\mathrm{H}^+],$$

and it can be seen that

$$\mathrm{pH} + \mathrm{pOH} = 14$$

and neutralization may sometimes be described as the situation when the concentration of $[\mathrm{H}^+]$ and $[\mathrm{OH}^-]$ are equal, or pH is 7.

For the case of the dissociation of a weak acid,

$$\mathrm{pH} = \mathrm{p}K_\mathrm{a} + \log(c_{\mathrm{A}-}/c_{\mathrm{HA}})$$

where $c_{\mathrm{A}-}$ and c_{HA} are the concentrations and $\mathrm{p}K_\mathrm{a}$ the dissociation constant for the acid, i.e. the pH at which the acid is half dissociated.

From the Nernst equation, we can see that for a monovalent ion, the cell e.m.f. (E) is related to its standard e.m.f. (E^θ) by:

$$E = E^\theta + RT/F \ln(a_{\alpha/\beta})$$

where a is the relative activity of the ion in phases α and β.

If an interface, such as a membrane, between two electrolyte phases (α and β) is created such that the membrane is perfectly *ion selective*, and the concentration of the ion is held constant in one phase, then a potential difference will be established between the two phases—i.e. a *membrane potential*, which responds in Nernstian fashion to the concentration of ion in the second phase.

Ion-selective Interfaces

The glass pH electrode is just one example where an interface has been created between two electrolyte phases, across which only a single ionic species can pass—in this case H^+. A selectively permeable membrane could, in principle, be used to respond to any ionic species, the membrane potential reflecting the ion's activity in the sample phase in a Nernstian fashion.

The effectiveness of any such membrane system is largely governed by the degree to which the target species will dominate the charge transport in the membrane. For the glass electrode for example, a membrane potential develops due to the affinity of the silicate network for certain cations.

The performance of an ion-selective membrane electrode is a functional relation between electrode potential and ion activity of the ion for which the electrode is selective (Fig. 3.1). An ideal ion-selective electrode for the ion 'i', produces a potential E according to the Nernst equation:

$$E = E^\circ \pm RT/nF \ln a_\mathrm{i}$$

so that the sensitivity, expressed in mV decade^{-1} activity, is represented by the slope of the curve shown in Fig. 3.1, and deviations from this linear dependence

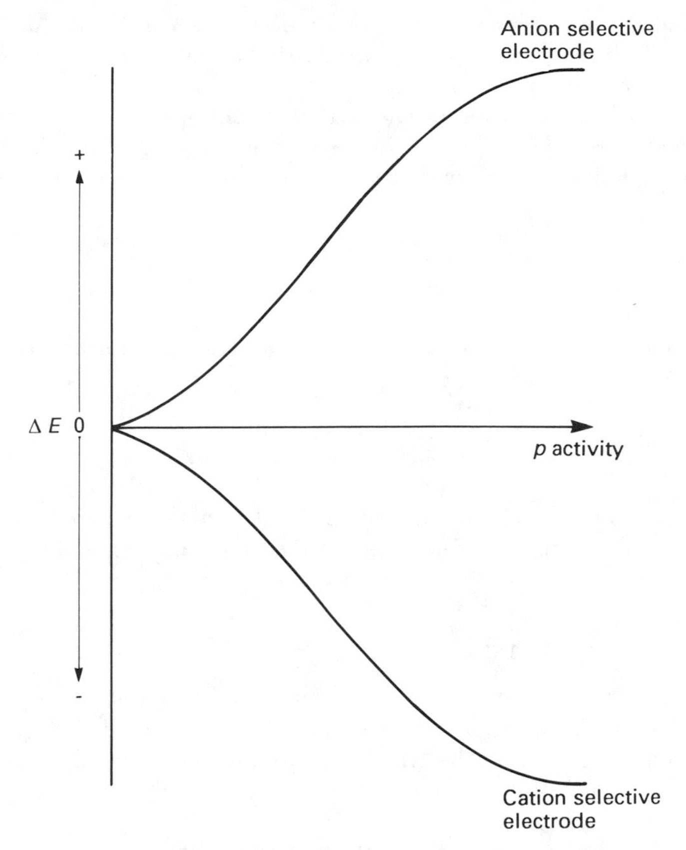

Fig. 3.1 Typical membrane electrode response.

occur at very low activities. The concentration where this alinearity begins is known as the limit of detection and is dependent on such factors as surface contamination, nature of the membrane, membrane defects, interference by other electrolytes and selectivity.

Considerable effort has been devoted to assessing and improving the selectivity of a given ion-selective electrode (ISE). The selectivity of a membrane for an ion 'a' in the presence of interfering ions 'b' is given by the general equation for electrode potential:

$$E = E^{\circ} \pm RT/\mathrm{a}F \ln (a_{\text{'a'}} + \Sigma k_{\mathrm{ab}} a_{\text{'b'}}{}^{\mathrm{a/b}})$$

where k_{ab} is the selectivity constant and a and b are charges on 'a' and 'b', respectively.

This equation is applicable to nearly all ISEs. When an electrode is very selective for ion 'a' in the presence of 'b', $k_{ab} \ll 1$. Similarly if the ISE shows a greater selectivity for an interfering ion 'b' than the target ion 'a', then $k_{ab} > 1$. Selectivity and thus k_{ab} is itself a function of the relative activities of the ions 'a' and 'b' and is not a constant. These variations in k_{ab} are associated with the mechanism of the electrode response, and with the changing environment for the ions with respect to one another.

It is obvious from these considerations that the performance of a particular ISE (or indeed any probe) in an analytical environment must be critically assessed. It is not only selectivity however that is of importance but also the following:

- selectivity
- slope and range of response
- stability and reproducibility
- response time
- sensitivity to ambient conditions (e.g. temperature, pressure, etc.)
- ease of maintenance
- ease of use
- availability
- cost and lifetime

The above points are all interdependent and will be related to the desired analytical environment and the nature of the ion-selective membrane.

CLASSIFICATION OF ISEs

Many different classifications of ISEs have been presented, but they will be considered here in terms of the type of membrane material. The classes may be identified as follows.

Glass electrodes These originate with the hydrogen-ion selective electrode, which particularly due to the high mobility of hydrogen ions, is very well behaved. Glasses with $k_{H/Na} \approx 10^{-14}$ and a wide response range (pH 0–14) are available. Subsequent electrodes have been developed for other cations, such as Na^+, K^+, NH_4^+.

Electrodes based on inorganic salts The common feature of these electrodes is that they are based on inorganic halides and sulphides, for example silver salts, lanthanum fluoride and heavy-metal sulphides. These membranes have been produced from preparations ranging from whole crystals to dispersions in an inert matrix, such as polythene or silicon rubber. They are targeted at ions such as halides, CN^-, S^{2-}, Ag^+, Cu^{2+}, and Pb^{2+}.

Electrodes based on organic ion exchangers and neutral carriers The electrodes based on neutral carriers generally have the highest selectivity in this class. However, cation exchangers or complexing agents or anion exchangers have been successfully employed in electrodes with liquid or solid membranes, selective to cations or anions, respectively.

Gas-sensing electrodes These electrodes represent an extension of ion-selective measurements to detection of gaseous analytes. Gas-sensing probes are complete electrochemical cells, incorporating both the ion-selective electrode and a reference electrode within the sensor. Assay of the target gaseous sample is not performed directly, but is related to a changing parameter (most usually pH) which can be monitored by an ion-selective electrode.

Each of these ISEs will now be discussed more fully.

Ion-selective Electrodes

GLASS ELECTRODES

The special ion selectivity demonstrated by *glass/electrolyte interfaces* (Dole, 1941; Garrels, 1967; Bates, 1973) has been recognized since the early part of this century. Glass electrodes which capitalized on this finding were developed and have been in use ever since. Measurements are made by immersing the electrode, so that the thin glass membrane is in full contact with the test solution, and recording the potential with respect to a reference electrode.

The cell construction becomes:

$$\mathrm{Hg/Hg_2Cl/KCl_{(sat)}/test\ solution/glass\ membrane/HCl\ (0.1\ M)/AgCl/Ag}$$

(internal reference)

reference electrode (SCE) | **glass electrode**

This is represented schematically for the glass electrode shown in Fig. 1.10b. In fact it can be seen from the cell construction above that the overall potential difference that will be measured is a sum of two junction potentials. The first, due to the liquid junction between the reference electrode and the test solution, should be small and constant, so that for a glass membrane sensitive to the species 'i', the cell potential is given by

$$E = \text{constant} + RT/nF \ln a_{\text{i solution}}$$

and that for a glass electrode sensitive to H^+ activity,

$$E = \text{constant} + 0.059\,\text{pH (volts)}$$

showing that the cell potential varies at the rate of 59 mV per decade change in hydrogen-ion activity.

For the titration of a weak acid with a weak base, the pH of the equivalence point will be given by:

$$\text{pH} = (\text{p}K_w/2) + (\text{p}K_a/2) - (\text{p}K_b/2)$$

so that unlike the strong acid–strong base system, the equivalence point is not at pH 7 but is influenced by the acidic and basic dissociation constants, K_a and K_b, respectively.

By comparison, acid–base indicator dyes are weak acids or bases which exhibit different colours depending on their ionic state,

$$\underset{\text{(acidic form)}}{HIn} \rightleftharpoons \underset{\text{(basic form)}}{In^-} + H^+$$

so that,

$$\begin{aligned} pH &= pK + \log(c_{In^-})/c_{HIn}) \\ &= pK + \log(\text{intensity colour } In^-)/(\text{intensity colour } HIn) \end{aligned}$$

and the ratio of the intensity of the colours of the two forms will give the pH, provided that K is known. In practice, the useful pH range of the indicator is limited to $pK_a \pm 1$. However, indicator dyes have been identified to cover the complete pH range, so that, in principle, the dye can be tuned to a particular analytical application (see p. 151).

Selectivity

In the glass electrode probe, the origin of the response is the exchange of ions in the layer at the surface of the glass. When the glass is immersed in an aqueous solution, a hydrated layer is formed at the surface whose thickness is dependent on the environment and the nature of the glass. Soda glasses for example have a hydrated layer $< 10^{-2}$ μm thick, whereas in a lithia-based glass, this layer can be as great as 10 μm.

At the interface between the glass and the sample, monovalent ions from the glass (Na^+ or Li^+) are exchanged in an equilibrium process with the determinand ion. The potential across this boundary varies according to the Nernst equation with the activity of the determinand. The electrode potential, however, can also be slightly influenced by factors largely independent of this activity. The flux of monovalent cations from the dry glass, behind this surface layer, causes a very slow drift in potential, for example, and an asymmetric potential error between the inner and outer surface of the glass may arise due to method of manufacture or dissimilar ageing processes at each interface.

In well-behaved commercial probes such errors have essentially been reduced to constants so that the variation is entirely due to the determinand ion and for a pH probe with optimized selectivity the performance may be described as already discussed, by the selectivity constant $k_{H/Na}$ in the form of the Nernst equation known as the *Nicolsky equation* (Nicolsky, 1937):

$$E = E^\circ + RT/F \ln(a_{H^+} + k_{H/Na} a_{Na^+})$$

In fact, as suggested earlier, $k_{H/Na}$ is not a constant, but Buck (1974) has shown generally that a plot of k_{ab} versus $\ln(a_a/a_b)$ is often linear and can provide useful information about the response.

Ion-selective glasses have also been developed for species other than H^+, of which the sodium-sensitive glass is the most successful. Nernstian responses in these electrodes extend down to 10^{-4}–10^{-5} M, and some probes respond to concentrations as low as 10^{-8} M, when used in flow systems where interference from H^+ has been eliminated by the use of a suitable buffer. Indeed interference by the H^+ ion is one of the major sources of error and its effect is to limit the pH range over which a Na^+-selective electrode may successfully be used. Conversely, the limit of detection for the sodium response will depend on the pH of the solution.

In comparison with $k_{Na/H}$ values of the order of 10^3, interference by K^+ ($k_{Na/K} < 3 \times 10^{-3}$) or NH_4^+ ($k_{Na/NH_4^+} < 10^{-6}$) is very small. All sodium sensitive glasses are however, very sensitive to silver ions ($k_{Na/Ag}$ 100–400!), this often representing the predominant response. This feature tends to direct the adoption of a calomel reference electrode in the sodium probe, rather than an Ag/AgCl reference, where leakage of Ag^+ across the liquid junction of the reference cell would cause serious interference.

Application of pH Measurement

pH measurement is arguably the most widely performed electrochemical assay. Applications are too numerous to consider in detail here, since its measurement is relevant in all situations where H^+ concentration might vary. In clinical environments for example, estimation of changes in the pH in body fluids provides a useful indication of the patient's state. The measurement is the total equilibrium situation which is the result of various individual equilibria involving protonation reactions. The transport of carbon dioxide as a waste product of respiration, for example, involves the dissolution of carbon dioxide according to:

$$CO_2 + H_2O \rightleftharpoons H_2CO_3 \rightleftharpoons H^+ + HCO_3^-$$

Amino acids, as we have already described are amphoteric compounds, containing an α-carboxyl group with a characteristic pK_a around 2 and an α-amino group

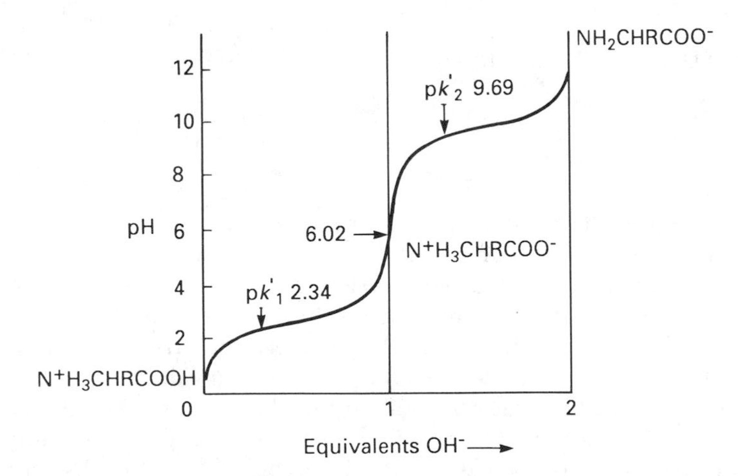

Fig. 3.2 Titration curve of alanine, with the pK_a values and the pH of zero net charge.

with pK_a about 10. Their acid–base behaviour can be formalized in terms of Brønsted–Lowry acids and follows the equations:

$$\underset{\text{(cation)}}{^{+}NH_3CHRCOOH} \underset{}{\overset{-H^+}{\rightleftharpoons}} \underset{\text{(zwitterion)}}{^{+}NH_3CHRCOO^-} \overset{-H^+}{\rightleftharpoons} \underset{\text{(anion)}}{NH_2CHRCOO^-}$$

The typical titration curve, therefore, is biphasic with two clearly defined steps, each with its own 'end point' corresponding to the two pK_a values (Fig. 3.2). Also evident in Fig. 3.2 is the isoelectric point, a point of inflection between the two phases, where there is no net charge on the molecule.

In the absence of an ionizable R group on the amino acid, the titration curves for the amino acids are all similar. The introduction of a side chain with additional ionizable groups further complicates the behaviour, as can be seen in Fig. 3.3 for lysine, histidine and glutamic acid. The last, for example, is described according to the dissociation sequence:

$$\begin{matrix} COOH \\ | \\ (CH_2)_2 \\ | \\ CHNH_3^+ \\ | \\ COOH \end{matrix} \underset{}{\overset{pK_a 2.1}{\rightleftharpoons}} \underset{\text{(zwitterion)}}{\begin{matrix} COOH \\ | \\ (CH_2)_2 \\ | \\ CHNH_3^+ \\ | \\ COO^- \end{matrix}} \overset{pK_a 4.1}{\rightleftharpoons} \begin{matrix} COO^- \\ | \\ (CH_2)_2 \\ | \\ CHNH_3^+ \\ | \\ COO^- \end{matrix} \overset{pK_a 9.5}{\rightleftharpoons} \begin{matrix} COO^- \\ | \\ (CH_2)_2 \\ | \\ CHNH_2 \\ | \\ COO^- \end{matrix}$$

while in lysine the 'basic' side group gives:

$$\begin{matrix} NH_3^+ \\ | \\ (CH_2)_4 \\ | \\ CHNH_3^+ \\ | \\ COOH \end{matrix} \overset{pK_a 2.2}{\rightleftharpoons} \begin{matrix} NH_3^+ \\ | \\ (CH_2)_4 \\ | \\ CHNH_3^+ \\ | \\ COO^- \end{matrix} \overset{pK_a 9.2}{\rightleftharpoons} \underset{\text{(zwitterion)}}{\begin{matrix} NH_3^+ \\ | \\ (CH_2)_4 \\ | \\ CHNH_2 \\ | \\ COO^- \end{matrix}} \overset{pK_a 10.8}{\rightleftharpoons} \begin{matrix} NH_2 \\ | \\ (CH_2)_4 \\ | \\ CHNH_2 \\ | \\ COO^- \end{matrix}$$

In each polypeptide there is still only one free terminal α-amino group and one α-carboxyl group. The titration curve is therefore primarily characteristic of the many side-chain R groups. As we have already discussed in Chapter 2, the major influence will come from those amino acid residues that are exposed on the surface of the polypeptide after it is folded in its three-dimensional structure; but since the pH will itself alter the non-covalent bonding between ionizable groups, this three-dimensional structure and thus the stability of the polypeptide is likely to be affected by the pH. The titration curve for a protein demonstrates its multivariate nature, with no distinct region easily identified as being due to the dissociation of any particular type of group. This polymer is best described only by its isoelectric point (pI),

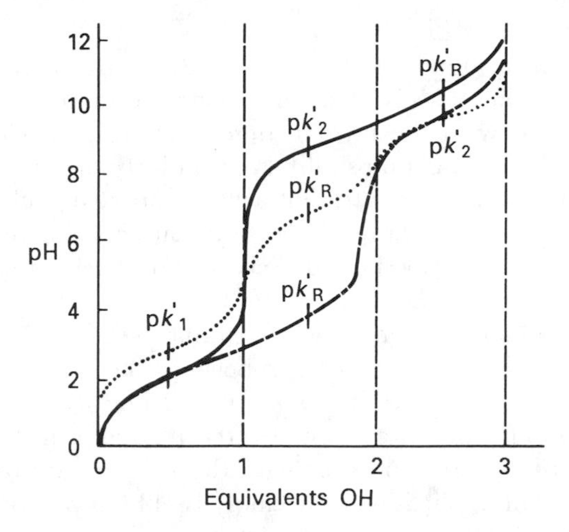

Fig. 3.3 Titration curves for glutamic acid (—•—•), lysine (——) and histidine (••••). In each case the pK_a of the R group is designated pK'_R.

$$\text{protein}^{+} \rightleftharpoons \text{protein}^{\pm} \rightleftharpoons \text{protein}^{-}$$
$$\xleftarrow{\text{acid}} \quad \text{p}I \quad \xrightarrow{\text{alkaline}}$$

whose value gives an indication of the net acidic or basic character of the side-chain residues.

These pH measurements, therefore, can provide only a general net estimate of protein acid/base behaviour. By reference to the requirement for a biosensor (see Chapter 1), where a unique changing and highly selective parameter must be identified as the transduction signal, they would have little analytical significance.

Enzyme-catalysed reactions can, however, include a change in pH; for example:

$$\text{penicillin} \xrightarrow{\textit{penicillinase}} \text{penicillolic acid}$$

or

$$\text{urea} \xrightarrow{\textit{urease}} NH_4^+ + HCO_3^-$$

so that in the presence of enzyme, the substrate (penicillin or urea) may be assayed by following the change in pH caused by the accumulation of products.

This then renders the measurement to be exclusive for a particular substrate, and thus satisfies the requirements of the model biosensor on specificity. This approach will be more fully investigated in Chapter 9.

Applications of Sodium Electrodes
Sodium electrodes have found a wide range of use at both macro and micro levels. One of the major applications is in the routine determination of sodium levels in biological fluids, and with the introduction of microelectrode devices, *in vivo* measurements are becoming a possibility, particularly in clinical environments.

Wide ranges of sodium concentration are measured in soil samples (Bower, 1959) and in natural and waste waters, the latter being monitored by an automatic method, in a development by Sekerka and Lechner (1974) for a range $0.1–100\ mg\ l^{-1}$.

More ingenious an application is the determination of salt in bacon (Halliday and Wood, 1966), where the sodium probe is pushed into macerated bacon samples, or placed in a hole bored into a bacon roll. The attraction of this method is the absence of sample preparation and extraction steps in the assay.

The low levels of detection possible with the sodium electrode mean that a particularly important application is the analysis of high-purity water. Very low levels of sodium must be detected in the monitoring of output from mixed-bed ion-exchange columns or from the water supply to boilers in, for example, power stations. Levels below 5×10^{-8} M can frequently be required here.

ELECTRODES BASED ON INORGANIC SALTS

These electrodes are based predominantly on silver salts, constructed in various forms, so that the membrane may be either homogeneous or heterogeneous. *Homogeneous membranes* consist entirely of the salt, either as a machined crystal or a pressed pellet of the appropriate shape. *Heterogeneous membranes*, on the other hand, consist of the active salt suspended in an inert matrix such as PVC, polythene or silicon rubber. Very often the performance of these two different electrode constructions are so similar that the method is immaterial to the response.

For a silver halide membrane contacted with elemental silver, the potential of the silver is developed as a result of the electron exchange between the silver metal and ions in the bulk of the metal and ion exchange between silver ions in the metal and silver ions at the silver solution interface. The solubility equilibrium between the halide ions and the silver ions at the electrode surface will be:

$$a_{Ag^+/int}\, a_{X^-/soln} = K_{(AgX)}$$

where $a_{Ag^+/int}$ is the Ag^+ activity at the Ag–solution interface, $a_{X^-/soln}$ is the X^- activity in the bulk solution and $K_{(AgX)}$ is the solubility product of AgX.

Assuming the activity of silver at the interface is unity, then the response to halide ions is given by:

$$E = E^{o}_{Ag^+/Ag} + RT/F \ln K_{(AgX)} - RT/F \ln a_{X^-/soln} - E_{ref}$$

and the analogous response of the electrode to Ag^+ ions is

$$E = E^{o}_{Ag^+/Ag} + RT/F \ln a_{Ag^+/int} - E_{ref}$$

This treatment is applicable to AgX membranes contacted with silver, so that an equilibrium between the elemental Ag and any excess halogen in the membrane is reached, resulting in unity activity of silver and halogen. For contacts made via materials other than silver, the Ag^+ activity of the membrane is fixed by the stoichiometry of the silver salt, and the standard potential will vary accordingly. Various theories have been considered in detail relating the standard potential of silver halide based electrodes to the stoichiometry of the active material (Marton and Pungor, 1971; Buck and Shephard, 1974).

The silver ion defect activity is particularly important when considering the detection limits of the membrane. A theory which appears to be generally supported by experimental data (Morf *et al.*, 1974), suggests that the response is affected by both the dissolution of the membrane and by the activity of the silver ion defects. If the latter amounts to z mmol g^{-1}, then for a membrane Ag_nX, at the membrane–solution interface:

$$a_{Ag^+/int} - a_{Ag^+/soln} - z = n(a_{X^{n-}/int} - a_{X^{n-}/soln})$$

Since

$$K_{(Ag_nX)} = a^n{}_{Ag^+/int} a_{X^{n-}}$$

then direct X^{n-} terms can be eliminated from this equation, assuming that $a_{X^{n-}/soln} = 0$:

$$a^{(n+1)}{}_{Ag^+/int} - a^n{}_{Ag^+/int}(a_{Ag^+/soln} + z) = nK_{(Ag_nX)}$$

Solving for $a_{Ag^+/int}$ when $n = 1$:

$$a_{Ag^+/int} = \{(a_{Ag^+/soln} + z) \pm \sqrt{[(a_{Ag^+/soln} + z)^2 - 4K_{(AgX)}]}\}/2$$

so that the potential of the electrode with respect to a given reference can be described by the two limiting cases:

where $z^2 \ll K_{(AgX)}$,

$$E = E_{Ag^+/Ag} + RT/F \ln\{a_{Ag^+/soln} + \sqrt{(a^2{}_{Ag^+/soln} - 4K_{(AgX)})}\}/2$$

and a Nernstian response is shown for $a^2{}_{Ag^+/soln} \gg 4K_{(AgX)}$, or at the other extreme,

where $z^2 \gg 4K_{(AgX)}$,

$$E = E_{Ag^+/Ag} + RT/F \ln(a_{Ag^+/soln} + z)$$

so that a Nernstian response is only shown when $a_{Ag^+/soln} \gg z$.

Similar calculations have been made for the response of Ag_nX membranes to various anions, based on the defect activity of the membrane (Morf *et al.*, 1974). Interference of the electrode response may also be treated in a similar way. Very few cations, for example, react with the silver compounds used for membrane preparations, although Hg^{2+} is an exception to this. A blocking film of mercuric sulphide forms on the surface of silver sulphide electrodes, while on the halide electrodes soluble mercuric halide complexes are formed, thus liberating Ag^+ ions, which will be sensed by the electrode:

$$Hg^{2+} + nAgI \rightleftharpoons HgI_n{}^{(2-n)} + nAg^+$$

The halide and sulphide electrodes can also suffer from interference by ligands or anions, forming a more insoluble silver salt:

$$AgX + mL^{n-} \rightleftharpoons AgL_m^{(1-mn)} + X^-$$
$$nAgX + Y^{n-} \rightleftharpoons Ag_nY + n^{X-}$$

so that the electrode senses the X^- ions generated at the membrane–solution interface. In these instances, the solubility product of the interfering complex that is formed must be added to the theoretical calculations, in addition to $K_{(AgX)}$. Within suitable limits however, a Nernstian response to these interferents is produced by the electrode, so that the electrodes may be successfully used to detect ions other than the determinands.

Where the interfering ion does not form a more soluble salt, the electrodes respond to sudden changes in the interfering ion activity with a transient 'overshoot' signal, before returning more slowly to an equilibrium level. Where $k_{ab} \ll 1$ and the interfering ion is in a large excess, Gratzl *et al.* (1985) have discussed this phenomenon in terms of desorption/adsorption and diffusion in the layer at the membrane interface. The potential overshoot is most prominent at low primary ion activities and high interfering ion activity, and becomes more limited at higher primary ion concentration, where the relative concentration change of interfering ion is less significant.

Applications of Electrodes Based on Inorganic Salts

The most widely applied of these ion-selective electrodes is the fluoride electrode. Of its many uses the most important is found in the analysis of water (natural, waste, potable, sea, etc.), where it is sufficiently stable to be employed in automatic monitoring systems in many instances (Erdmann, 1975).

The major application of chloride-ion selective electrodes is also in the analysis of waters, particularly in boiler water. A rather useful medical application, however, is in the screening for cystic fibrosis via the estimation of chloride in sweat. This measurement is sufficiently successful for Orion Research Inc. to have developed an electrode specifically for this purpose. Sweat is produced by electrically stimulating the sweat glands and the electrode is placed on the skin for a direct measurement of the chloride concentration (Kopito and Scwachman, 1969).

ELECTRODES BASED ON ORGANIC ION EXCHANGERS AND NEUTRAL CARRIERS

These electrodes have their roots in liquid ion-exchange electrodes, where the ion exchanger was in solution in an organic solvent. It has, however, generally proved more convenient to incorporate the ion exchanger in an inert matrix such as PVC, polythene or silicone rubber to produce a 'solid-state' membrane. In theory, ion exchangers could be developed for any ion. In practice many of the ion exchangers are unsatisfactory, with poor selectivities.

Of similar construction are the ion-selective electrodes involving neutral carriers as the active material. These neutral carriers are usually macrotetrolide

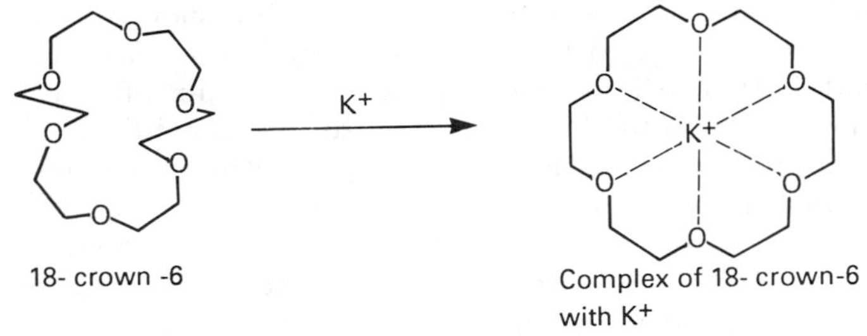

18- crown -6

K^+

Complex of 18- crown-6 with K^+

Fig. 3.4 Conformational rearrangement associated with complexation of 18-crown-6 with K^+.

antibiotics, depsipeptide antibiotics or more increasingly cyclic polyethers (crown compounds) (Izatt *et al.*, 1985). A common feature of these carriers is the central cavity which favours complex formation with the target ion. Neutral carriers have been tailored to give the correct cavity size for a particular determinand.

Ionophore Modelling

Initial ionophores based on crown ethers were derived from the 18-crown-6 ether (Fig. 3.4), targeted at the K^+ ion. Cram and Trueblood (1981) postulated that efficient binding of the target was influenced by (a) the ability to form stable complexes and (b) the attainment of optimal binding with minimal changes in the conformation of host, target and solvent. As may be seen in Fig. 3.4, the former criterion is satisfied by the 18-crown-6 ether, but with too much conformational reorganization to achieve really effective binding. Complexation has been improved by designing and synthesizing *spherands* and *hemispherands* (Fig. 3.5), the former often producing such highly stable complexes, that 'decomplexation' is too slow to be of application to ion-selective electrodes. Molecules with some degree of pre-organization such as the hemispherands prove to be more suitable for this analytical application, and the introduction of different substituents to adjust the 'locked' and fluxional features of the host macrocycle provides control over the kinetic performance of the ligand (Lockhart, 1986).

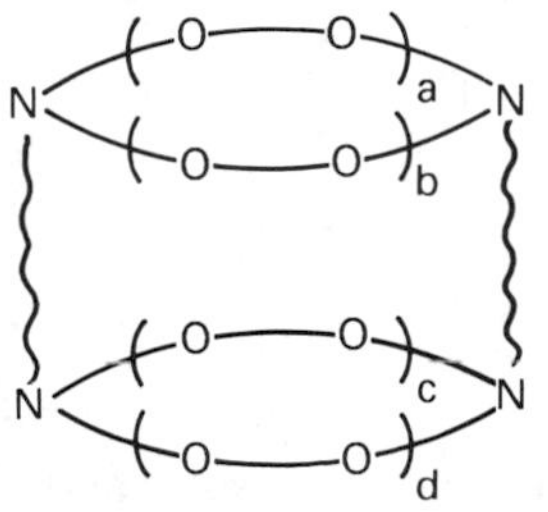

More three-dimensional host structures, achieved with bridged polycyclic receptors, give further steric or even chiral barriers to cation selectivity. Sutherland (1986) has analysed the role of tricyclic receptors of a general structure in their selectivity for the dication $NH_3^+(CH_2)_nNH_3^+$, and shown that selectivity is a function of host flexibility, with flexible macrocycles selecting a range rather than a unique dication. Selectivity is increased as separation of the two receptor sites more closely matches the dication characteristics, with distinction between a single CH_2 group achievable. The general problem of guest recognition can be extended to include the behaviour of natural biomolecules such as enzymes, antibodies, etc. The requirement by these proteins to achieve active site recognition of the substrate (see Fig. 2.4) is a more complex version of these ionophore hosts. The synthetic modelling of polycyclic receptor systems to replicate this unique protein specificity is being addressed largely with the aid of charge calculations and computer graphics.

Limits of Detection

One of the major factors effecting the lifetime of the electrode is the leaching of the ion exchanger or neutral carrier from the membrane. These ion selectors are very

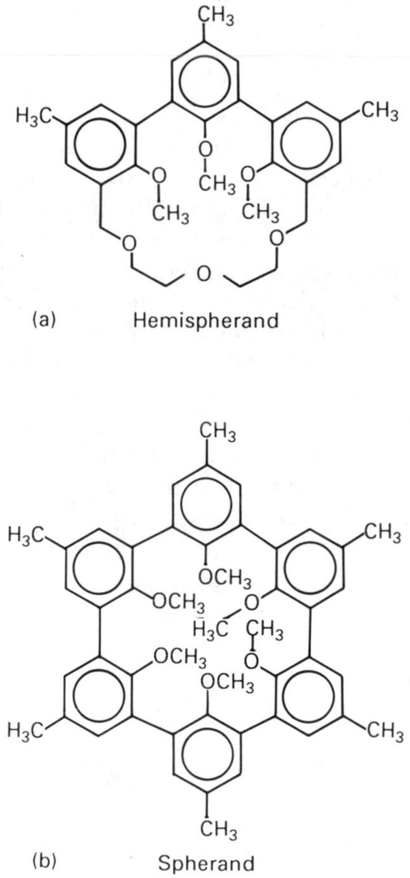

(a) Hemispherand

(b) Spherand

Fig. 3.5 Crown compounds: (a) a hemispherand and (b) a spherand.

slightly soluble in water and diffuse from the surface of the membrane into the solution, thus rendering the electrode useless.

Obviously, the detection limits of these electrodes will be set by the distribution coefficient for the determinand between the membrane and the sample and the equilibrium activity of the determinand in the membrane. The selectivity of the membrane will be a function of the relative stabilities of the complex formation of the membrane with the determinand and interferent. For determinand A^+ and interferent B^+ with neutral complexing ligand S:

$$B^+_{(soln)} + AS^+_{(memb)} \underset{}{\overset{k_{AB}}{\rightleftharpoons}} BS^+_{(memb)} + A^+_{(soln)}$$

and

$$k_{AB} = \frac{K_{BS^+} k_{BS^+}}{K_{AS^+} k_{AS^+}}$$

where K_{AS^+} and K_{BS^+} are the stability constants in solution and k_{AB^+} and k_{BS^+} are their respective partition coefficients.

Since the partition coefficient for monovalent cations is very similar for many antibiotic-based membranes, these systems show a close correlation between k_{AB} and K_{BS^+}/K_{AS^+}. This correlation is also shown by some crown compounds (Rechnitz and Eyal, 1972).

In general, selectivities can be achieved that are comparable with other ion-selective electrode systems. The major attraction of these types of membrane, however, is the potential ability to design a membrane selective to any cation and deposit it at many electrode geometries and sizes.

Applications

The greatest impact of the calcium electrode has been made in the analysis of serum and plasma. The physiologically active calcium is the ionized form and under normal conditions this forms only about 52% of the total calcium present (Fig. 3.6), its proportion to bound calcium being effected by the condition of the patient. For example, changes in carbon dioxide levels cause a change in pH upsetting the bound–unbound equilibrium, which will be reflected by a calcium-selective electrode, responding only to the unbound calcium.

Potassium ion-selective electrodes based on the antibiotic valinomycin or similar analogues have also been most successful in monitoring potassium uptake in body fluids.

The nitrate electrode is one of the most popular ion-exchange electrodes, particularly in the analysis of plant materials, soils, waters and effluents. The electrodes however suffer from interference primarily from chloride and bicarbonate, the former with a greater than 10-fold sensitivity over nitrate! Selectivity of the nitrate electrode is therefore not ideal and its adoption must be attributed, at least partly, to the absence of any more satisfactory method for nitrate.

A liquid membrane electrode based on a lipophilic derivative of vitamin B_{12} also shows a similar selectivity for Cl^- and NO_3^-. Sensitivity to NO_2^- is however $> 10^4$ over either of these ions, and is correlated with the reversible coordination

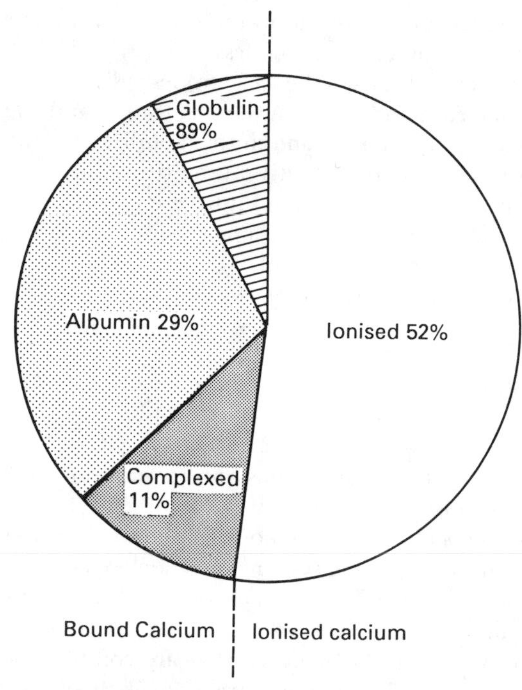

Fig. 3.6 Distribution of calcium plasma.

of the anion with the lipophilic Co(III) complex, SCN^- being the major interfering anion (Schulthess *et al.*, 1985). Under intensive use the lifetime of the electrode was estimated to exceed 6 months, and under test for NO_2^- in meat samples, a good correlation was obtained with existing methods.

GAS-SENSING ELECTRODES

The gas-sensing probes represent a development based on ion-selective electrodes and complementary to the ion-selective electrodes themselves. The probes are usually more selective than the systems already described, but they do not make the measurement directly, but estimate the partial pressure of a gas due to the resultant change in a parameter that can be estimated via an ion-selective device. For example, dissolution of carbon dioxide proceeds according to the equilibria:

$$CO_2 + H_2O \rightleftharpoons H_2CO_3 \rightleftharpoons H^+ + HCO_3^-$$

and

$$K = \frac{[H^+][HCO_3^-]}{[H_2CO_3]}$$

or

$$pH = pK + \log \frac{[HCO_3^-]}{[H_2CO_3]}$$

In practice, in equilibrium with gaseous carbon dioxide, the concentration $[CO_2]$ can be substituted for $[H_2CO_3]$ and

$$pH = pK' + \log \frac{[HCO_3^-]}{[CO_2]}$$

The concentration of dissolved carbon dioxide can therefore be related to pH and monitored as already described with the pH electrode.

The concept of measuring the partial pressure of a gas in this way was first set out by Stow *et al.* (1957). Improvements in the probe made by Severinghaus and Bradley (1958) gave rise to the *Severinghaus electrode* which in various forms is still commonly in use for the monitoring of blood carbon dioxide levels. The same principle can be applied to any acidic or basic gas. Indeed, the base ion-selective electrode is nearly always a pH electrode. In general the determinand gas diffuses from the sample into a thin electrolyte film on the surface of the ion-selective electrode until an equilibrium is reached between sample and electrolyte gas concentrations (Fig. 3.7). For carbon dioxide, ammonia, sulphur dioxide and nitrogen oxide probes, dissolution of the gas in the thin film causes a change in the H^+ concentration which is detected by the electrode.

As would be anticipated, the response time for these probes has been shown to be dependent on the thicknesses of the various phases and the membrane characteristics. This was determined by Ross *et al.* (1973): At $t=0$ the

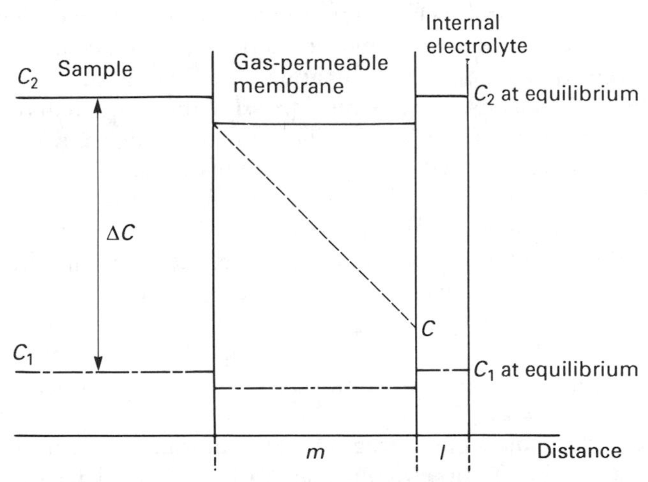

Fig. 3.7 Concentration profile for determinand gas in phases next to the electrode, following a step change in concentration.

concentration is changed from C_1 to C_2 and assuming that partition equilibrium at the membrane interface is very rapid, the concentration in the membrane interface, C_{memb} will rapidly change causing a concentration gradient across the membrane and thus a net flux of diffusing species across the membrane. For an electrode of area A, the flux will be (see below for abbreviations):

$$\text{Flux} = -AD\Delta C_{memb}/m$$

This flux results in a change in the total concentration of the diffusing species in the internal electrolyte. This species can exist in neutral (C) or ionic or complexed (C_B) forms, so that,

$$Al\left(\frac{dC_T}{dt}\right) = \text{Flux} \qquad C_T = C + C_B$$

giving

$$\frac{dC_T}{dt} = \frac{d(C+C_B)}{dt} = (1 + dC_B/dC)\frac{dC}{dt} = -(Dk/lm)(C_2 - C)$$

so that,

$$(1 + dC_B/dC)dC = -(Dk/lm)(C_2 - C)dt$$

Considering t for the fractional approach to equilibrium ε, Ross *et al.* (1973) have determined the response time by integration of this equation, where $C_2 - C_1$ is sufficiently small for dC_B/dC to be considered a constant or $dC_B/dC \ll 1$:

$$t = lm/Dk(1 + dC_B/dC) \ln \Delta C/C_2\varepsilon$$

where l = thickness of thin film, m = thickness of membrane, k = partition coefficient between thin film and membrane, D = diffusion coefficient in membrane, C = concentration of neutral species in internal electrolyte, C_B = concentration of ionic and complexed forms in internal electrolyte, C_T = total concentration of all forms, $\Delta C = C_2 - C_1$, ΔC_{memb} = concentration gradient across membrane and $\varepsilon = |(C_2 - C)/C_2|$. It can be seen from this expression that a different response time is predicted for step changes in concentration from low to high and for high to low. For $C_2 > C_1$ the response time will be independent of the size of the change and will be shorter than for the equivalent change in concentration where $C_2 < C_1$. These predictions have been substantiated in practice.

In general the potential response for a gas-sensing electrode utilizing a pH probe can be described by a Nernst-type equation of the form:

$$E = E^{o}_{gas} \pm k/n \ln C_{gas}$$

where E^{o}_{gas} contains all the constant terms and (+) is valid for acidic gases and (−) is valid for basic gases.

Membrane performance is most frequently the limiting factor in the effectiveness of these gas-sensing probes. Membrane coatings or breakage can terminate the life of an electrode, although the base pH probe may remain fully functional. Indeed the role of the membrane is particularly paramount to the selectivity of the probe. This gas-permeable layer provides a barrier to other sample species that might otherwise cause a response by the underlying electrode. Even if the ion-selective electrode is a pH probe, the possibility of interference by hydrogen ions may be safely ignored unless they cause a change in the activity of the species that is to be detected. A gas-sensing probe can, therefore, only suffer direct interference from dissolved gaseous species that can also diffuse across the membrane, and will cause a pH response of comparable magnitude to that of the determinand.

These compound electrode systems represent a concept in probe design akin to that of the biosensor. A general base-sensing element is employed with a wide range response to sample species from multiple origins. Further elements are added to the base element, whose characteristics induce a specificity to the transduced signal. These additional elements act as a selective barrier to all but the target species. In the ion-selective electrodes this barrier is the ion-selective membrane, while in the case of these gas-sensing probes, the construction elements are extended to include a gas-permeable membrane, to give the final target analyte selection. Future chapters will discuss both synthetic and naturally occurring target recognition barriers with various physical, chemical and biological properties.

References

Bates, R.G. (1973). *Determination of pH*, 2nd edn. Interscience. New York, Wiley.
Bower, C.A. (1959). *Soil Sci. Soc. Am. Proc.* **23**, p. 29.
Buck, R.P. (1974). *Anal. Chim. Acta* **73**, p. 321.
Buck, R.P. and Shephard, V.R. (1974). *Anal. Chem.* **46**, p. 2097.
Cram, D.J. and Trueblood, K.N. (1981). *Top. Curr. Chem.* **98**, p. 43.
Dole, M. (1941). *The Glass Electrode*, New York, Wiley.
Erdmann, D.E. (1975). *Env. Sci. Technol.* **9**, p. 252.
Garrels, R.M. (1967). *Glass Electrodes for Hydrogen and Other Cations*, Ed. Eisenman, G. New York, Marcel Dekker.
Gratzl, M., Linder, E. and Pungor, E. (1985). *Anal. Chem.* **57**, p. 1506.
Halliday, J.H. and Wood, F.W. (1966). *Analyst* **91**, p. 802.
Izatt, R.M., Bradshaw, J.S., Nielsen, S.A., Lamb, J.D. and Christensen, J.J. (1985). *Chem. Rev.* **85**, p. 271.
Kopito, L. and Scwachman, H. (1969). *Pediatrics* **43**, p. 794.
Lockhart, J.C. (1986). *J. Chem. Soc.* **82**, pp. 1161–1167.
Marton, A. and Pungor, E. (1971). *Anal. Chim. Acta* **54**, p. 209.
Morf, W.E., Kahr, G. and Simon, W. (1974). *Anal. Chem.* **46**, p. 1538.
Nicolsky, B.P. (1937). *Acta Physicochim. URSS* **7**, p. 597.
Rechnitz, G.A. and Eyal, E. (1972). *Anal. Chem.* **44**, p. 370.
Ross, J.W., Riseman, J.H. and Krueger, J.A. (1973). *Pure Appl. Chem.* **36**, p. 473.
Schulthess, P., Arumann, D., Kräutler, B., Carderas, C., Stepanek, R. and Simon, W. (1985). *Anal. Chem.* **57**, p. 1397.
Sekerka, I. and Lechner, J.F. (1974). *Anal. Lett.* **7**, p. 463.
Severinghaus, W. and Bradley, A.F. (1958). *J. Appl. Physiol.* **13**, p. 515.
Stow, R.W., Baer, R.F. and Randall, B.F. (1957). *Arch. Phys. Med. Rehabil.* **38**, p. 646.
Sutherland, I.O. (1986). *J. Chem. Soc.* **82**, pp. 1145–1159.

Chapter 4

Semiconductor Electrodes

Semiconductor Electrodes: Introduction

The electronic properties of solids are usually described by the molecular orbital theory in terms of the 'band model', where the 'molecular' orbitals are so closely spaced that they are essentially continuous bands. The *bonding* energy levels are filled and form the *valence band*, while the unfilled *anti-bonding* energy levels form the *conduction band*. Separation of these two bands, the *forbidden region*, gives a band gap of E_g (Fig. 4.1).

In a metal lattice the bonding and anti-bonding molecular orbitals overlap (Fig. 4.2) so that the conduction band is also partially filled and the loosely bound electrons form a mobile 'electron gas', where an electron can move between filled and vacant electronic energy levels at virtually the same energy. This property imparts high conductivity to the solid.

For larger values of E_g the bands do not overlap and the valence band is almost filled and the conduction band almost empty. Promotion of an electron into the conduction band can occur here by thermal excitation, leaving a hole behind in the valence band (Fig. 4.3). The number of conductance band electrons (n_i) is approximated by the expression:

$$n_i = p_i \approx 2.5 \times 10^{19} \exp(-E_g/2kT)\ \mathrm{cm}^{-3}\ (25°\mathrm{C})$$

where p_i = number of valence band holes. Materials which will produce such holes and electrons by thermal excitation at room temperature are called *intrinsic semiconductors*. Silicon for example, with $E_g = 1.1$ eV has $n_i \approx 1.4 \times 10^{10}$ cm^{-3}. However, for $E_g > 1.5$ eV there are so few such carriers produced that the solids are insulators.

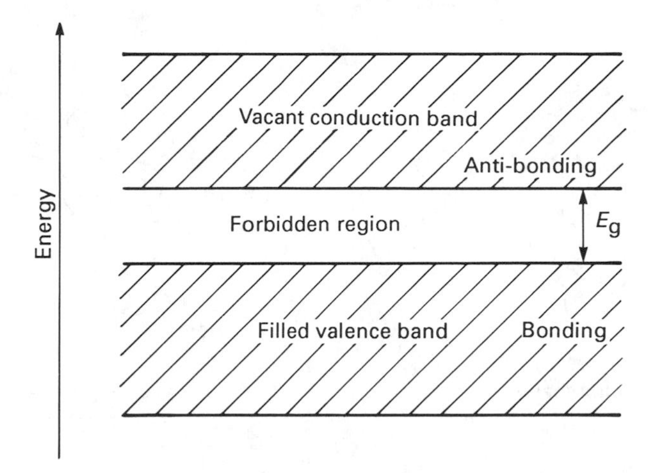

Fig. 4.1 Semiconductor band gap energy model.

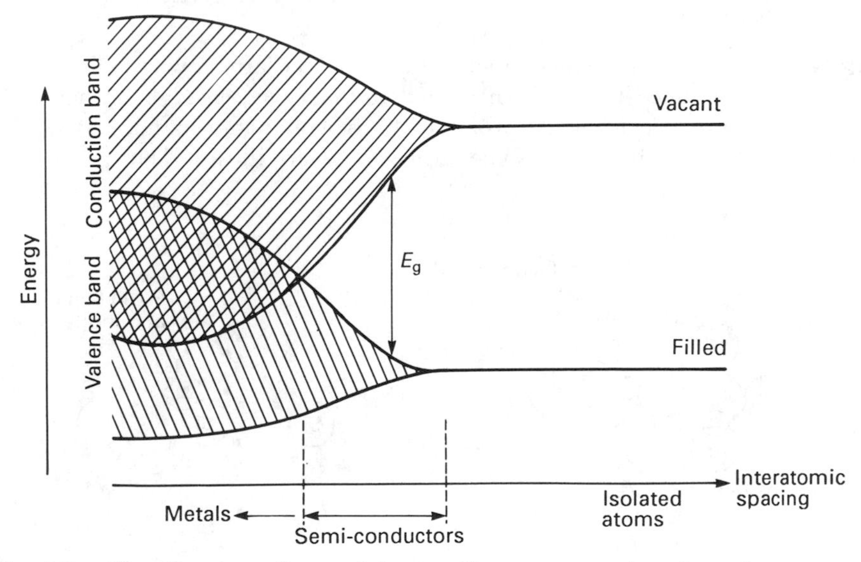

Fig. 4.2 Classification of material according to energy bands and interatomic spacing.

n-*Type Semiconductors*

Electrons can be artificially introduced in the conduction band, or holes created in the valence band by the use of small amounts of dopants. Arsenic behaves as an electron donor when added to silicon, since it introduces a further energy level E_D, very close to the bottom of the conduction band of the pure silicon (Fig. 4.4a). At room temperature most of the arsenic will be ionized, and the free electron will be

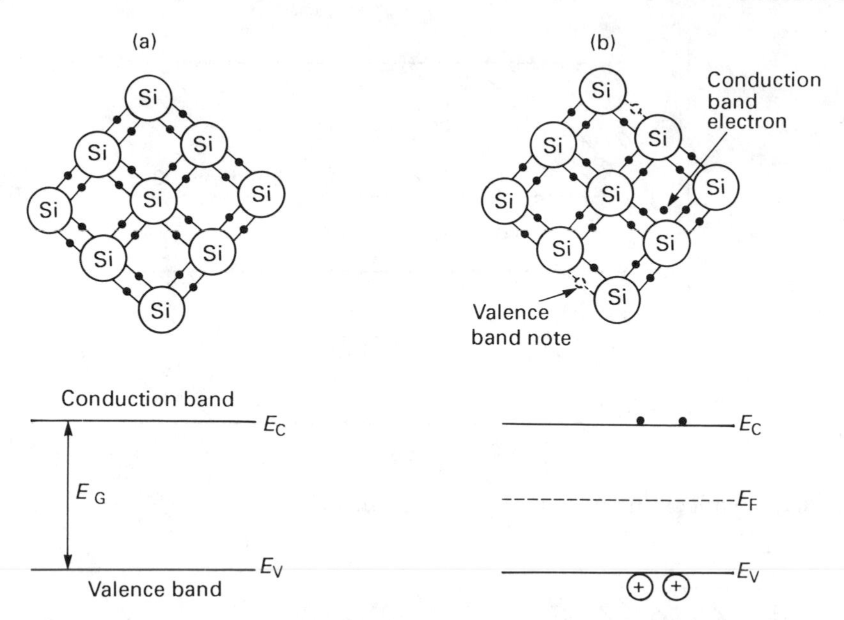

Fig. 4.3 The intrinsic semiconductor. (a) A perfect lattice, where all electrons are confined to the valence band and no holes exist; (b) at higher temperature some electrons are promoted to the conduction band, leaving holes in the balance band and some lattice bonds are thus broken.

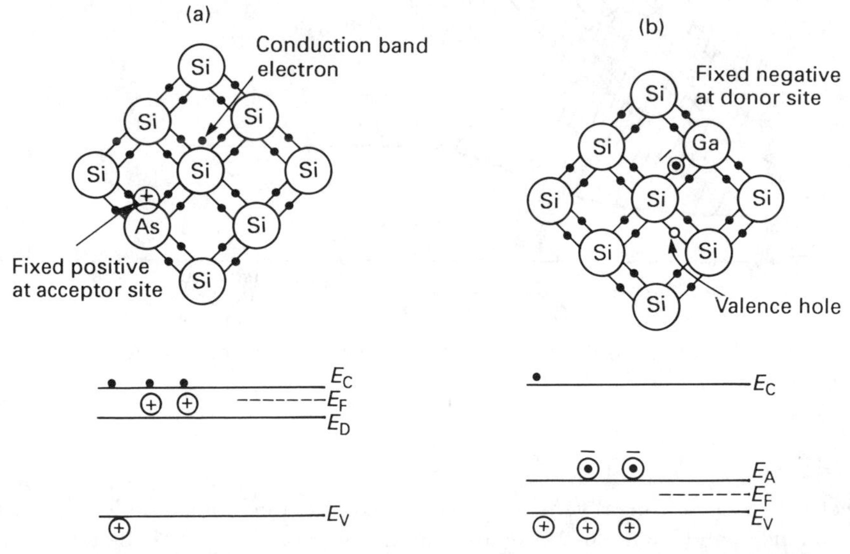

Fig. 4.4 Energy bands and representation of the extrinsic semiconductor lattice for (a) doped *n*-type semiconductor and (b) doped *p*-type ~~se~~miconductor.

introduced into the closely lying conduction band. In this instance, therefore, the electron density (n) in the conduction band does not reflect the hole density (p) in the valence band, since the remaining As^+ are fixed in the energy level E_D and not E_V. Conductivity in this material can therefore be attributed mainly to the E_C electrons, and a material such as this doped with donor atoms is called an *n-type semiconductor*.

p-*Type Semiconductors*

Conversely, when gallium is the dopant for silicon an energy level E_A is created close to the top of the valence band (Fig. 4.4b), thus allowing electrons to be thermally excited from the valence band into this new lower energy level E_A, leaving holes in the valence band. Conductivity in this instance can be attributed to the hole density and so such a material is termed *p-type semiconductor*.

Fermi Levels

The energy level where the probability of filling is 0.5 is called the *Fermi level*, E_F. For an intrinsic semiconductor this level will be mid-way between the conductance and valence bands, but for doped materials E_F depends on the doping level. For p-doped materials, E_F will lie nearer the valence band, while for n-doped materials it will be nearer the conduction band.

MIS Structures

When materials with different conduction properties are placed next to one another, the energy band diagram must adjust to accommodate each interface characteristics. For example, consider the structure: metal–insulator–semiconductor (MIS). Assuming ideal behaviour (Fig. 4.5a), the work function for the electrons in the metal is equivalent to the semiconductor. If no voltage is applied across the structure the system is in equilibrium and the Fermi levels across the metal and semiconductor are equal.

When a voltage is applied between metal and semiconductor (with respect to the semiconductor at 0 V), the Fermi levels of the two materials are separated by an amount equal to the potential difference, and the equilibrium situation is disturbed, creating a capacitor-like system, with the metal–insulator and semiconductor–insulator interfaces becoming charged.

With p-type silicon as semiconductor and a negative applied potential, the field that is created will cause the positively charged holes to migrate to the semiconductor–insulator interface and similarly an accumulation of electrons at the metal–insulator interface. The excess charge in the semiconductor does not reside at the surface, as it would do in a metal, but is distributed back away from the interface in a *space charge region*. This is analogous to the diffuse double layer that forms in solution at an electrode–solution interface.

The increase in the hole concentration at the semiconductor surface would imply that the Fermi level must move nearer the valence band at the interface. However, if the system is in thermal equilibrium, the Fermi level must remain flat,

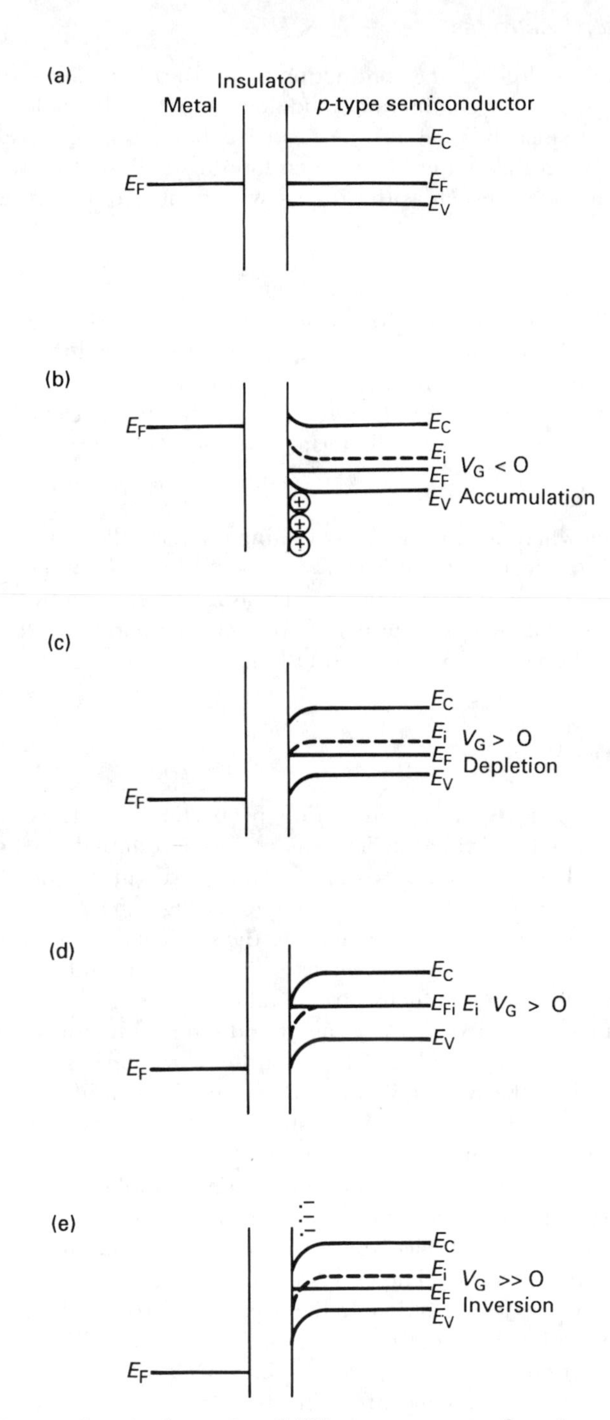

Fig. 4.5 Energy bands through a MIS structure as a function of applied voltage V_G.

so that the same effect is achieved by the bending of the valence band towards the Fermi level together with an analogous shift in the conduction band (Fig. 4.5b). Conversely, if a positive potential is applied between metal and semiconductor, then the positive holes are repelled from the semiconductor–insulator interface, there is a depletion in the interfacial region and the valence and conduction bands bend accordingly (Fig. 4.5c).

Increasing the positive potential that is applied eventually leads to the situation where, even for *p*-type silicon, the hole and electron concentrations near the interface are equal. At this point the Fermi level will once more lie mid-way between valence and conduction bands (i.e. equivalent to the intrinsic level E_i) (Fig. 4.5d). Further increases in potential will take the interface region into a situation with an excess electron concentration, thus causing *inversion* to *n*-type (Fig. 4.5e).

In reality, the work functions for the electrons in metal and semiconductor are not equivalent and so a net voltage must be applied in order to simulate the initial equilibrium condition. Other deviations from ideal behaviour in the three phases are all collectively accommodated by the voltage that must be applied in order to achieve a flat band condition. This voltage is thus termed the 'flat-band' voltage, V_{FB}.

Semiconductor–Solution Interface

For a semiconductor in contact with a solution containing the radox couple O/R, the Fermi level in solution is calculated in terms of the redox potential E^o (Fig. 4.6). Consider an *n*-type semiconductor interface with the O/R redox couple in solution. If the E_F of the semiconductor lies above that of the solution there will be a net flow of electrons across the semiconductor–solution interface into the solution. By analogy with the MIS model, the excess positive charge at the semiconductor surface is distributed in a space charge region, causing the conduction band to curve away from the Fermi level. The electric field in this region is analogous to the diffuse double layer, so that any excess holes created here would move to the interface and any excess electrons into the bulk semiconductor. If this interface region is irradiated with light of energy greater than the band gap E_g, then the absorption of photons causes electron-hole pairs to be separated, the electrons moving into the bulk semiconductor and to an external circuit, and the holes migrating to the interface, with an effective potential equivalent to the valence band. These holes will thus cause the oxidation of R to O (Fig. 4.6c). Similarly, an interface between a *p*-type semiconductor and the redox couple O/R in solution (Fig. 4.7) can be photo-activated to cause the reduction of O to R. However, since a hole accumulation layer (a depletion layer) would form at a *p*-type/redox couple interface for redox couples positive of E_{FB}, these photoeffects are confined to more negative potentials. This also applies to *n*-type semiconductors, where photo-assisted electrode reactions must be confined to redox couples more positive than E_{FB}.

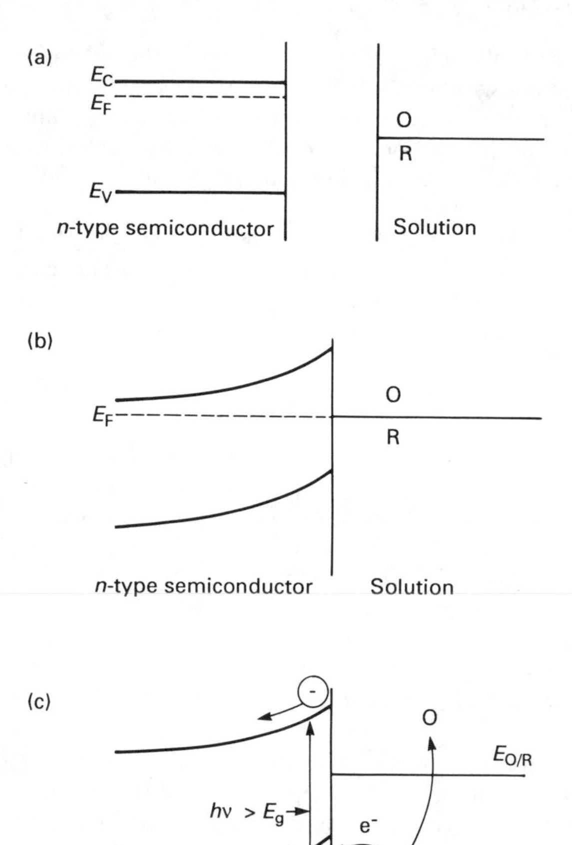

Fig. 4.6 Formation of a junction between an *n*-type semiconductor and a solution containing a redox couple O/R. (a) Before contact; (b) at equilibrium in the dark; (c) after irradiation, where $h\nu > E_g$.

Field Effect Transistor (FET)

A field effect transistor is arranged so that changes in the MIS interfaces already described can be monitored and controlled. For example, inversion at the *p*-type semiconductor–insulator interface could be monitored by two *n*-type 'sensors' placed at either side of the *p*-type layer (Fig. 4.8; IGFET, insulated gate FET). Normally there would be no connection between these two 'sensors', but under inversion conditions for the *p*-type layer, the *n*-type inversion channel at the interface would give a connection channel between the two *n*-type sensor areas. In constructing a FET device, these two *n*-type sensor regions are known as the source

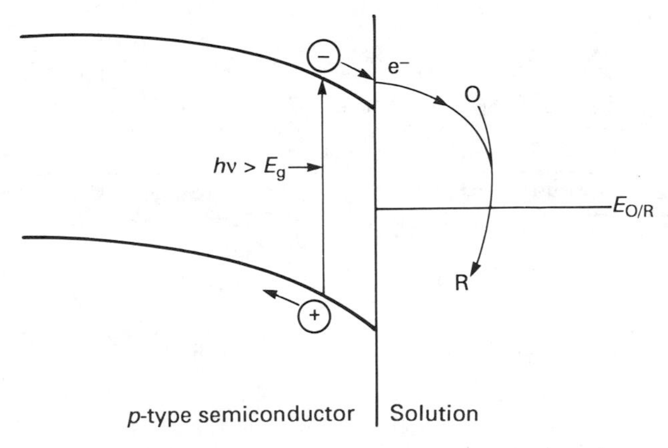

Fig. 4.7 Formation of a junction between a *p*-type semiconductor and a solution containing a redox couple O/R after irradiation, where $h\nu > E_g$.

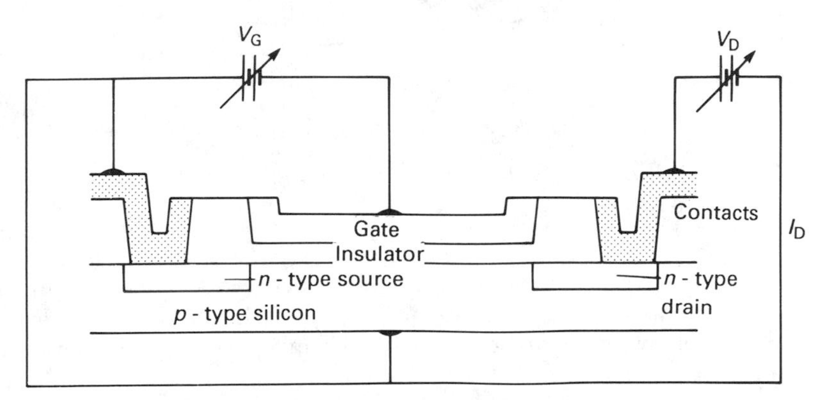

Fig. 4.8 Diagrammatic representation of an IGFET.

and drain. The drain is maintained at a voltage V_D, with respect to the source at 0 V, but no current flows between source and drain until the gate voltage V_G, between metal and semiconductor, reaches the threshold voltage V_T, where inversion is induced at the semiconductor–insulator interface. Then, the current I_D, which flows between source and drain will be determined by the electrical resistance of the surface inversion layer and the magnitude of V_D.

Current–voltage relationships describe the density of mobile electrons in the interface inversion layer as a function of V_G and V_D, and the position 'y', along the channel (Fig. 4.9), since V_y must change from 0 V at the source to V_D at the drain. Above the threshold voltage V_T, V_D modulates the number of electrons in the inversion layer, thus altering the effective conductance. Three limiting models of the inversion channel can be considered:

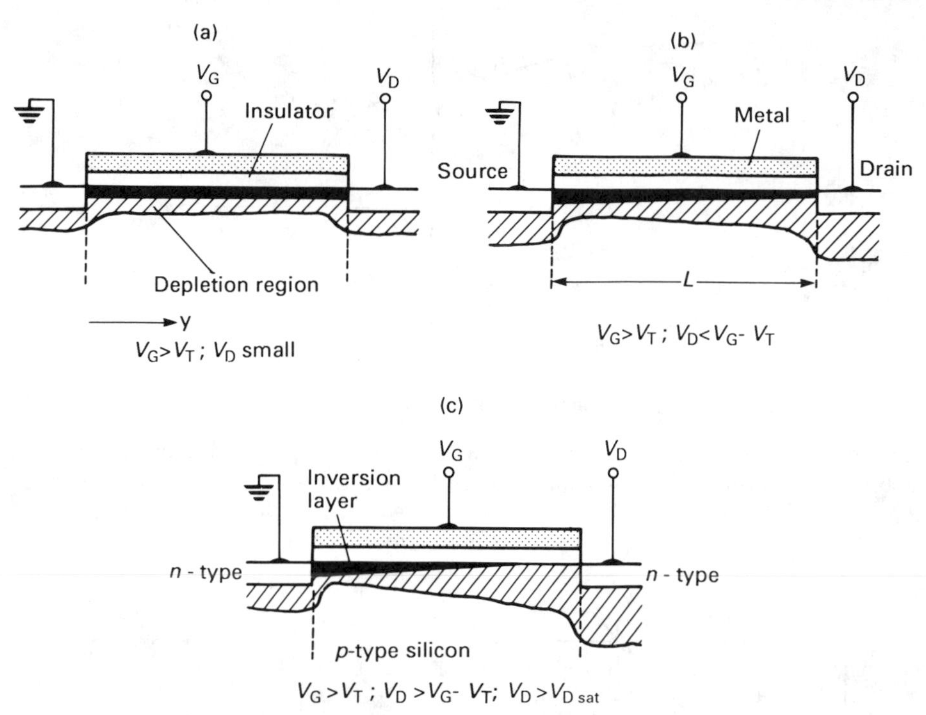

Fig. 4.9 The IGFET channel as a function of V_D for conditions of $V_G > V_T$.

(1) V_D is small so that V_y does not vary greatly between source and drain (Fig. 4.9a). In this instance the charge in the channel due to the mobile electrons Q_n, can be considered to be approximately independent of position and for a channel of length L and width W:

$$Q_n = -C_o(V_G - V_T)WL$$

where C_o is the capacitance of the insulator. For a given charge Q_n, the current flowing through this inversion layer will be related to the path length L, the electron mobility μ_n, and the field to which the electrons are exposed V_D/L.

$$I_D = \frac{-Q_n(\mu_n V_D)}{L \quad L}$$

Substituting for Q_n,

$$I_D = \frac{C_o W \mu_n}{L}(V_G - V_T)V_D$$

(2) If V_D is not negligible compared with V_G, then V_y alters between source and drain (Fig. 4.9b). The effective channel bias near the drain will be much

smaller than near the source, where $V_y \rightarrow 0$. This will cause the electron density to decrease from source to drain, but due to V_D at the drain the *depletion layer* will increase in this region. This model can be approximated by an average value for V_y of $V_D/2$, giving for $V_D < V_{Dsat}$:

$$I_D = \frac{C_o W \mu_n}{L} [V_G - (V_D/2) - V_T] V_D$$

(3) When $V_D > V_G - V_T$ the electron density near the drain will be reduced to zero and a depletion will exist at the semiconductor–insulator interface (Fig. 4.9c). In this model the inversion layer does not extend across the whole length L, but the high electric field of the depletion zone accelerates electrons between the end of the channel and the drain, at a velocity which is independent of the drain voltage. This is in the saturated region of operation, i.e. $V_D \geqslant V_{Dsat}$, and it follows that

$$V_{Dsat} = (V_G - V_T)$$

I_D therefore becomes

$$I_D = \frac{C_o W \mu_n}{L} \frac{(V_G - V_T)^2}{2}$$

Figure 4.10 shows the saturated and unsaturated regions for V_D, as a function of the applied gate voltage V_G, and although accurate quantitative measurements cannot be achieved with the above simplified treatment, the interdependence of V_G, V_T and V_D is demonstrated.

This introduction to semiconductor structures highlights analogies with double-layer construction in electrolyte solutions, and indicates their potential suitability as electrode materials.

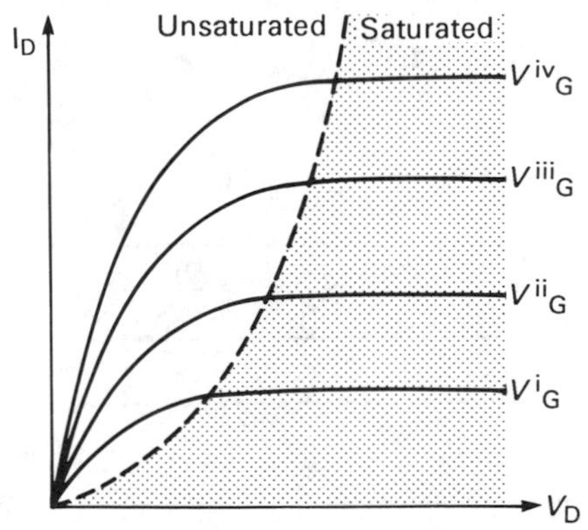

Fig. 4.10 V_D as a function of applied gate voltage V_G, showing the saturated and unsaturated regions for V_D. $V_G^I < V_G^{II} < V_G^{III} < V_G^{IV}$.

Chemically Sensitive Field Effect Transistors (CHEMFETs)

Transformation of the IGFET already described into a sensor element requires the identification of a component in the FET that can act as a sensing surface. An obvious choice is the metal unit in a MIS sandwich. Selective adsorption of gaseous species at a catalytic metal surface is observed for many transition-metal surfaces. If the catalytic metal is sufficiently thin and forms the gate of the IGFET, then gas molecules dissociated on the metal gas surface will diffuse through the metal layer to the metal–insulator interface, where they will be polarized, forming a dipole layer and causing a voltage drop across the interface (Fig. 4.11), which adds to the applied gate voltage V_G. For Langmuir adsorption the voltage shift ΔV will be given by

$$\Delta V = \frac{\Delta V_{max} k p^{\frac{1}{2}}}{(1 + k p^{\frac{1}{2}})}$$

where p is gas pressure, ΔV_{max} is maximum observed shift and k is a constant for a given atmosphere.

Alternatively, if the metal gate is replaced with a chemically sensitive layer, whose charge distribution properties respond to a series in the sample medium, then a potential V_G, is established across the insulator with respect to a reference electrode placed in the solution (Fig. 4.12), so that equations derived for an IGFET must be corrected for the reference electrode interface potential, E_{ref}, and the membrane–solution interface potential, $\phi_{mem-sol}$. For example, for a chemically sensitive membrane that responds to the ion 'i',

$$\phi_{mem-sol} = \frac{1}{z^i F}(\mu^i_{sol} - \mu^i_{mem})$$

where μ^i is the chemical potential in a given phase and z^i is the number of elementary charges. From the Nernst equation this can be related to the solution ion activity a^i:

$$\frac{1}{z^i F}(\mu^i_{sol} - \mu^i_{mem}) = E^o + \frac{RT}{z^i F} \ln a^i$$

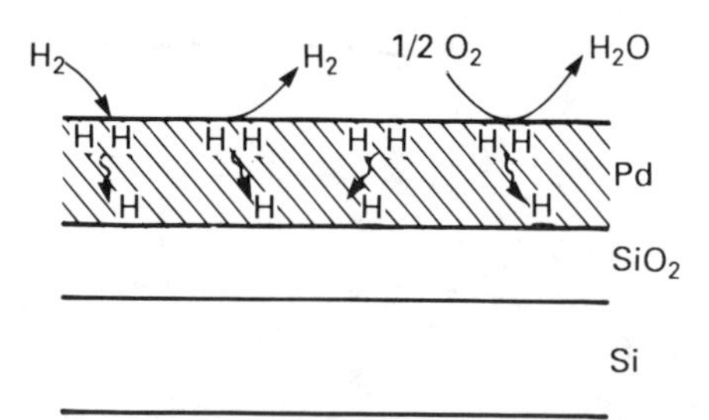

Fig. 4.11 Migration of H atoms after the catalytic dissociation of H_2 at the surface of a metal gate and their surface-catalysed reactions.

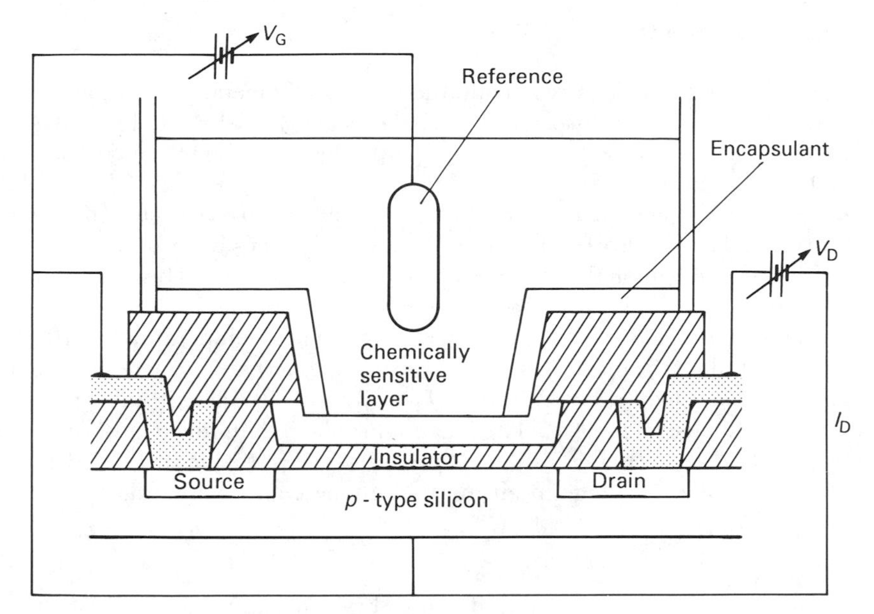

Fig. 4.12 Diagrammatic representation of the construction of an ISFET.

so that the ISFET (ion-selective FET) response to a change in the activity of the ion 'i', would be for $V_D < V_{D\,sat}$:

$$I_D = \frac{C_o W \mu_D}{L} (V_G - V_{T*} - E_{ref} - E^o + RT/z^i F \ln a^i - V_D/2) V_D$$

V_{T*} is used here rather than V_T, which accounted for work function differences between metal and semiconductor; in this case,

$$\Phi_{ms} = \Phi_{m-sol} + \Phi_{sol-mem} + \Phi_{mem-s}$$

but the interfaces 'm–sol' and 'sol–mem' are already accounted for by E_{ref} and $\phi_{sol-mem}$, respectively, so V_{T*} normally refers to the 'mem–s' (membrane–semiconductor) difference, although the term E^o is often included in its value, so that the above relationship for I_D reduces to:

$$I_D = \frac{C_o W \mu_n}{L} (V_G - V_{T*} - E_{ref} \pm RT/z^i F \ln a^i - V_D/2) V_D$$

and for $V_D > V_{D\,sat}$

$$I_D = \frac{C_o W \mu_n}{2L} (V_G - V_{T*} - E_{ref} + RT/z^i F \ln a^i)^2$$

Gas-sensitive Metal Gate (IGFET)

The first FET sensor of this type contained a catalytic metal gate of palladium (Pd), which catalysed the dissociation of H_2 molecules at its surface (Lundström *et al.*, 1975). Some of the hydrogen atoms so produced diffuse through the thin (<300 nm) Pd layer to adsorb at the Pd–SiO_2 interface (see Fig. 4.11). In the presence of oxygen, chemical reactions occurring at the surface producing water, deplete the surface of adsorbed hydrogen atoms, so that sensitivity of the device to hydrogen is dependent on the presence of oxidizing gases and will be greatest in an inert atmosphere.

Normally, this type of device is operated at an elevated temperature (>100°C) in order to speed up the reaction at the surface. In an inert atmosphere the limit of sensitivity is reported to be 0.01 ppm H_2 or 3×10^{-5} Pa and in air 1 ppm or 5×10^{-4} Pa (Lundström, 1981). The maximum voltage shift due to adsorption of hydrogen atoms at the metal–insulator interface was reported to be around 0.5 V, with a typical rate of change in an inert atmosphere approximating to

$$\Delta V \approx K p^{\frac{1}{2}}$$

At low levels of H_2, $K \approx 27$ mV/ppm.

These Pd-gate structures are manufactured by evaporating a layer of Pd on a *p*-silicon chip, which is covered with a 100 nm thick oxide layer. Two configurations are possible, either as a capacitor (Fig. 4.13a) or as a FET (Fig. 4.13b).

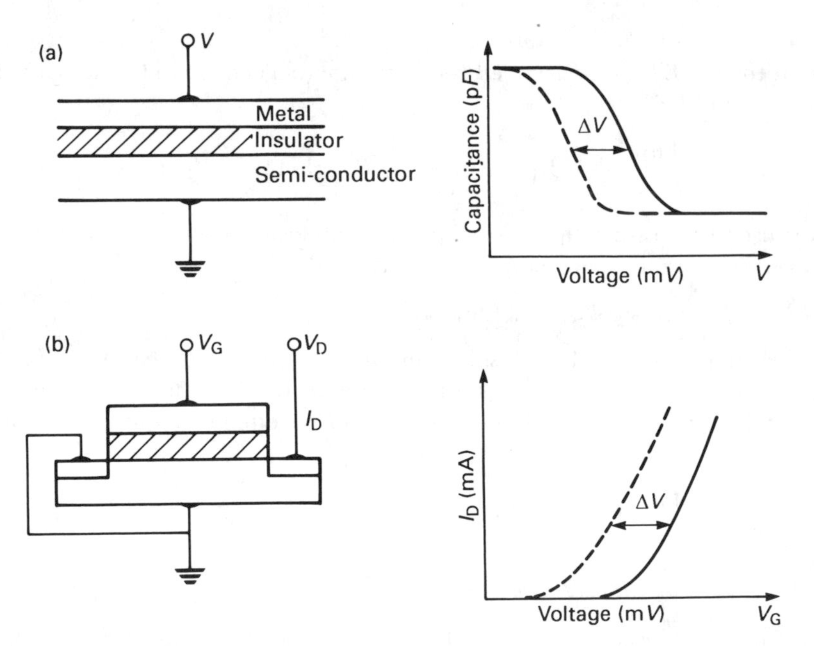

Fig. 4.13 Configurations for a gas sensitive metal gate IGFET: (a) as a capacitor, giving a gas induced shift in the capacitance–voltage curve; (b) as a FET, giving a gas-induced change in the I_D–voltage characteristics.

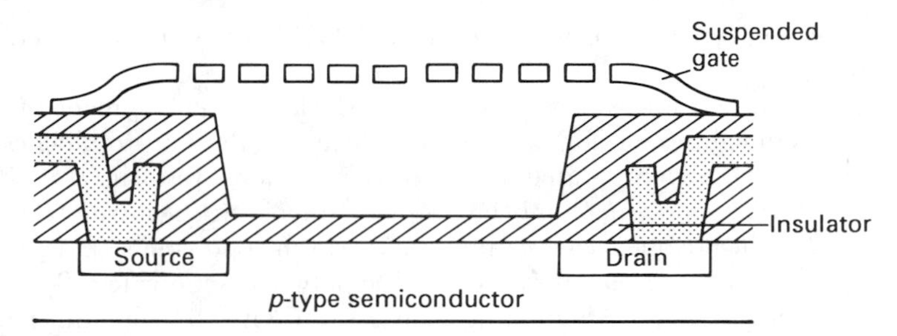

Fig. 4.14 Diagrammatic representation of a suspended gate FET.

The Pd catalyst also shows some sensitivity to other gases that might be present in the ambient atmosphere such as CO, NH_3, H_2S, CH_4 and C_4H_{10}. For ultra-thin Pd (25–40 Å), Jordan Maclay *et al.* (1988) report a linear response to CO in air, in the 1000–10 000 ppm CO range. Ammonia also has a low sensitivity which may be advantageously enhanced by incorporating a second thin layer (3 nm) of a catalytic metal such as iridium. The resulting hybrid metal gate shows a high NH_3 sensitivity and low H_2 sensitivity (Spetz *et al.*, 1984, 1987; Lundström and Danielsson, 1985). Details of the mechanism whereby NH_3 is able to cause a voltage shift have not been fully characterized, although it is clear that the mechanism differs to that proposed for hydrogen.

Suspended Gate Field Effect Transistor (SGFET)

The aforementioned H_2-sensitive FET relies for its operation on a change in the electron work function of the metal gate, due to the presence of H_2 dissolved in the metal layer. Applications of such a device are limited by the sensitivity that can be achieved by the choice of hybrid metal gates, where the target species can diffuse *through* the gate layer. The range could be extended by the introduction of a chemically sensitive layer within the transistor structure. Various geometries have been proposed to achieve this. In the SGFET (Fig. 4.14) the metal gate is not in direct contact with the insulator but forms a mesh structure above a gap (typically 2000 Å). Gaseous molecules with a dipole moment may diffuse through the mesh and become adsorbed on the inner surface of the gate or else on the insulator surface where they create a dipole potential, which may be considered to be in series with the applied voltage V_G.

The sensitivity of this type of FET device is not limited by the catalytic properties of the gate material. Indeed, it is sensitive to most polar species! Selectivity may be induced by the modification of the gate structure. For example, chemically selective layers may be electrochemically deposited on the metal gate.

Polypyrrole may be electrochemically deposited to give a layer on the gate which responds to lower aliphatic alcohols (Josowicz and Janata, 1986), producing a change in the work function which causes a positive shift in V_T. The response which is possibly associated with a hydrogen-bonding type interaction between the polypyrrole film and the alcohol–OH groups, is near Nernstian (60 mV/decade).

Polypyrrole in fact forms a convenient matrix for the entrapment of other chemical functionalities or for surface modification. Both nitrotoluenes and nitroarenes have been incorporated in the polypyrrole films (Josowicz *et al.*, 1987). In the former case a copolymer has been proposed, involving the nitrotolyl radical (Fig. 4.15), but the nitrobenzenes cannot readily form copolymers, and a charge transfer complex between the aromatic and the polymer is indicated. By 'finger-printing' the polymers by this modification a sensitivity to aromatics may be induced in both these films. An unmodified polypyrrole film shows no sensitivity to toluene, but a copolymer with nitrotoluene or a polymer including nitrobenzene responds to toluene.

Selectivity via Pattern Recognition

The use of multidimensional analysis in order to separate overlapping signals is becoming more widely adopted as a means of increasing selectivity. The technique requires that a measurement be made with respect to more than one variable, such that an appropriate set of simultaneous equations reveal both the nature of the response and its magnitude in terms of concentration of reacting species.

Müller and Lange (1986) have investigated this approach for zeolite-covered MOS sensors, sensitive to hydrogen and a range of hydrocarbons. On exposure to the test gas the voltage V_G changes with an initial rate $d(\Delta V_G)/dt$, to reach a final steady-state value V'_G. A simple model (Lundström, 1981) suggests that the initial rate of change of $V_G(S_i)$ is proportional to concentration and gives a final value $V'_G(S_f)$ which is approximately proportional to the square root of the concentration. The value $S_i/(S_f)^2$ should therefore be characteristic of the test gas and independent of concentration.

Modulation of this characteristic by the introduction of additional parameters such as pore size of the zeolite and/or temperature further increases the amount of information and selectivity that can be obtained from a single sensor material.

This example demonstrates one use of data processing to induce a selectivity to a signal response. Slight modifications in sampling environment can produce significant changes in the signal that is transduced. These changes can be employed to advantage in multidimensional analyses to impart both selectivity and sensitivity. This is an approach which is likely to become increasingly useful.

Ion-selective Field Effect Transistor (ISFET)

Deposition of the chemically sensitive film directly on the insulator and recording changes in the film with respect to a reference electrode placed in solution is the technique which has been most widely investigated as suitable for the detection of

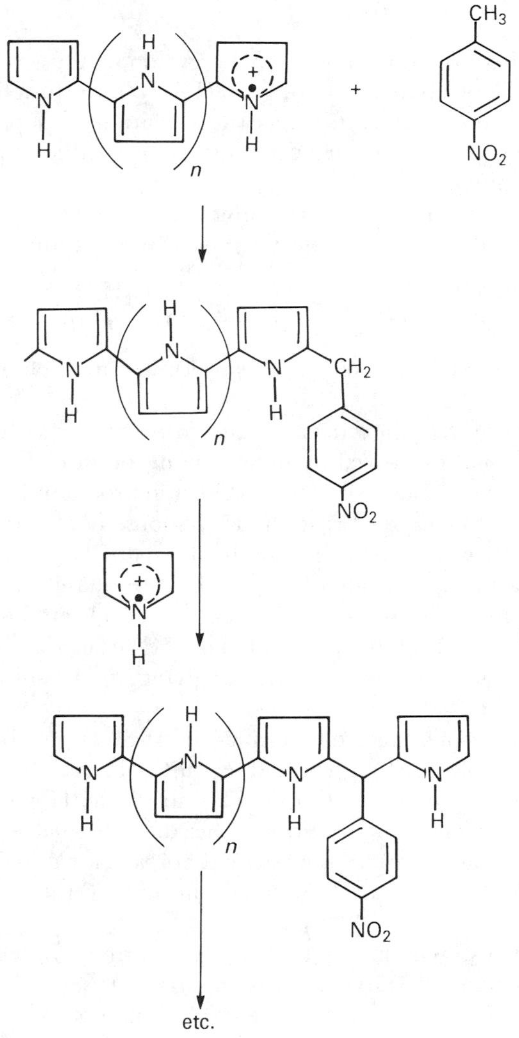

Fig. 4.15 Proposed incorporation of nitrotoluene in the pyrrole polymer.

ionic species. As with ISEs, ionophores are the obvious choice of a chemically sensitive layer for this application.

The ISFET would appear to offer a solution to the problem of noise 'pick-up' frequently encountered in the use of a conventional ISE, since the 'on-chip' sensing and control construction eliminates the wire connections between high-impedance components.

The membranes usually used in ISFETs are identical to those encountered with ISEs with the addition of a few unique examples.

SOLID-STATE MEMBRANES

The first pH-sensitive ISFET reported by Bergveld (1970), was based on the ion-exchange properties of a bare insulator gate. However, the practical value of a pH ISFET based on a SiO_2 gate is little, since the insulation properties of the layer are lost on immersion in solution for even short periods, due to hydration of the thermally grown SiO_2.

The generally accepted model for H^+ interaction involves the reversible surface reactions (Yates *et al.*, 1974; Siu and Cobbold, 1979) at amphoteric sites,

$$S\text{–}OH_2^+ \rightleftharpoons S\text{–}OH + H^+ \qquad (K_1 = [S\text{–}OH][H^+]/[S\text{–}OH_2^+])$$
$$S\text{–}OH \rightleftharpoons S\text{–}O^- + H^+ \qquad (K_2 = [S\text{–}O^-][H^+]/[S\text{–}OH])$$

so that for low values of K_1 and K_2, the surface will not behave in a Nernstian manner towards H^+.

Silicon nitride (Si_3N_4) gate devices can be prepared with good insulating properties that are not destroyed by hydration in solution (Esashi and Matsuo, 1978). Si_3N_4 can be deposited with excellent reproducibility to give a pH-sensitive interface showing a Nernstian H^+ response of 50–60 mV/decade. The wide range of pH sensitivity of this surface has been attributed to silanol groups at different positions in the hydrated layer at the interphase.

Other semiconductor–insulator structures have also been found to show a pH response where H^+ bonding sites are present on the surface. TiO_2–electrolyte and Ge–electrolyte are such examples, the latter giving an H^+ shift in the flat band potential of 60 mV/decade.

The deposition of solid-state membranes for ISFETs can be achieved using techniques compatible with integrated circuit fabrication. It thus offers a particularly attractive solution to the ISFET structure. However, the development of solid-state membranes for species other than H^+ has so far met with only limited success, although aluminosilicate and borosilicate glass can be deposited to produce small Na^+ sensors, fabricated entirely by integrated circuit processes.

Unfortunately Ag_2S which, apart from its use in silver- and sulphide-sensitive electrodes (see Chapter 3), is a basic matrix for several other solid-state membranes for ISEs, cannot be evaporated. In common with many other ion-selective materials, dissociation or some other chemical change cannot be avoided in these high vacuum techniques.

POLYMER MEMBRANES

Polymer membranes are usually prepared by casting over the gate region of the FET from a volatile organic solvent. This is the same procedure as that required for the ISEs already described (see Chapter 3); in fact any of the polymeric ionophore membranes that have been developed for ISEs would be seemingly suitable for use directly with ISFETs.

Membranes that are thinner than 50 μm have been found to suffer from pin-hole defects, which will fill with electrolyte, eventually shorting out the membrane. The best results seem to have been achieved with multi-stage castings to give a membrane thickness $> 100\,\mu m$.

The most successful ISFETs employing these polymer membranes have been directed at such ions as K^+ and Ca^{2+}, the latter employing as ionophore, a phosphoric acid-based derivative such as *p*-(1,1,3,3-tetramethylbutyl-phenyl) phosphoric acid with calcium ion sensitivity (Griffiths *et al.*, 1972) and the former, valinomycin (Band *et al.*, 1977, 1978) or similar synthetic crown ether. In both these cases membranes thinner than $40\,\mu m$ produced substantially reduced responses ($< 40\,mV/pIon^+$) due to pin-hole defects. With membranes 80–150 μm thick, this pin-hole phenomenon was not observed, and while gradual deterioration of the calcium membrane reduced its lifetime, the usable life expectancy for the K^+ sensor can extend beyond a month.

Leaching of the ionophore from the polymer matrix is a serious limitation in the application of a sensor device. Not only does its loss reduce the effective lifetime of the device, but it renders it potentially unsuitable for *in vivo* use. Chemical modification of valinomycin so as to allow attachment to the membrane polymer is not easily achieved without simultaneous modification of its otherwise unrivalled selectivity for K^+. Various synthetic hemispherands and calix crowns which might be suitable for membrane attachment, have therefore been tested for comparable selectivity. Table 4.1 shows, for example, the structure of some such synthetic ionophores with comparable (Moody and Thomas, 1972; Dijkstra *et al.*, 1988; Reinhoudt *et al.*, 1988; Van den Berg, 1988) and reduced selectivity, and in particular Van den Berg reports that no selectivity is lost for the 21-membered hemispherand when functionalized prior to attachment to a membrane.

MULTIPLE FUNCTION ISFETs

A quadruple function ISFET for pH, Na^+, K^+ and Ca^{2+}, developed primarily for clinical applications (Sibbald *et al.*, 1985), employed the glass, phosphoric acid based and valinomycin membranes for Na^+ and K^+ respectively, while pH sensitivity was achieved at a bare gate Si_3N_4 structure or at a pH-sensitive glass sputtered over the Si_3N_4 gate region.

This particular device provides an excellent demonstration of the need to adapt specific sensors for a particular application. In this example the multi-ISFET was aimed towards continuous monitoring of whole blood at the bedside or during surgery. Tested under laboratory conditions the device performed well for all four functions. However, under clinical situations in whole blood, where fluctuations in blood electrolyte composition would occur (Sibbald *et al.*, 1983, 1984), considerable difficulties in the analysis of Na^+ were encountered, concerned primarily with the selectivity and sensitivity of the Na^+-selective membrane. While this membrane is adequate for many Na^+ estimations, the range and environment of the Na^+ in whole blood renders it unsuitable there.

In clinical applications it is not always the absolute concentrations of the analytes that are required, but rather a change in concentration. Ideally, 0.1 mV resolution is desirable, a goal which is difficult—if not sometimes impossible—to

Table 4.1

Compound	k^+/Na^+ (approx. value)
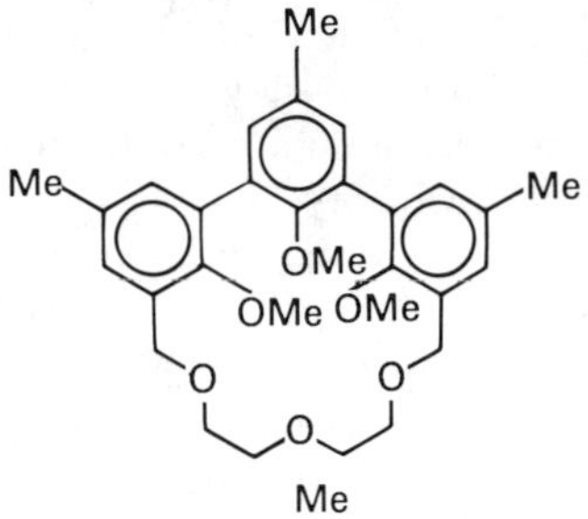 18-membered hemispherand	5×10^{-1}
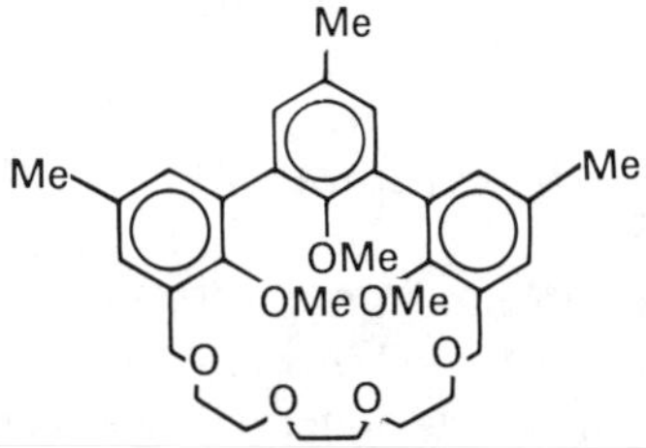 21-membered hemispherand	3.5×10^{-3}
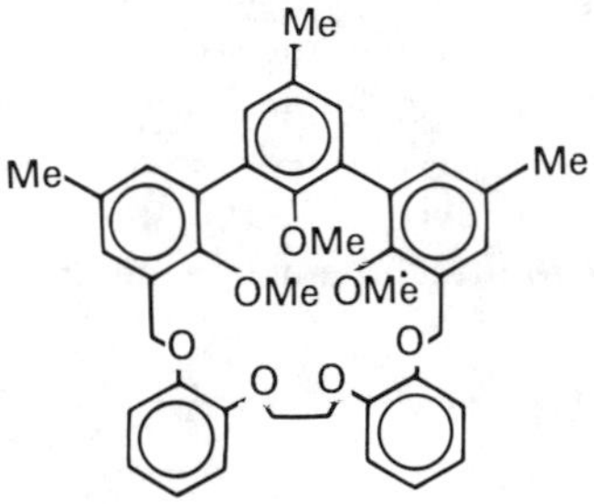 21-membered dibenzo hemispherand	1.5×10^{-3}
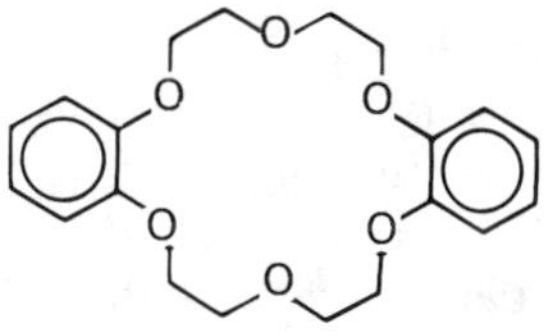 Dibenzo-18-crown-6	1×10^{-1}
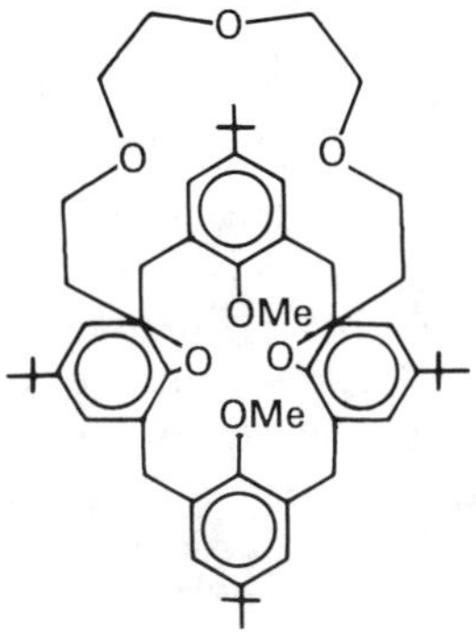 Dimethyl calix [4]-arene-crown 5	5×10^{-4}
Valinomycin	

achieve with ISEs. Even with an ISFET, hydration of the electro-active gate during the first 12 h or so of operation can give a drift >0.2 mV/h.

HETEROGENEOUS MEMBRANES

Heterogeneous membranes are a solution to the fabrication of solid-state membranes of semi conducting salts. In a procedure pioneered by Pungor (1967) and co-workers, an inorganic salt of low solubility is usually dispersed in a suitable polymer. Various polymers have been investigated, the resulting membranes showing very sub-Nernstian (5–27 mV/decade) to near Nernstian (>50 mV/decade) responses. Silicone rubber, for example, has been generally discarded as a support matrix due to its poor Nernstian response, whereas the elastomeric polymer, polyfluorinated phosphazene (PNF), which can be readily dissolved in ketones and other common organic solvents, is employed successfully in the fabrication of near Nernstian membranes.

Finely divided silver chloride powder (75%) and PNF (25%) cast from methylisobutylketone gave a membrane which responded to Cl^- with a slope of 52 mV/decade (Shiramizu *et al.*, 1979). A better selectivity for Cl^- could be achieved by partially replacing the AgCl with silver sulphide in the ratio of 4 : 1. The selectivity of the membrane prepared from 75% of this silver salt mixture and 25% PNF is compared with the previous membrane in Table 4.2.

Indeed adjustments in the silver salt mixture can be successfully made to 'tune' for selectivity for particular anions. An AgI–Ag_2S (6 : 1) preparation, for example, gives sensitivity for I^- of the order of 60 mV/decade and for CN^- of the same order with selectivity constants $K_{I^-/Cl^-} = 1.14 \times 10^{-5}$ and $K_{CN^-Cl^-} = 2 \times 10^{-4}$.

Reference FET

The modification of the IGFET to produce an ISFET requires, like its ISE counterpart, a reference electrode. Although operation without a reference has been proposed, it would now appear to be generally accepted that a reference

Table 4.2 Selectivity constants for heterogeneous membranes

Ion	*Concentration (M)*	*Selectivity constant $K_{Cl/i}$*	
		AgCl : PNF (3 : 1)	AgCl : Ag_2S : *PNF* (12 : 3 : 5)
Br^-	7.5×10^{-5}	1.87	5.4
I^-	7.5×10^{-6}	poisoned	0.42
SO_4^{2-}	0.01	0.01	4.5×10^{-3}
NO_3^-	0.1	1.3×10^{-4}	5.1×10^{-4}

electrode is required. In common with the ISE, the Ag/AgCl electrode is usually employed, connected with the sample via a liquid junction.

Comte and Janata (1978) tested an alternative approach where a differential measurement of pH was made between two pH FETs. A pH FET in a buffered solution at constant pH was connected via a liquid junction to a second sample pH FET. They suggested that the most attractive feature of this reference gate was its ability to compensate automatically for temperature change and noise. Such elimination of ambient effects is obviously only directly possible between 'matched' components such as these ISFETs and is less feasible between unmatched systems such as an IS gate and an Ag/AgCl reference electrode.

An alternative solution to the reference FET is to design a gate material with *no* ion sensitivity. Matsuo and Esashi (1978) have proposed a parylene gate FET, fabricated using integrated circuit technology to deposit 1000 Å of Parylene C (polymonochloro-*p*-xylene) on Si_3N_4. The device shows little ion sensitivity (e.g. about 0.5 mV/pH), and under conditions where little sensitivity to any other interfering sample species is also indicated the usefulness of such a FET is apparent as a reference system.

CHEMFET Operation

It may be deduced from the earlier examination of FET structures, that the fundamental response is a change in the drain current due to the activity at the gate region. Under conditions of strong inversion a change in the activity of the ions in solution causes a change in the number of mobile carriers in the channel, and thus for a given constant drain voltage, a change in the drain current.

Obviously two modes of operation are possible where:

(i) V_G and V_D constant; I_D monitored
(ii) I_D and V_D constant; V_G monitored

(i) *Operation with Constant* V_G

In this mode of operation all externally applied voltages (V_D and V_G) are held constant and I_D measured via an operational amplifier in the current to voltage converter mode (Fig. 4.16a), with gain control of the output signal provided with the feedback resistor R such that:

$$V_{out} = -I_D R$$

Unfortunately, in practice, V_{out} is not a direct reflection of I_D but includes series resistance in the source and drain, causing deviations from linear behaviour. The device must therefore be calibrated across the entire range in order to account for these deviations.

(ii) *Operation with Constant* I_D

The above circuit may be extended (Fig. 4.16b), so that control of V_G is provided by the output V_{out}, to hold I_D constant. V_{out} is fed through a voltage divider into a differential amplifier, where the inverting input is tied to 0 V, such that 0 V at the

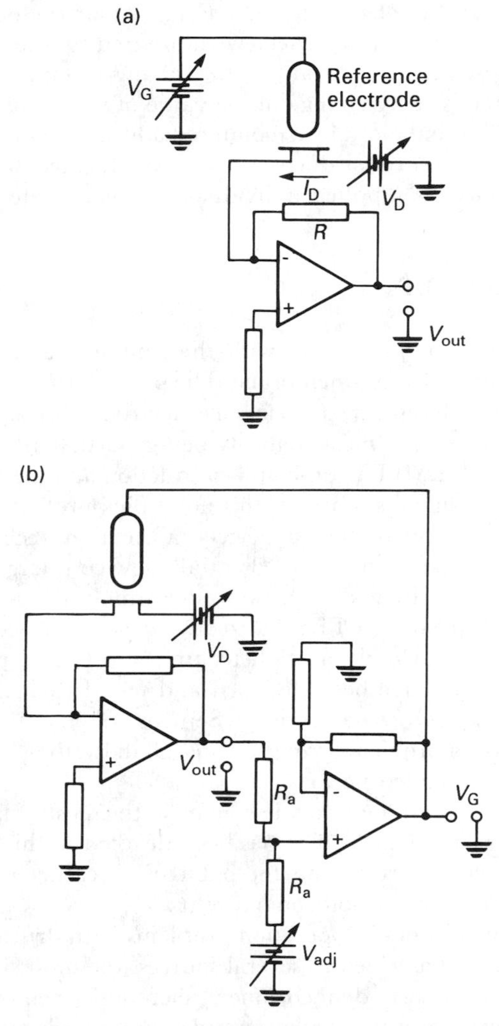

Fig. 4.16 Mode of operation for CHEMFETs. (a) Basic circuit for operation with constant V_G; (b) basic circuit for operation with constant I_D.

non-inverting input must be achieved by:

$$V_{out} + V_{adj} = 0$$

and since

$$I_D = -V_{out}/R$$

then

$$I_D = V_{adj}/R$$

and the required I_D can be set by adjusting V_{adj}, so that the output from the differential amplifier will control V_G to achieve the desired constant drain current.

This mode of operation has the advantage that changes in interfacial potential can be monitored directly. Any change in the value of the term $\pm RT/z/F \ln a^i$ contributing to I_D, will be balanced by a monitored adjustment of V_G. In multiple function FET devices, however, the circuitry is complicated by this mode of operation, since V_G must be supplied individually for each gate function.

Assessment of CHEMFETs

In an ISE the electrode is in contact with the sample via a buffer solution contained behind an ion-selective membrane. The potential difference generated across this membrane with respect to a reference electrode is measured by remote high-impedance circuitry, the measurements being particularly susceptible to noise artifacts. In the CHEMFET, on-chip transduction and the low impedance of the processed output signal suppresses this noise pick-up feature.

The use of existing semiconductor device fabrication technology in the manufacture of CHEMFETs provides a potentially low cost, large-scale production, where multiple-gate single-chip arrays offer simultaneous monitoring of several target species. Indeed these FET-based devices can now be created in such small sizes (10–100 μm) that multiple catheter tip sensors are a real possibility.

pH catheter tip sensors have been designed and tested (Schepel *et al.*, 1984; Kohama *et al.*, 1984) and some devices (Sentron, Holland) are available commercially. However a problem facing this and all indwelling sensors is that of biocompatibility over extended periods.

Adequate protection must be provided against the host's natural defence mechanisms. Although significant effort has been devoted to this problem, only partial success has been reported, and the pH ISFET catheter tip sensor has occasionally failed due to lack of biocompatibility.

Under more ideal conditions of operation problems with drift have also been encountered. This may be ascribed to several sources, the most significant error probably arising from a poorly defined inner 'reference' (membrane–insulator interface) and imperfections in encapsulation and membrane deposition. Leakage of fluid at encapsulation boundaries and through defects in the membrane structure or gate insulator are problems that remain to be fully solved.

Fogt *et al.* (1985) ascribed the existence of a poorly defined inner membrane–insulator interface to signal interference caused by CO_2 gas in the sample solution. The observation was explained by an interaction of the CO_2 with traces of H_2O at the interface, causing changes in H^+ concentration.

One of the roots of this problem appears to lie in the poor adhesion properties between the membrane and the FET, so that gradual detachment of the membrane leads to electrolytic shunts around its edges. Blackburn and Janata (1982) have developed a suspended mesh ISFET with an inherent means of anchoring the ion-selective membrane. This design is an apparent precursor to the SGFET described earlier, since a suspended mesh polymer film with an array of

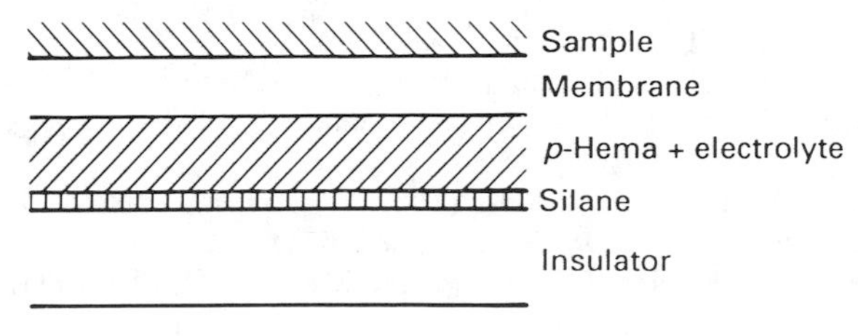

Fig. 4.17 Sandwiching structure proposed by Sudhölter *et al.* (1986), involving covalently attached poly(hydroxyethyl methacrylate) (p-HEMA). The resulting ISFET has reported improved stability.

10 μm square holes is fabricated 1 μm above the gate insulator and the polymer ion-selective membrane cast over the gate area. The liquid polymer flows under the mesh, displacing the air and filling the gap and the mesh becomes an integral part of the membrane. The authors reported drift-free operation for devices prepared in this way over a 60-day period, compared with 24 h for the conventional ISFET.

Alternatively, a more clearly defined interface has been provided by an intermediate layer of Ag/AgCl (Dror *et al.*, 1987) between insulator and membrane, but this requires free Cl^- at the membrane interface to participate in the equilibrium.

Sudhölter *et al.* (1986) have proposed a hydrophobic intermediate layer of poly(hydroxyethyl methacrylate) (p-HEMA), formed by covalent attachment of the monomer to the insulator surface via a silane, followed by photopolymerization and 'filling' with appropriate electrolyte. With the overlaying membrane, the resulting ISFET structure (Fig. 4.17) showed improved stability and providing that the 'filling' buffer had $pH < 4$, where there would be no influence from CO_2-induced proton generation, the signal was stable to varying CO_2 levels. This construction, with its 'internal' filling electrolyte is more closely related to the traditional ISE and allows the use of the more 'noisy' acrylate and silicon polymer matrix membranes.

Probably due to the availability of ion-selective membranes, the greatest development effort appears to have been made in this direction. However, the use of chemically sensitive layers whose work function varies due to an interaction with a target species is a potentially vast field ideally suited for inclusion in FET devices.

References

Band, D.M., Kratochvil, J. and Treasure, T. (1977). *J. Physiol.* **265**, 5P.

Band, D.M., Kratochvil, J., Poole Wilson, P.A. and Treasure, T. (1978). *Analyst* **103**, p. 246.

Bergveld, P. (1970). *IEEE Trans. Biomed. Eng.*, BME-17, p. 70.

Blackburn, G.F. and Janata, J. (1982). *J. Electochem. Soc.* **129**, p. 2580.

Comte, P.A. and Janata, J. (1978). *Anal. Chim. Acta* **101**, p. 247.
Dijkstra, P.J., den Hertog, Jr., H.J., van Steen, B.J., Zijlstra, S., Skowronska-Ptasinska, M., Reinhoudt, D.N., Eerden, J. van and Harkema, S. (1988). *J. Org. Chem.* **53**(2), pp. 374–382.
Dror, M., Bergs, E.A. and Rhodes, R.K. (1987). *Sensors and Actuators* **11**, p. 23.
Esashi, M. and Matsuo, T. (1978). *IEEE Trans.*, BME-25, p. 184.
Fogt, E.J., Untereker, D.F., Norenberg, M.S. and Meyerhoff, M.E. (1985). *Anal. Chem.* **57**, p. 1995.
Griffiths, G.H., Moody, G.J. and Thomas, J.D.R. (1972). *Analyst* **97**, p. 420.
Jordan Maclay, G., Jelly, K.W., Nowroozi-Esfahani, S. and Formosa, M. (1988). *Sensors and Actuators* **14**, p. 331.
Josowicz, M. and Janata, J. (1986). *Anal. Chem.* **58**, p. 514.
Josowicz, M., Janata, J., Ashley, K. and Pons, S. (1987). *Anal. Chem.* **59**, p. 253.
Kohama, A., Nakamura, Y., Nakamura, M, Yano, M. and Shibatani, K. (1984). *Crit. Care Med.* **12**, p. 940.
Lundström, I. (1981). *Sensors and Actuators* **1**, p. 403.
Lundström, I., Shivaraman, M.S. and Svensson, C. (1975). *J. Appl. Phys.* **46**, p. 3876.
Lundström, I. and Danielsson, B. (1985). *Sensors and Actuators* **8**, p. 3876.
Matsuo, T. and Esashi, M. (1978). *Proc. 153rd Annual Meeting of Electrochemical Society* **78**(83), p. 202.
Moody, G.J. and Thomas, J.D.R. (1972). *Talanta* **19**, p. 623.
Müller, R. and Lange, E. (1986). *Sensors and Actuators* **9**, pp. 39–48.
Pungor, E. (1967). *Anal. Chem.* **39**, p. 28A.
Reinhoudt, D.N., Dijkstra, P.J., In't Veld, P.J.A., Bügge, K.E., Harkema, S., Ungaro, R. and Ghidini, E. (1988). *Pure and Appl. Chem.* **60**(4), pp. 477–482.
Schepel, S.J., de Rooij, N.F., Koning, G., Oeseburg, B. and Zijlstra, W.G. (1984). *Med. Biol. Eng. Comput.* **22**, p. 6.
Shiramizu, B.T., Janata, J. and Moss, S.D. (1979). *Anal. Chim. Acta* **108**, p. 161.
Sibbald, A. (1983). *IEEE Proc.* **130**, p. 233.
Sibbald, A., Whalley, P.D. and Covington, A.K. (1984). *Anal. Chim. Acta* **159**, p. 47.
Sibbald, A., Covington, A.K. and Carter, R.F. (1985). *Med. Biol. Eng. Comput.* **23**, p. 329.
Siu, W.M. and Cobbold, R.S.C. (1979). *IEEE Trans. Electron. Devices*, ED-26, p. 1805.
Spetz, A., Lundström, I. and Danielsson, B. (1984). *Anal. Chim. Acta* **163**, p. 143.
Spetz, A., Armgarth, M. and Lundström, I. (1987). *Sensors and Actuators* **11**(4), p. 349.
Sudhölter, E.J.R., Skowronska-Ptasinska, M., van der Wal, P.D., van der Berg, A. and Reinhoudt, D.N. (1986). Patent, *Ned. Oktrooiaanvrage*, NL 8602242.
Van den Berg, A. (1988). 'Ion Sensors Based on ISFETs with Synthetic Ionophores'. Thesis of the University of Twente.
Yates, D.E., Levine, S. and Healy, T.W. (1974). *J. Chem. Soc. Faraday Trans.* **70**, p. 1807.

Chapter 5

Amperometric Assay Techniques

Analysis of Charge Transfer

Change in oxidation state, or charge transfer, is a feature common to many chemical and biochemical reactions. The transfer of charge may be described generally by

$$O + ne^- \rightleftharpoons R$$

where n is the number of electrons (e) transferred between oxidant (O) and reductant (R). This equation can describe a simple charge-transfer electrode reaction, where electrons are transferred between electrode and electroactive species at the surface of an electrode.

As with any chemical process, the thermodynamics and the kinetics must be considered, and the equilibrium situation, when there is no net charge transfer, may be described by the *Nernst equation*,

$$E = E^{\theta} + RT/F \ln \frac{[\text{electron acceptor}]}{[\text{electron donor}]}$$

As we have seen in Chapter 3 the equilibrium potential (E) may be related to the standard electrode potential (E^{θ}) and the concentrations of the oxidized and reduced species.

If the electrode potential is altered from the equilibrium potential then equilibrium can only be re-established by adjustment of the concentrations of oxidant and reductant. This will require a net current flow or charge transfer, the magnitude of which will depend on the kinetics of the electron transfer, and will

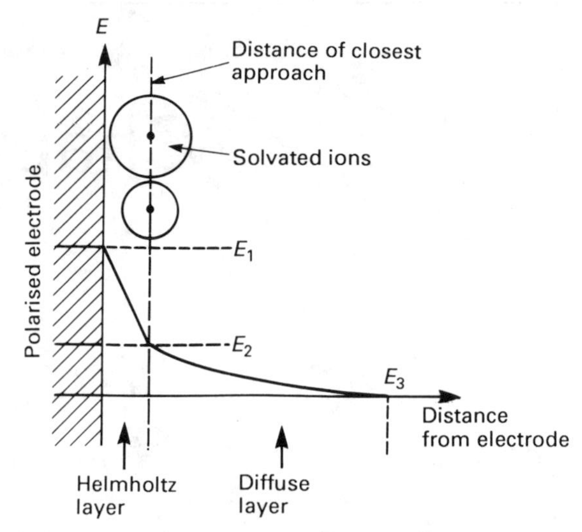

Fig. 5.1 Potential profile through double layer next to the surface of an electrode.

establish a concentration gradient for the oxidant and reductant at the electrode surface.

In a simple model of heterogeneous electron transfer, it may be considered that transfer takes place not at the electrode surface, but in the Helmholtz layer. This may be rationalized by consideration of the double layer. When a positive electrode at potential E_1 is in a solution of a supporting electrolyte, solvated anions from the electrolyte will collect in an orderly layer immediately in front of the electrode, in the plane of closest approach, i.e. the Helmholtz layer (Fig. 5.1). On the solution side of this layer is the diffuse layer, in which there is a predominance of anions. This organization at the electrode causes the potential to vary from E_1 at the electrode to E_s in the solution, with a discontinuity in the variation represented by the potential of the Helmholtz layer, E_2.

The potential difference across the double layer at the electrode surface will assist the transfer of an ion through the double layer in one direction, but will inhibit its transfer in the opposite direction. The result of this is to raise the potential energy of the electroactive reactants by the amount nEF, where n is the number of electrons transferred. However, when the activated complex is reached at the maximum in the free-energy barrier, this potential energy difference is only some fraction, β, of the energy difference nEF, since E varies through the double layer (Fig. 5.2). The rate of electron transfer is thus governed by the expression

$$k_e = kT/h \exp\{-[\Delta G^{\ddagger} - \beta nEF]/RT\}$$

where k_e is the intrinsic charge-transfer rate constant (compare with Arrhenius equation, $k = A\exp\{-E_{act}/RT\}$).

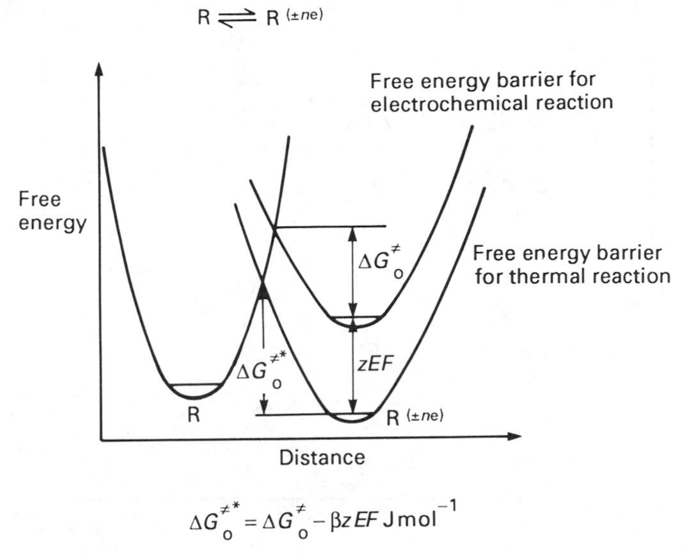

Fig. 5.2 Potential energy barriers for the thermal and electrochemical reactions of $R \rightleftharpoons R^{(\pm ne)}$.

Voltammetric Techniques

CYCLIC VOLTAMMETRY

Physical methods that are employed to study electrochemical charge transfer reactions may be defined generally under the title of voltammetry and involve the measurement of current–voltage relationships.

If the potential is varied linearly with time at a microelectrode in quiescent solution, the current–potential curve is governed by the rate of transfer of electrons given by the expression for k_e. Linear variation of potential along a triangular waveform with time and recording of the current is a technique known as *cyclic voltammetry* (CV) (Fig. 5.3).

The simplest case involves the reversible reduction of a species, without accompanying chemical reactions:

$$O \underset{-ne}{\overset{+ne}{\rightleftharpoons}} R$$

As the electrode potential (E) is swept through E^o (the redox potential), the surface concentration of O will change according to the Nernst equation:

$$[O]/[R] = \exp[nF/RT(E - E^o)]$$

If the reaction is under diffusion control then a current proportional to concentration of O will be obtained and, as O is reduced to R, a layer will develop next to the electrode in which O has been depleted (diffusion layer), causing a

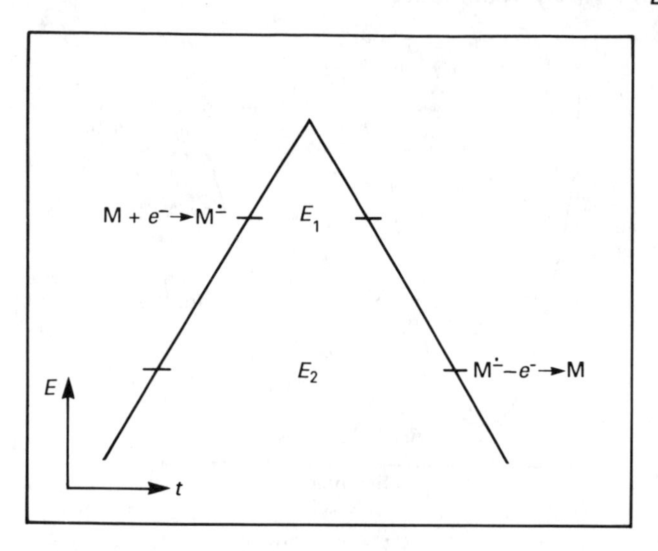

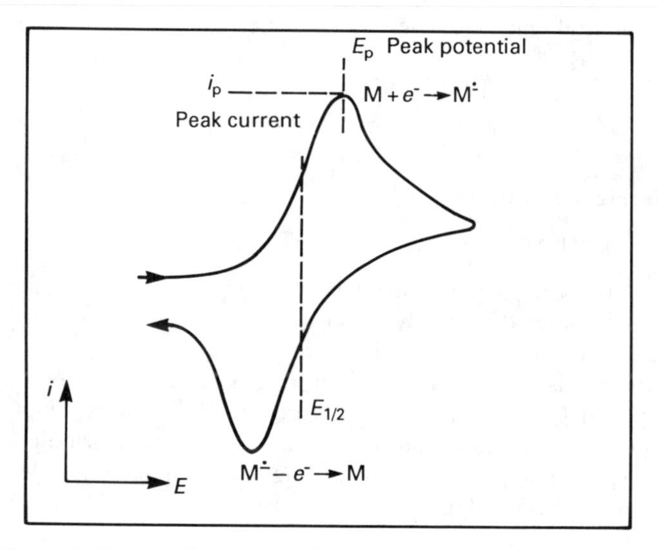

Fig. 5.3 *E–t* and *E–i* profiles for cyclic voltammetry.

peak in the observed CV. On the reverse scan, the re-oxidation of R will be obtained and the maximum current for the two peaks should be equal and for planar diffusion given by

$$i_p = -0.4463nF(nF/RT)^{1/2}C_oD^{1/2}v^{1/2}$$

This is the *Randles–Sevčik* equation and at room temperature (26°C) reduces to

$$i_p = -2.69 \times 10^5 n^{3/2}C_oD^{1/2}v^{1/2}$$

and it can be seen that a plot of $v^{1/2}$ against peak current would be linear for this diffusion-controlled process, and at a given scan speed peak current is directly proportional to concentration. The peak separation for a totally reversible redox couple is $57/n$ mV.

ELECTRON-TRANSFER RATE CONSTANTS

In fact the magnitude of the current at any given potential will depend on the heterogeneous rate constants (electron transfer rates at the electrode interface) for the forward (k_f) and backward (k_b) reactions at that potential. The rate constants may thus be expressed in terms of electrode potential (E)

$$k_f = k_f^o \exp[-\alpha nFE/RT]$$
$$k_b = k_b^o \exp[(1-\alpha)nFE/RT]$$

where α is known as the transfer coefficient for the forward reaction. This leads to the current density, i (A cm^{-2}),

$$i = nF\{k_f^o[\mathrm{O}]\exp[-\alpha nFE/RT] - k_b^o[\mathrm{R}]\exp[(1-\alpha)nFE/RT]\}$$

At the equilibrium potential, E_e, there is no net electrochemical reaction occurring and i_0, the *exchange current density* is equal in both directions, giving a net current, i, of zero.

OVERPOTENTIAL

The situation changes as the applied potential deviates from this equilibrium value. This potential difference is termed the *overpotential*, η

$$\eta = E - E^o$$

and i may be written as

$$i = i_0[\exp(\alpha nF\eta/RT) - \exp((1-\alpha)nF\eta/RT)] \qquad \textit{(Butler–Volmer equation)}$$

and at large positive and negative values of η, the anodic and cathodic reactions respectively will be favoured. However, the changes that occur in the exchange current for small changes in η ($\pm RT/\alpha nF$) are indicative of the ease at which a given electrode reaction occurs, so that an overpotential of $RT/\alpha nF$ would result in a current density of the order of

$$i \approx i_0 nF\eta/RT$$

High values for i_0 indicate a 'fast' reversible reaction, while low values are seen for irreversible systems. Exchange current values vary between different electroactive species and *different electrode materials*. The latter property has obvious implications in the choice of electrode as an analytical base-sensor.

As the overpotential is increased in the anodic or cathodic direction, the total current becomes dominated by that particular potential branch, and increases exponentially until the electrode kinetics are matched by, and then overtaken, by

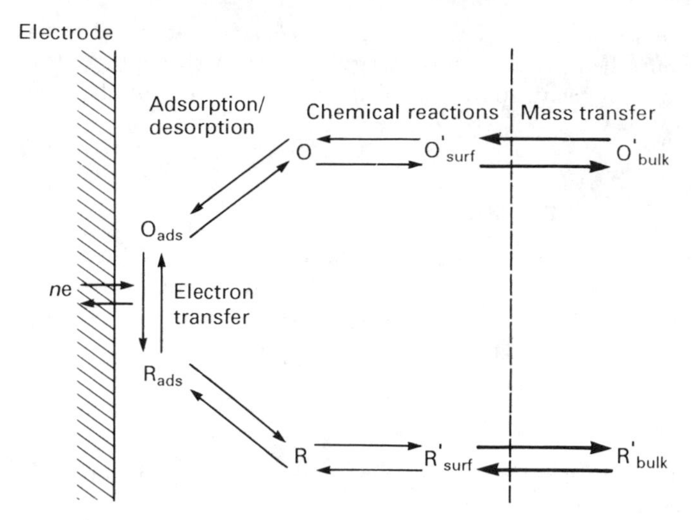

Fig. 5.4 Pathway of surface processes involved in a general electrode reaction.

diffusional limitations of transport of the reactive species from solution to the electrode surface.

In general, therefore, the current due to the overall electrode reaction is determined by (Fig. 5.4):

(1) Mass transfer from bulk solution to the electrode surface
(2) Electron transfer at the electrode surface
(3) Preceding or following chemical reactions
(4) Surface reactions such as adsorption

CV provides a powerful mechanistic tool to examine these reactions.

ADSORBED SPECIES

If the reactants or products are not freely soluble in solution, but are adsorbed or in some way immobilized at the electrode surface, then mass-transfer effects cease to play any role and the major difference compared with a solution redox couple, is that the peaks are sharper and symmetrical, with little or no peak separation. Since a fixed amount of reactant is present at the beginning of the scan, the current returns to zero after the peak (rather than assuming a diffusion-controlled value as for the solution species). In this instance the current, peak width and potential depend on the type of surface immobilization. However, many can be approximated by a Langmuir isotherm-type model where $E_f = E_b$ and the peak current density is given by

$$i_p = n^2 F^2 C_{ads} v / 4RT$$

where C_{ads} is the concentration of adsorbed species, i.e. the current is directly proportional to sweep rate. The area under the cathodic peak corresponds to the charge associated with the reduction of adsorbed O and similarly the area under the anodic peak is due to the oxidation of adsorbed species.

Systems where an electroactive species is attached to an electrode by chemical means can be compared with these adsorbed species since the number of reactant sites on the surface is fixed. All these electrodes, known as *modified electrodes*, make a significant contribution to electrochemical biosensors, since it is through the modification of the electrode to facilitate electron transfer, or by immobilization of a biorecognition molecule, that specificity for a chosen analyte is often induced in the electrode. Modified electrodes will be discussed in greater detail in later sections.

COUPLED CATALYTIC CHEMICAL REACTIONS

In the situation where there is a coupled chemical reaction of the type

$$O + ne^- \rightleftharpoons R$$
$$R + A \xrightarrow{k} O + B$$

i.e. the reactant (O) can be regenerated by the chemical reaction of R with A, then a competition will exist between the two reactions. At fast scan speed (v) and/or small k then the chemical reaction may have no effect, and the CV will then show the typical reversible behaviour shown in Fig. 5.3. However, for larger values of k, and/or as v is decreased then more O will be produced chemically, and the CV will show a decrease in the current due to

$$R \rightarrow O + ne$$

and since effectively more reactant is produced, an increase in the current

$$O + ne \rightarrow R$$

and deviation from the $i_p/v^{1/2}$ behaviour described above. At very slow scan speeds, the reverse peak may disappear altogether, and the cathodic current reaches a limiting plateau which is independent of sweep speed, and given by

$$i = -nFC_o(DkC_A)^{1/2}$$

Identification of a catalytic mechanism is readily made through analysis of the $i_p/v^{1/2}$ value which shows a characteristic increase with decrease in v. As indicated in the above equation, values of k may be obtained from the current plateau (i). However, an alternative method comparing the peak currents in the presence (i_k) and absence (i_d) of A has been solved by Nicholson and Shain (1964) to provide tables from which k can be estimated.

Examination of the above relationships for i_p for the two limiting cases

(i) where i_p approachcs diffusion control and is given by the Randles–Sevčik equation, and
(ii) where i_p is dominated by the catalytic reaction and is independent of v,

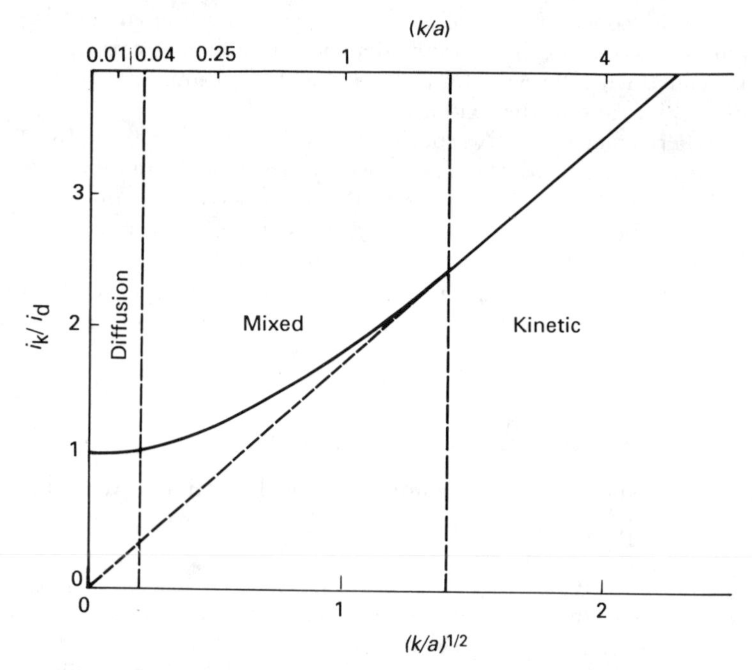

Fig. 5.5 Theoretical curve of i_k/i_d versus $(k/a)^{1/2}$, showing areas of kinetic, diffusion and mixed control.

shows that for a fixed concentration of A, i_k is related to i_d by the factor

$$(kRT/nFv)^{1/2}$$

This reduces to

$$(k/a)^{1/2}$$

where $a = nFv/RT$ so that a plot of i_k/i_d for various values of v can lead to the kinetic parameter k/a by comparison of theoretical and experimental values. A theoretical curve of i_k/i_d against $(k/a)^{1/2}$ may be compiled (Fig. 5.5). This i_k/i_d relationship reveals areas of kinetic, diffusion and mixed control.

For values of $(k/a) > 1$, i.e. when

$$k > nFv/RT$$

the plot is essentially linear and dominated by the catalytic process. In this region $(k/a)^{1/2}$ can be accurately obtained from the i_k/i_d ratio for different values of v. It can also be seen from the above relationships that a plot of k/a against $1/v$ should give a straight line of slope kRT/nF allowing k to be solved.

In general, the problem

$$O + ne^- \rightleftharpoons R$$
$$R + A \xrightarrow{k} O + B$$

would involve consideration of a second-order reaction and the diffusion of a species A. However, if working conditions can assume that A is present in excess, then its concentration remains essentially unchanged, and the experiment can be considered as pseudo-first order. The kinetic parameter, k, is therefore the pseudo-first-order rate constant. A second-order rate constant, k_s, may be estimated from values of k obtained at different concentrations of A:

$$k_s = k/[\mathrm{A}]$$

A full treatment of the kinetics of this redox catalysis has been described by Savèant and co-workers (Savèant and Su, 1984, 1985; Nadjo *et al.*, 1985).

Potential Step Techniques

In the model so far described, linear diffusion to a planar electrode has been assumed, which can be characterized by application of Fick's laws concerning diffusion in one dimension. For the simple electron transfer,

$$\mathrm{O} + n\mathrm{e}^- \rightarrow \mathrm{R}$$

Fick's second law, discussing the change of concentration with time can be solved for the planar electrode to give the current response on applying a potential step from a potential where the electrode process is negligible to one where it is diffusion controlled:

$$I_t = \frac{nFD^{1/2}C_o}{\pi^{1/2}t^{1/2}} \qquad \text{(Cottrell equation)}$$

That is, current decays with respect to $t^{1/2}$ so that a plot of I_t against $t^{1/2}$ would be linear and for given t, current will be proportional to bulk concentration. With increase in time the concentration gradient for the reactant extends further away from the electrode surface, as shown in Fig. 5.6. At long time, a steady-state current is reached, which is independent of time, and is given by

$$I = \frac{-nFDC_o}{\delta}$$

where δ defines the boundary layer thickness of the concentration gradient

$$\delta \propto (Dt)^{1/2}$$

and the current is directly proportional to concentration of the reactant O.

For the redox catalysis,

$$\mathrm{O} + n\mathrm{e}^- \rightarrow \mathrm{R}$$
$$\mathrm{R} + \mathrm{A} \xrightarrow{k} \mathrm{O} + \mathrm{B}$$

the thickness of the boundary layer will depend on the rate at which A reacts with the electrochemically produced R, and for large values of k it is very thin. The

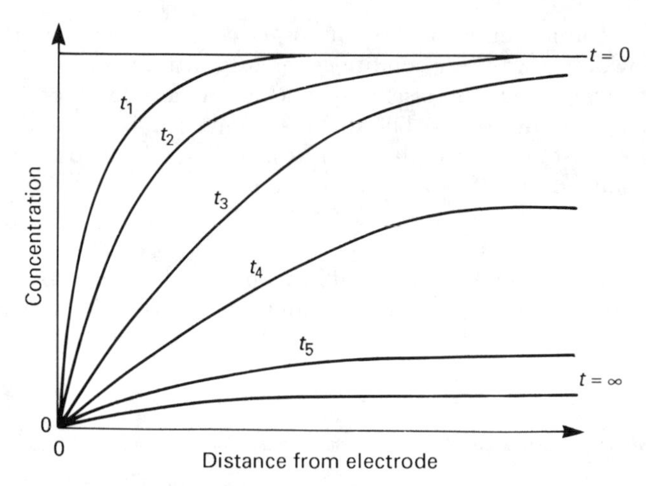

Fig. 5.6 Concentration profiles for an electroactive species, at different times after applying a potential step from an inert potential to a potential of diffusion control: $t_1 < t_2 < t_3 < t_4 < t_5$.

steady-state current then depends on the rate of the homogeneous chemical reaction, and is given by

$$I = -nFD^{1/2}k^{1/2}C_o$$

Non-steady-state Measurement

The steady-state current is always less than the transient time-dependent current which is recorded early after a potential step (Fig. 5.7). It is possible to infer, therefore, that this transient current is likely to have better signal-to-noise characteristics than the steady-state current. Exploitation of this transient current for amperometric assay can lead to increased resolution and limit of detection.

As will be seen in following sections, the nature of the potential step—i.e. the coordinates of the waveform—can be carefully adjusted to increase the selectivity and sensitivity of this base amperometric sensor. For example, the potential step may be of sufficient duration for steady state to have been achieved or so short that the concentration gradient has only penetrated a short distance from the electrode surface. The voltage may be stepped to a potential where the electrode is under the kinetic control of charge transfer between electrode and electroactive species, or else limited by mass transport to the electrode surface—or else it may be under mixed control.

The simplest case is found where the applied potential corresponds to the region where a limiting current is observed—i.e. where all of the electroactive species that approaches the electrode is electrolysed and the surface concentration is zero, so

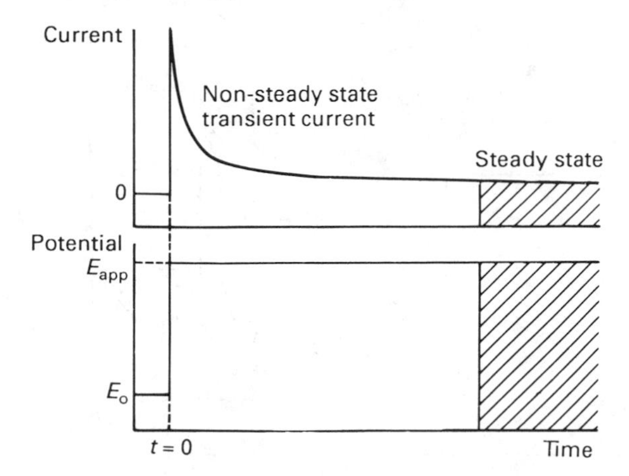

Fig. 5.7 The *i–t* profile following the application of a step change in potential.

that the concentration gradient and the diffusion layer thickness, δ, are time dependent.

In many, although not all, applications of amperometric measurement in sensors, the electrode is retained behind a membrane, whose function it is to allow the *selective* diffusion of analyte (or an electroactive species due to the analyte) to the electrode surface, while prohibiting the passage of interfering species from the sample solution. As will be seen in subsequent sections, the nature of this membrane can be very varied, providing, for example, merely a gas-permeable membrane for the oxygen electrode, or containing an immobilized biorecognition molecule such as an enzyme for the selective assay of the enzyme's substrate.

However, regardless of the exact nature of the membrane, it results in discrete diffusional barriers being established at the electrode. For example, if the membrane-covered electrode system shown in Fig. 5.8 is considered, where the membrane of thickness b is separated from the electrode surface by an electrolyte layer of thickness a, then assuming equilibrium conditions at the membrane–solution interfaces, the concentration of analyte in the solution sample phase (C_s) is related to the concentration in the membrane phase (C_m) by the distribution coefficient, K:

$$C_m = KC_s$$

During steady-state operation of the electrode, so that the electroactive species [O] is reduced under diffusion control, the concentration profiles are continually modified until a profile is reached at $t \rightarrow \infty$ (Fig. 5.8b), for a system in contact with a *stirred* sample solution).

The current monitored by the electrode is determined by the rate of diffusion of electroactive species to the electrode surface; therefore for the system here:

$$i = nFD_e(dC_e/dx)_{x=-a}$$

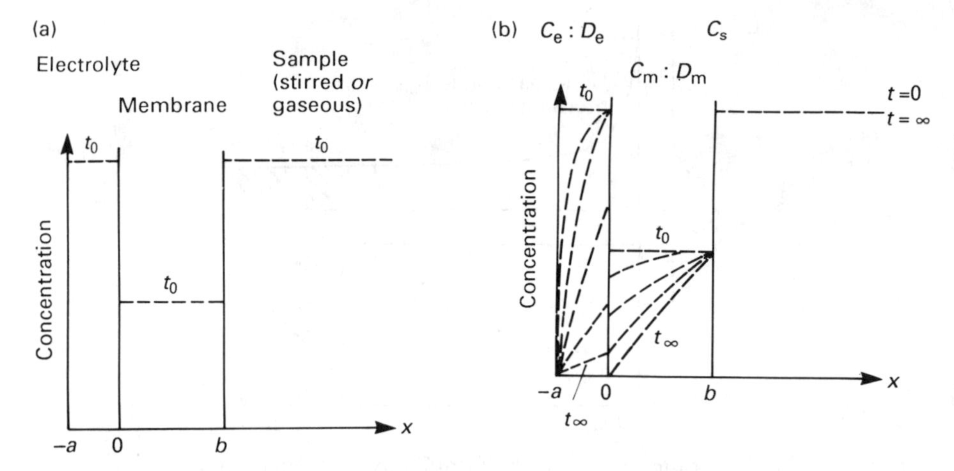

Fig. 5.8 Concentration profiles for a membrane-covered electrode. (a) The initial profile; (b) the development of concentration gradients through the different phases during attainment of steady state.

By imposing suitable boundary solutions the current density may be solved into three time zones.

(1) When the diffusion layer is retained within the electrolyte,

$$i \approx nFC_s(D_e/\pi t)^{1/2}$$

(2) When the diffusion layer has penetrated the membrane and the analyte concentration on the inner membrane surface is less than C_s,

$$i \approx nFC_s(P_m/\pi t)^{1/2}$$

$D_m K_m = P_m$ where P_m is the permeability coefficient and K_m is the ratio C_s/C_m.

(3) When steady-state current is reached,

$$i = nFC_s P_m/b$$

In aqueous electrolyte of a typical 10 μm thickness, the first time zone is < 100 ms. The membrane diffusional barrier will then retain the diffusional layer within this membrane phase for a period depending on its thickness and diffusion coefficient.

Since, in the simplest case, diffusion through the membrane is likely to be the slowest contribution to the overall process, the response time (t_r) for a membrane electrode can be approximated by

$$t_r \approx b^2/D_m$$

where b is the membrane thickness. It follows therefore that small values of b will result in a faster response time. It is often an advantage, however, to increase b so that the concentration gradient does not penetrate beyond the outer membrane interface, thus creating a local concentration gradient of analyte in an *unstirred* sample. In many cases, therefore, these two opposing effects must be balanced to achieve a compromise.

In addition to the advantages associated with increased current signal in non-steady-state operation, it is apparent from the foregoing discussion that in this mode, the duration of the potential step can be chosen so that the diffusional layer is *always* contained within the membrane (*note*: sufficient time must elapse between the application of each potential step for the concentration profile to re-establish at its $t=0$ value). Manipulation of these parameters for non-steady-state operation will be discussed for particular systems.

Applications of Charge-transfer Measurement—the Oxygen Electrode

An electrochemical assay method for oxygen, based on a membrane-covered electrode has been available for three decades (Clark, 1956), but nevertheless it is still being constantly modified and perfected.

It is particularly relevant in the context of biosensors, since not only is the measurement of gaseous or dissolved oxygen important in its own right, but linked to bacteria or tissue slices (Chapter 7) or enzymes (Chapter 8) it provides an indication of the state of dynamic equilibrium existing between oxygen utilization by the system and re-oxygenation. This is itself dependent on the presence or absence of a particular substrate, essential to that biorecognition system and can as such be applied in the assay of that substrate.

One of the essential requirements and indeed characteristics of a biosensor is its analyte selectivity, i.e. lack of response to other similar species which may also be present in the assay sample. The biosensor aims to achieve this mandate by selection of a biorecognition system which will utilize the analyte of interest, and through the manipulation of the control and data-capture parameters of the 'base sensor'. In the case of the oxygen electrode, selectivity for oxygen over other electroactive species, which can diffuse through the membrane cover, is achieved entirely through control of the electrode parameters.

Applications of oxygen assay are numerous, but a particularly good example of selectivity through sensor control is found in the measurement of dissolved oxygen in body tissue in a medical environment. An enormous effort has been devoted towards perfecting the oxygen electrode and achieving optimal electrode performance under all clinical conditions. The techniques and manipulations involved here, and the treatment of interfering signals produced by other electroactive species present in the sample, will therefore serve as examples for detailed examination.

The reduction of oxygen is a complex process which can occur via a number of mechanisms depending on reaction conditions (i.e. pH, solvent, etc.). In aprotic solvents, the reaction proceeds via a one-electron reduction to the superoxide

$$O_2 + e^- \rightarrow O_2^-$$

whereas in aqueous acidic media, two- or four-electron processes are recorded

$$O_2 + 2H^+ + 2e^- \rightarrow H_2O_2$$
$$O_2 + 4H^+ + 4e^- \rightarrow 2H_2O$$

Similarly at neutral or alkaline pH, two or four electron, one- or two-step mechanisms are proposed, which also depend on electrode material and electrode potential:

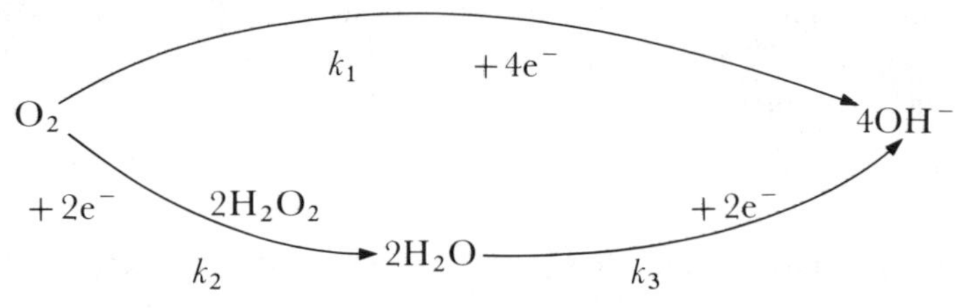

Obviously, for purposes of oxygen *assay*, the four-electron transfer, giving a higher current density, will provide a more favourable signal.

A major feature that should be recognized in all these mechanistic routes is that the reaction is essentially irreversible, so that amperometric assay of the O_2 is non-passive and will lead to depletion of O_2 in the sample. Errors concerned with this

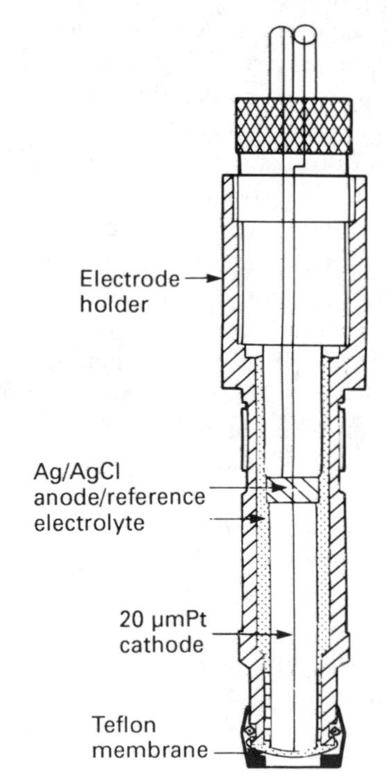

Fig. 5.9 Section through the radiometer Clark-type electrode.

depletion can be a major source of problems in biosensors involving an O_2 base sensor.

The Clark oxygen electrode was a membrane-covered electrode, where the electrolyte retained behind the membrane maintained a pH environment for the electrode independent of the sample, and where the oxygen-permeable membrane barrier prevented electroactive or surface-active impurities from the sample giving an erroneous signal or poisoning the electrode. The platinum electrode was polarized on the diffusion-controlled plateau of the oxygen reduction, and the current recorded was proportional to the concentration of oxygen in the sample solution. The oxygen from the sample diffuses through the membrane to the electrolyte and is reduced at the electrode surface.

As has already been established, the steady-state current

$$i = nF\alpha_{m} pP_{s}(D_{m}/b)$$

where pP_s is the partial pressure of oxygen in the sample and α_m is the solubility of oxygen in the membrane phase, indicates that the thinner the membrane, the more sensitive the measurement. In principle, a range of membranes could be employed with different physical characteristics and thicknesses. In practice however, the most frequently applied membranes are Teflon or silicon < 25 μm thick, fixed by the mechanical means of an O-ring or sleeve over the end of the electrode body (Fig. 5.9).

Sources of Error—Depletion of Sample

In consideration of the theoretical treatment of the membrane-covered oxygen electrode, it has been assumed that the concentration at the outer membrane edge (membrane–sample interface) remains constant, and is not depleted by the electrode reaction. In unstirred sample solution this condition is not fulfilled and the diffusion layer thickness extends beyond the interface into the sample and the recorded electrode current reflects the oxygen depletion at this interface. The effect of stirring can be simulated in a flow-through cell, where controlling the flow rate can alter the accuracy of the reading, the effect decreasing with increase in membrane thickness. In general, a compromise between a fast response time (thin membrane) and sensitivity to flow rate must be made.

For practical applications, for example *in vivo* monitoring of oxygen in blood, this depletion effect can have serious implications, since not only can a local oxygen depletion be detrimental to the surrounding tissue but in a working environment where electrode calibration is usually carried out in the gaseous phase (equivalent to a stirred sample), and operation is required in the liquid blood phase, a significant reading error would be obtained. This error is known as the liquid–gas difference and for a Clark-type Pt electrode of 20 μm diameter covered with a 12 μm Teflon membrane, it is of the order of 5%.

Various attempts have been made to reduce this error, but as already discussed the obvious solution to be found by increasing the membrane thickness also leads to an undesirable increase in response time and reduction in sensitivity.

Mancy and co-workers (Mancy *et al.*, 1962; Wise *et al.*, 1980) presented the theory for non-steady-state measurement and showed its application to membrane-covered oxygen electrodes. Subsequently the technique has been exploited, expanded and further perfected.

Previously, the steady-state equation has been applied

$$i = nF\alpha_m(D_m/b)pP_s$$

but if the measurement were made while the diffusion layer was retained within the membrane,

$$i \approx nF\alpha_m(D_m/\pi t)^{1/2}pP_s$$

or better still the electrolyte

$$i \approx nF\alpha_e(D_e/\pi t)^{1/2}pP_s$$

then a number of advantages can be gained:

(1) Higher current density at shorter times giving increased sensitivity (Fig. 5.7).
(2) Diffusion layer not penetrated into sample, therefore no liquid–gas difference and no depletion of sample.
(3) Independent of membrane (for very short times).
(4) Electrode life increased due to decreased 'on' time.

This non-steady-state operation is achieved by applying the polarizing voltage to the electrode as a rectangular wave (Fig. 5.10a), stepping between the polarizing voltage for diffusional controlled reduction of oxygen (pulse) and a 'rest' potential where the O_2 is not electroactive (space).

Brooks (1980) showed that the liquid–gas difference was not only dependent on the pulse length but also on the space length and that for a given space length the difference increased with increasing pulse length. This interrelationship could be explained by the time-dependent relaxation of the concentration gradients that had been established during the measuring pulse. In general terms it is necessary to make the space sufficiently long for the electrode to have no 'memory' of the previous pulse.

Non-Faradaic Current Error

Various sampling techniques have been employed for this non-steady-state operation (Fig. 5.10b, c). Rather than recording the entire current profile, a single sample is taken at a predetermined time (t_1) (Fig. 5.10b) after the beginning of the potential step. However, a better signal-to-noise ratio is obtained if the charge is recorded between t_2 and t_3 (Fig. 5.10c), that is to say the current signal integrated between two time limits,

$$\int_{t_1}^{t_2} i\mathrm{d}t = 2nFC_s(D/\pi)^{1/2}[t^{1/2}]_{t_1}^{t_2}$$

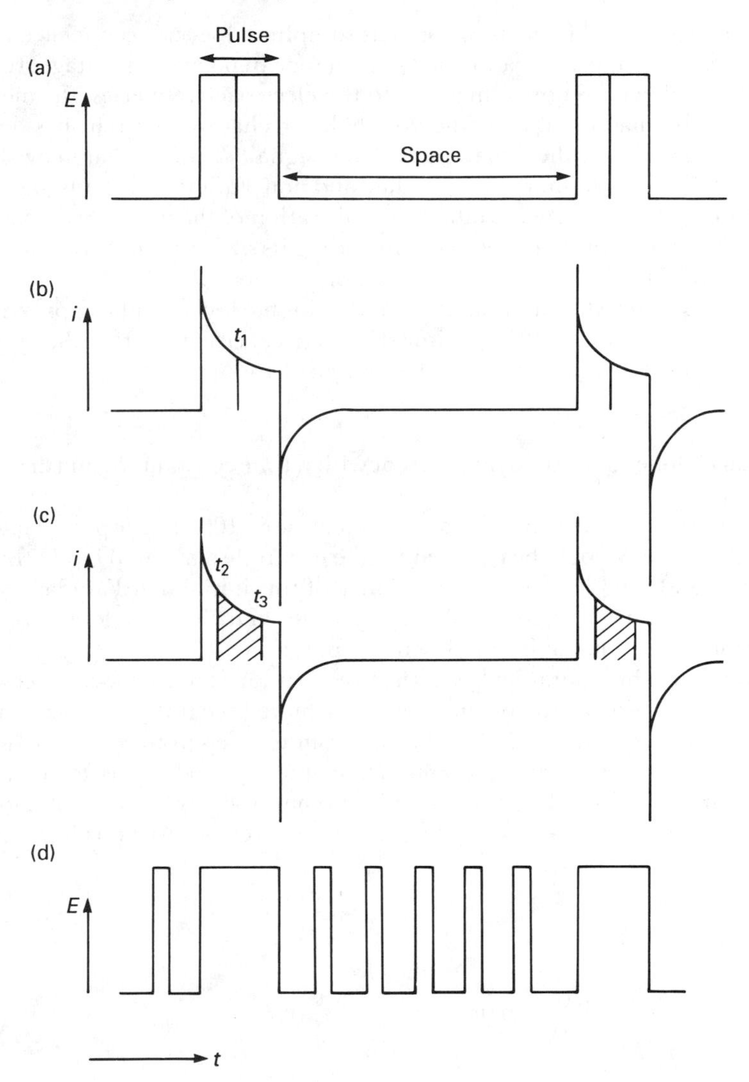

Fig. 5.10 Non-steady-state sampling techniques. (a) Rectangular wave polarizing voltage, described by V_p (pulse voltage), V_s (space voltage), t_p (length of pulse) and t_s (length of space); (b) and (c) current response to potential profile in (a) with (b) single-point current sampling at time t_1 and (c) integrated current sampling between times t_2 and t_3; (d) complex potential control waveform proposed to overcome non-Faradaic error.

As can be seen from Fig. 5.10(c), current sampling does not commence at $t=0$ immediately after the application of a step change in polarizing voltage. In order to ensure reliable current readings due to the electroactive species, the measurements must be made well after the double-layer charging current has decayed essentially to zero and the remaining current signal is entirely Faradaic. Figure 5.11 shows the contribution from Faradaic and non-Faradaic currents to the total current after a step change in voltage. The duration of the non-Faradaic current depends on the nature of the electrode (including its size) and the 'structure' of the solution giving the electrode interface double layer.

In a greatly simplified approximation, the double layer can be approximately represented by a resistor (R)/capacitor (C) electrical circuit, so that the result of a potential step of magnitude E, can be described by

$$i = E/R \exp(-t/RC)$$

that is, an exponentially decaying current with a time constant RC and dependent on potential.

It has been suggested that for a microelectrode (100 μm), a pulse length no greater than 50 mS would be required for the electrode current to be *independent* of membrane and sample phase, *but* avoidance of this initial non-Faradaic current error would require a considerably longer pulse and this therefore imposes an undesirable time constraint on the sensor operation.

The origin of this limitation lies in the time constant for double-layer charging after the application of the potential step. A more recent development in non-steady-state control mode (Hall, 1986) has been to attempt to retain the double-layer structure, even during the non-active space period, so as to reduce the charging time required. The more complex potential-control waveform shown in Fig. 5.10(d) appears to achieve this mandate. Previously control during the

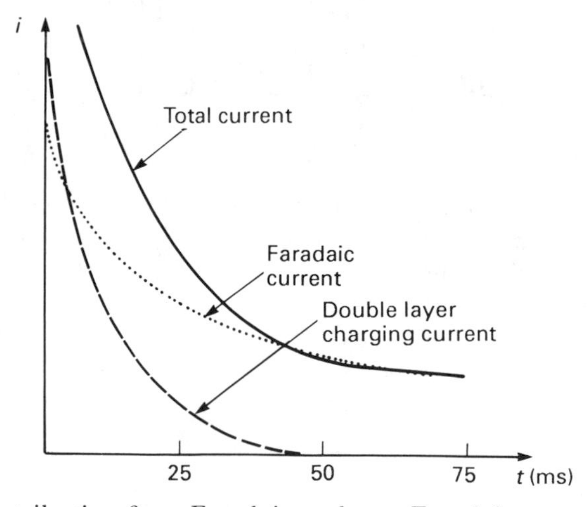

Fig. 5.11 Contribution from Faradaic and non-Faradaic currents to the total electrode current.

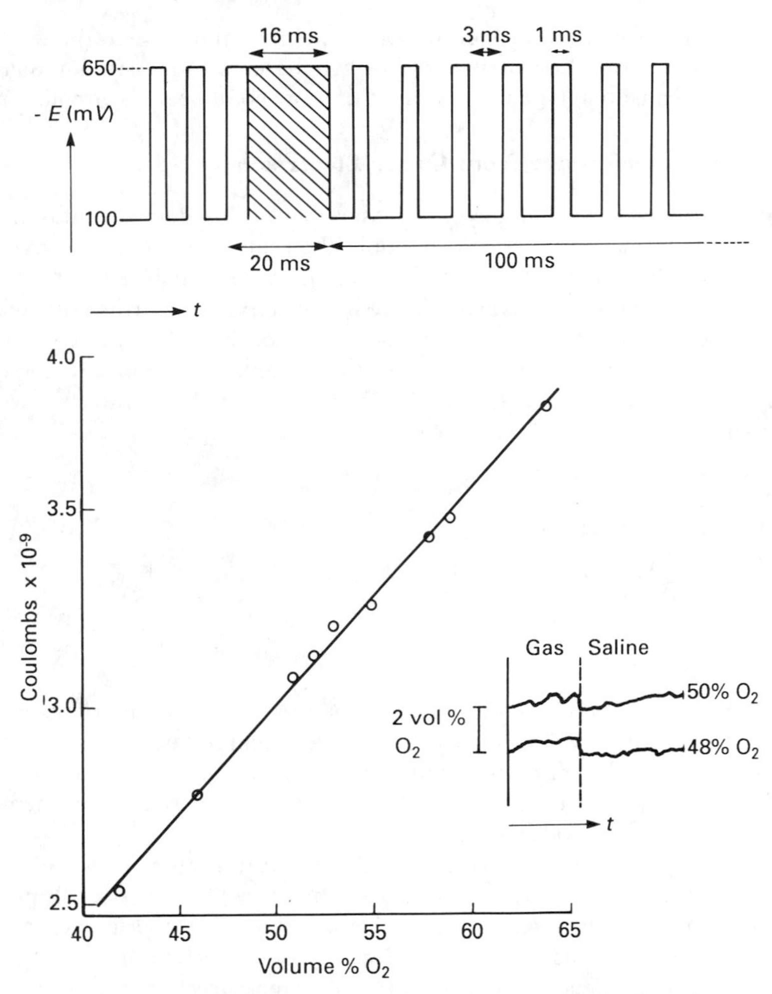

Fig. 5.12 Demonstration (below) of the elimination of the liquid–gas difference for oxygen measurement at a 125 μm Ag electrode covered with a 6 μm Teflon, when employing the complex waveform shown above. The current was sampled for 16 ms from $t=4$ ms to $t=20$ ms over a 20 ms measurement pulse. The liquid–gas difference (inset) was $<1\%$, compared with 5% for 20 μm Pt covered with 12 μm Teflon.

'space' had been at a constant potential where there is no electroactivity. In this more complex waveform, the non-Faradaic conditioning is maintained by a series of 1 mS pulses to the electroactive measuring potential—this pulse is probably too short to initiate the electrochemistry, and should avoid depletion of electroactive species at the electrode surface or in the electrolyte layer. This treatment results in the realization of considerably shorter measuring pulses (20 mS) and the report of a minimal liquid–gas difference for a membrane oxygen electrode (Fig. 5.12).

The benefits of this mode of operation are evident for pulses up to about 100 mS, but for applications where longer pulses can be employed without incurring errors, there is no advantage in applying this more complex measurement mode.

Selectivity: Interference from Other Electroactive Species

An important application of oxygen measurement is in environmental monitoring. In natural or waste waters, for example, fluctuations in dissolved oxygen can drastically modify the level of pollution over a very short time scale.

In the amperometric oxygen electrode, selectivity is primarily achieved through choice of electrode material and control of the polarizing voltage. SO_2 therefore, which is frequently encountered in environmental monitoring and has a redox couple $\approx 750\,mV$, more cathodic than oxygen, will not interfere with the measurement:

$$H_2SO_3 + 4H^+ + 4e^- \rightarrow S + 3H_2O$$

However such gases as NO

$$2NO + 2H^+ + 2e^- \rightarrow N_2O + H_2O$$

or Cl_2

$$Cl_2 + 2e^- \rightarrow 2Cl^-$$

which have a redox potential more anodic than oxygen could interfere, although usually these compounds are only present in trace amounts. In cases where the presence of interfering gases is significant, the membrane is chosen to act as a selective filter usually aiming to allow the O_2 molecule to permeate the membrane more easily than the 'pollutants'.

In some applications, however, even the membrane filter is an inadequate solution. The catheter oxygen electrode is an obvious solution to the monitoring of blood–oxygen levels in medical care. The possibility of interference by other blood-gases, in particular anaesthetic agents, has been emphasized by many groups. The consequence of erroneous oxygen measurement could lead to an incorrect and possibly fatal estimate of the patient's state.

Identification of anaesthetic errors in oxygen monitoring followed the work of Severinghaus *et al.* (1971) on the anaesthetic agent halothane. Nitrous oxide has also been identified as the source of a significant error in the oxygen reading—but not on all commercially available electrodes!

NITROUS OXIDE

Examination of the current–voltage plots for nitrous oxide at different electrode materials (Fig. 5.13a) shows the influence of electrode in the reduction:

$$N_2O + H_2O + 2e^- \rightarrow N_2 + 2OH^-$$

While the original Clark electrode utilized Pt, commercially available oxygen sensors for medical applications have employed both silver and gold. Polarized at

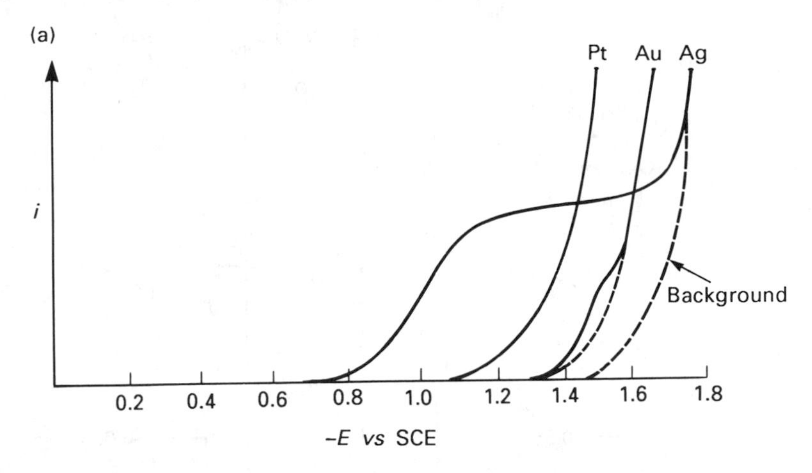

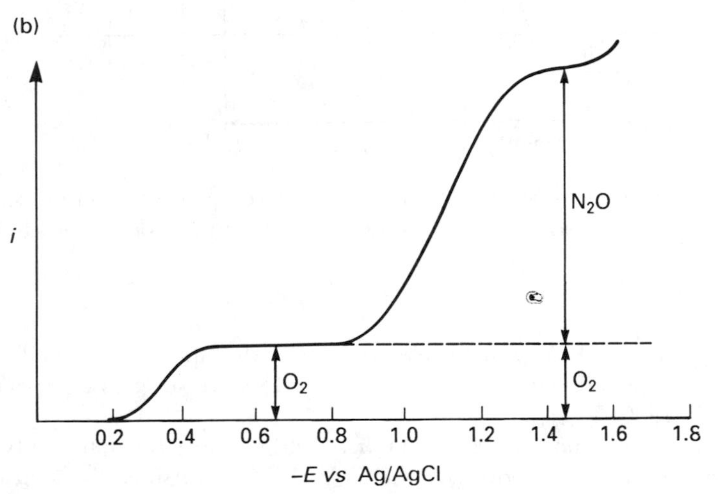

Fig. 5.13 (a) Reduction of nitrous oxide on different electrode materials; (b) reduction of a mixture of oxygen–nitrous oxide at silver electrode.

−1 V versus Ag/AgCl it is not surprising therefore that the silver sensor gives a response for N_2O. Evidence for N_2O sensitivity on gold electrodes after long periods of use is not substantiated by these *i–V* plots, but this was attributed to contamination of the electrode by silver migration from the Ag/AgCl reference/secondary electrode (Albery *et al.*, 1978), which is employed in almost all commercial oxygen electrodes. (The presence of migrating Ag^+ ions from the reference electrode was also shown to cause interference at potentiometric sodium-sensitive glass electrodes; see Chapter 3.)

As may be seen in Fig. 5.13(b) efficient control of the polarizing voltage to be more anodic than −0.8 V versus SCE provides a simple solution to this problem of N_2O error. In fact potential selection also reveals a method for the simultaneous

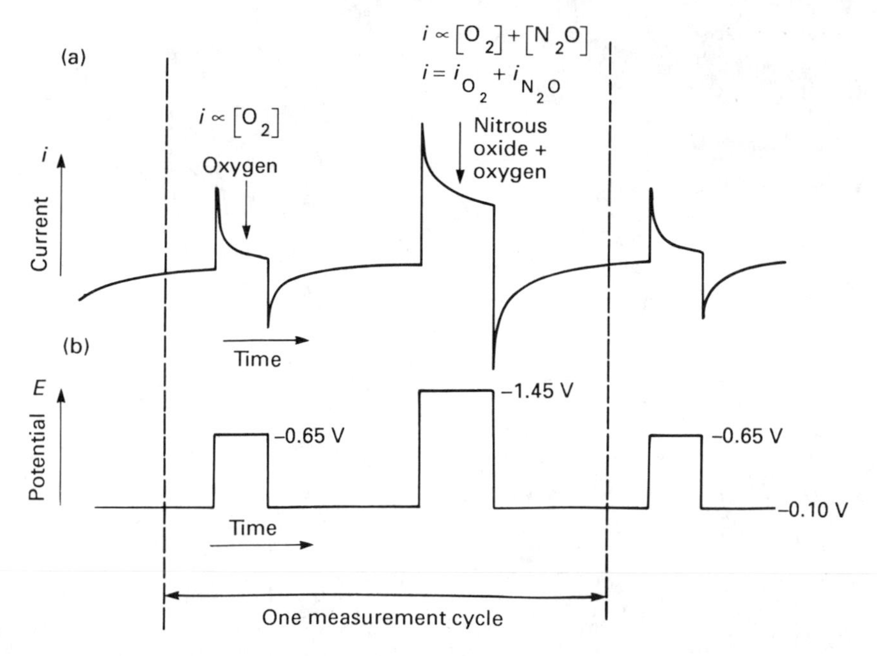

Fig. 5.14 Measurement regime for the consecutive determination of oxygen and nitrous oxide at a single Ag electrode. (a) Current response following application of the potential waveform shown in (b).

assay of O_2 and N_2O at a single electrode, via the application of alternate polarizing voltage pulses to -0.65 V and -1.45 V versus SCE, respectively (Brooks *et al.*, 1980).

The technique is demonstrated in Fig. 5.14 and employs the non-steady-state operation mode discussed above. At -0.65 V the current signal is exclusive to oxygen and can be calibrated for oxygen concentration:

$$i_{-0.65} = I_{O_2}$$
$$i_{-0.65} \propto [O_2]$$

At -1.45 V versus SCE, on the limiting plateau for N_2O, the current response can be due to both N_2O and O_2:

$$i_{-1.45} = i_{O_2} + i_{N_2O}$$

but the N_2O current here may be calculated by subtraction of the O_2 current which was calculated in the previous pulse at -0.65 V.

HALOTHANE

Halothane reduction at 'oxygen electrodes' presents a problem very similar to that of nitrous oxide. Figure 5.15 shows the reduction of halothane on various metals.

It suggests, however, that separation of the oxygen and halothane signals on silver is very unlikely since their half-wave potentials are too close (−0.43 V versus SCE and −0.56 V versus SCE respectively). Even at a gold electrode, where the reduction potentials *are* separated, the presence of halothane has been shown to inhibit the oxygen reduction current at −0.65 V versus SCE, thus causing a serious underestimation if oxygen is assayed at this potential (Hall *et al.*, 1988). The O_2 signal at −1.45 V appeared unaffected, but at this potential halothane is also reduced and in steady-state measurements, the two signals cannot be separated.

The adoption of multidimensional analysis was highlighted in Chapter 4 as a means of separating overlapping signals. Some considerable effort has been devoted towards improving oxygen/halothane selectivity on gold; a preliminary solution has been found by the introduction of the time dimension into the analysis and manipulation of the electrode kinetics on the time axis.

If the rate constant for the electron transfer (k) is very slow, then the electrode reaction does not achieve diffusion control. Normally as the overpotential is increased, the electrode kinetics become matched by the diffusional limitations, and at higher overpotentials mass-transfer effects dominate.

As discussed above, Fick's second law can be solved for a diffusion-controlled process to give the time-dependent relationship for the current response—the Cottrell equation, but for an *irreversible* electrode reaction which is not under diffusion control, possibly of the type (Albery *et al.*, 1981b)

$$\underset{\text{(halothane)}}{Ag + e^- + CHClBrCF_3} \xrightarrow{\text{slow}} (Ag \ldots Br \ldots CHClCF_3)^- \xrightarrow[\text{fast}]{H_2O,\ e^-} Ag + Br^- + CH_2ClCF_3 + OH^-$$

then the solution has two limiting forms. At short times

$$i_k \approx -nFkC_s[(1 - 2kt^{1/2})/(\pi^{1/2}D^{1/2})]$$

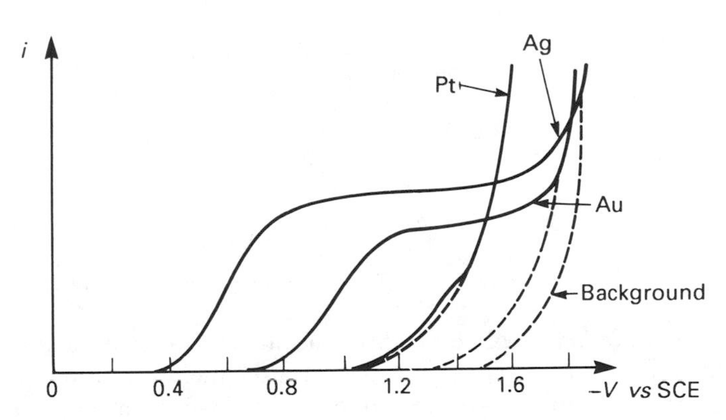

Fig. 5.15 The reduction of halothane on different electrode materials.

and at longer times, the solution reduces to the Cottrell equation:

$$i_d \approx (-nFC_sD^{1/2})/(\pi^{1/2}t^{1/2})$$

Empirically, this means that at short times the current i_k is only some fraction of the expected diffusion-controlled current i_d, solved for the same value of t.

How can this knowledge assist in separating two superimposed amperometric responses, in this case the oxygen and halothane signals? In a previous example, separation of the O_2 and N_2O signals was achieved by identification of two polarizing voltage zones:

zone 1 O_2 reduced $(-0.65\ V)$
zone 2 O_2+N_2O reduced $(-1.45\ V)$

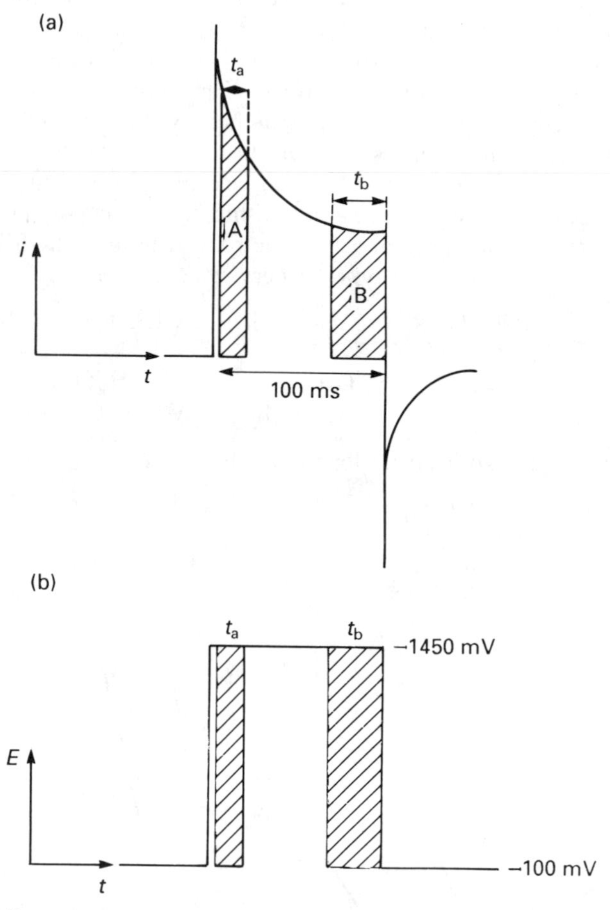

Fig. 5.16 Measurement regime for the simultaneous measurement of oxygen and halothane on gold. (a) Current recorded after the application of the potential steps shown in (b); (b) Two integration zones t_a and t_b give $\{i_{k(hal)}+i_{d(O_2)}\}$ and $\{i_{d(hal)}+i_{d(O_2)}\}$ respectively.

In this example the signal can be separated into two time zones:

zone 1 O_2 diffusional control: halothane kinetic control ($\leqslant 20\,\text{mS}$)
zone 2 O_2 and halothane under diffusion control ($>50\,\text{mS}$)

In practice, this assay can be performed on a *single* pulse (Fig. 5.16) where current is sampled in the two time zones, and the data solved by considering the two limiting forms i_d and i_k (Hall, 1986).

The detailed treatment of some specific examples of sources of error which can occur in the use of an amperometric oxygen electrode has served here as a means of investigating the role of electrode control parameters in achieving enhanced stability, sensitivity and selectivity. For oxygen electrodes linked to biological recognition systems such as enzymes or whole cells (discussed in Chapters 7, 8), these parameters may display a different degree of importance, but their manipulation could drastically alter the output, range and selectivity of the biosensor.

Amperometric Electrodes for Estimation of Ion Concentration

Ion-selective electrodes have come to be defined as potentiometric electrodes, but the introduction of an indicator species, which shows changes in its electrochemical behaviour due to the presence of a particular ionic species, can allow indirect ion concentration estimation at an amperometric electrode. The principle is similar to that required for the development of ion-selective membranes for potentiometric measurements, in so far as the electrode surface is modified with a polymer film, which shows high affinity and selectivity for the target ion. For this application, however, it must also show additional redox activity, which can serve as the internal standard. Introduction of the affinity characteristics can be achieved either as part of the polymer backbone or by incorporation of affinity species via ion exchange.

The dihydroxyazo dyes will form metalloderivatives. Antipyrylazo-III (AP-III), for example,

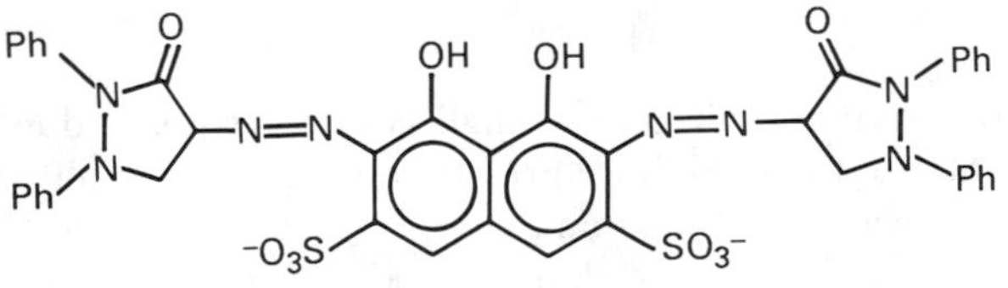

has a high affinity towards calcium, without interference from other divalent ions (Hurrell and Abruña, 1988). The dye shows an oxidation peak at +0.55 V versus SCE, which is sensitive to the presence of Ca^{2+}. Incorporated by ion exchange into a viologen polymer, deposited at a Pt electrode, the resultant electrode, monitored at +0.55 V, responds to Ca^{2+} with a limit of detection at around 10^{-7} M. This is comparable with a potentiometric Ca^{2+} sensor.

Polymer films of $[Ni^{II}Fe^{II/III}(CN_6)]^{2-/-}$ may be electrochemically deposited on nickel electrodes from solutions containing hexacyanoferrate ions. The redox potential of the $Fe^{II/III}$ couple varies with the lattice structure, i.e. the crystal field surrounding the iron centre, caused by the nature of the electrolyte cation.

This property can be exploited in the determination of ion concentration, since a difference in surface affinity exists between the various alkali cations, Cs^+ having the highest affinity (Amos *et al.*, 1988). A shift of 0.65 V is seen for the $Fe^{II/III}$ couple, between Na^+ and Cs^+ solutions. The Cs^+-related peak current for the $Fe^{II/III}$ couple shows a complicated relationship to Cs^+ concentration, but for a given Na^+ concentration, the Na^+-related peak is decreased in proportion to $\log[Cs^+]$, with a detection limit of 10^{-8} M. This offers an improvement over potentiometric methods, which only allow detection down to 10^{-5} M Cs^+, with considerable interference from other alkali cations. In this amperometric method, in fact, Na^+ is necessary but concentrations >0.1 M Na^+ are insensitive to changes in Na^+, so that under operation in these 'Na^+ saturation' conditions serious cross-reactivity with other ionic species does not occur.

Many other examples also exist of amperometric linked assays of ionic species. Their comparability with potentiometric methods is largely a function of the particular application.

Macromolecular Systems

Electron transfer to biological macromolecules will be dealt with more specifically in Chapter 8, but the background and the limitations are introduced here. On the whole, electrochemical analysis of macromolecules is made difficult by their low diffusion coefficients and frequent inaccessibility of the electroactive groups (usually contained in the coenzyme element) for electrode processes.

Although in non-aqueous solutions the carboxyl group of the amino acids may be electrochemically reduced, their electroactivity in aqueous solution is usually confined to cystine. The proposed mechanism for the reaction involves the reduction of adsorbed cystine (Stankovich and Bard, 1977):

$$(RSSR)_{ads} + 2H^+ + 2e^- \rightarrow 2(RSH)_{soln}$$

Proteins containing this residue will give a small current response due to the electrochemical reduction of their sulphide bridges, but this is generally of little analytical significance (Cecil and Weitzman, 1964).

More useful is the catalytic current that is produced in cobalt or nickel solutions. This, known as the *Brdička current*, has been widely exploited in clinical analysis (Brdička, 1933a, b).

A similar picture emerges for the nucleic acids, since only cytosine and adenine are reducible in aqueous solution (Janik and Elving, 1968), and the electrochemical behaviour of the nucleosides and nucleotides become a feature of their base content and adsorbability. Polynucleotides containing adenine and guanine may also be oxidized at carbon electrodes (Brabec and Dryhurst, 1978) and both

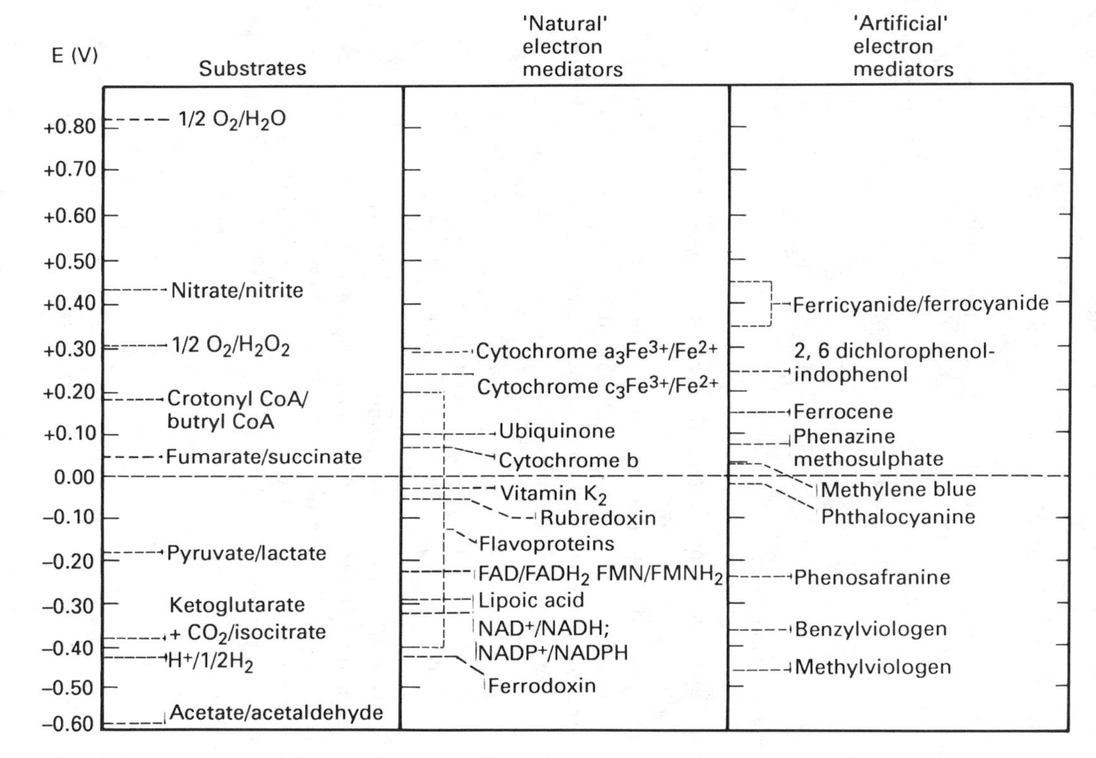

Fig. 5.17 Values of E at pH 7 and 298 K for several redox couples of interest.

RNA and DNA produce two peaks, the more flexible single-stranded polynucleotides giving a higher peak than the ordered double-stranded polynucleotide. This is also the case with their electrochemical reduction. Although this latter feature may suggest a method for the transduction of the interaction between a DNA single-stranded template sequence and its complement, the measurement is not straightforward, since the polynucleotides are strongly adsorbed at the electrode surface and a major contribution to the total signal would come from electrode fouling and non-specific adsorption and electrochemistry.

It is the poor charge transfer to macromolecules and 'random' adsorption of both electroactive and non-electroactive species at the electrode surface that inhibits their use in electrochemical analysis and has resulted in the use of the more indirect approach that will be discussed subsequently.

Redox Enzymes

The class of enzymes known as oxido-reductases are electron-transfer agents participating in a chain of *charge-transfer* reactions, the oxidases for example culminating in molecular oxygen. These charge-transfer enzymes in particular would appear to be ideal candidates for utilization in the indirect analysis of their substrates. Three *major* classes of these enzymes are most usually involved, as defined by their prosthetic group (or redox centre):

(1) *Pyridine-linked dehydrogenases* requiring NAD or NADP
(2) *Flavin-linked dehydrogenases* containing FAD or FMN
(3) *Cytochromes* containing porphyrin ring system

Other classes are also found, among which the NAD-independent quinoproteins can sometimes offer an advantage of being independent of oxygen.

The standard electrode potentials for these systems are shown in Fig. 5.17, where they are compared with those of some 'artificial' redox catalysts and also E^θ for some of the enzyme substrates. The analogy with the catalytic system already discussed is obvious:

$$\mathrm{O} + ne^- \rightleftharpoons \mathrm{R}$$
$$\mathrm{R} + \mathrm{A} \rightarrow \mathrm{O} + \mathrm{B} \qquad \text{or} \qquad \mathrm{O} + \mathrm{A} \rightarrow \mathrm{R} + \mathrm{B}$$

where, if O and R now refer to the oxidized and reduced forms of the enzyme, and A is the substrate (S) and B the product (P), the analogous enzyme–substrate reaction is:

$$\mathrm{R}_{(\mathrm{enz})} + \mathrm{S} \rightarrow \mathrm{O}_{(\mathrm{enz})} + \mathrm{P} \qquad \text{or} \qquad \mathrm{O}_{(\mathrm{enz})} + \mathrm{S} \rightarrow \mathrm{R}_{(\mathrm{enz})} + \mathrm{P}$$

The electrochemistry of the redox group of these enzymes is shown by examining the voltammetric curves of their respective prosthetic groups. FMN for example, shows a pH-dependent redox couple (Fig. 5.18):

The process may occur in two steps via a semiquinoid free-radical intermediate. In neutral solution, the curves obtained for FMN by cyclic voltammetry show two peaks, thus indicating the two-step process.

However, although electron transfer to the isolated prosthetic groups has been achieved at potentials at or close to the standard electrode potential, attempts to achieve reversible charge transfer between electrode and enzyme have met with varying degrees of success and have generally required a large overpotential. In the enzyme, where the electroactive cofactor is embedded in a protein coating, there is only limited accessibility between the redox centre and the electrode, so that electron transfer is inhibited and a limiting current is not normally observed. At the renewable surface of a dropping mercury electrode, a peak can be obtained at a potential only about 100 mV more anodic than the well-behaved electrochemistry of FAD, but entrapped in a polyacrylamide gel at a Pt gauze electrode or incorporated in a carbon paste electrode, only a very low and poorly defined

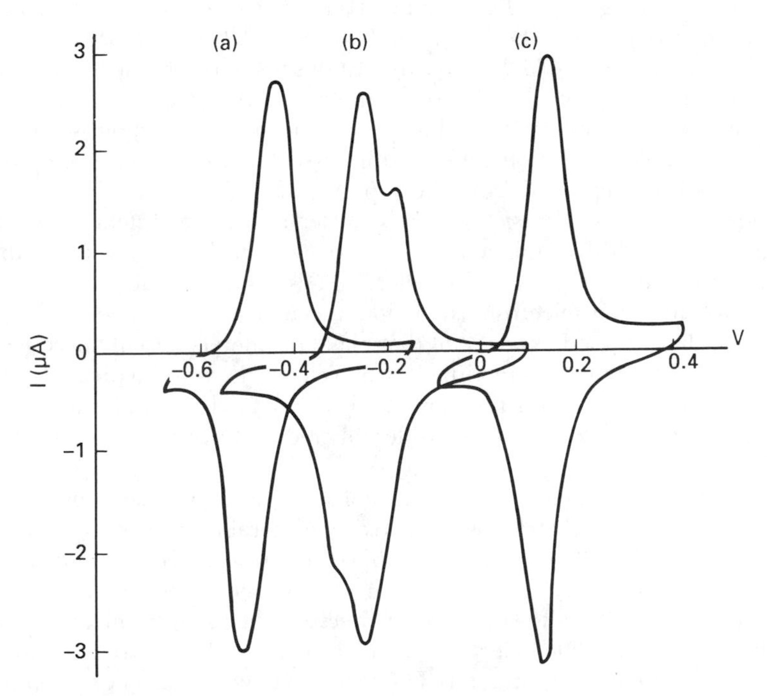

Fig. 5.18 Cyclic voltammetric curves for FMN at: (a) pH 13.8; (b) pH 8.4; (c) pH 0.94.

current density can be attributed to electron transfer between electrode and enzyme. In fact with glucose oxidase covalently grafted to a Pt electrode *no* exchange current could be obtained with the protein prosthetic group (Bourdillon *et al.*, 1979; Kamin and Wilson, 1980). These problems obviously prevent the use of a simple electrode material to provide the function of *direct* re-oxidation (and thus regeneration) of enzyme, following the enzymic oxidation of its substrate. This presents a limitation in the design of enzyme-linked assays which monitor charge transfer.

The use of enzymes in electrocatalysis requires the solution to this problem through modification of the electron transport at the electrode. The main approaches that have been made are:

(1) Adsorption of a 'modifier' on the electrode
(2) Covalent attachment of an activating reagent on the electrode
(3) Deposition of polymer film
(4) Use of low-molecular-weight mediators

Modified Electrodes

Much of the pioneering work on modified electrodes involved the adsorption of organic functional groups. For example, enhanced redox activity has been recorded for cytochrome C (Eddowes and Hill, 1979; Albery *et al.*, 1981a) at a gold electrode in the presence of 4,4′-bipyridyl. In this instance, the electron promoter is adsorbed reversibly on the electrode, in equilibrium with its bulk solution. Irreversible adsorption can be achieved by the use of promoters which are insoluble in aqueous solution and are dip-coated onto the electrode from an organic solvent or deposited from the vapour phase.

A common feature of *electron promoters* is an electron-rich π-system where charge delocalization is readily stabilized. In the same way, carbon electrodes with their extended π-systems are particularly effective adsorption surfaces.

The two- and four-electron reduction mechanisms of oxygen have been discussed earlier. A good example of the role of promoters for enhanced electron transfer is in the creation of a modified carbon electrode, where a dicobalt cofacial porphyrin was adsorbed on the surface (Fig. 5.19), thus catalysing the four-electron reduction of oxygen at a considerably reduced overpotential (Collman *et al.*, 1980).

Because of its importance as a cofactor for many enzymic reactions, the redox chemistry of NAD^+/NADH has attracted considerable interest. Oxidation of NADH has been achieved at unmodified electrodes, but usually with a considerable overpotential and giving a product which adsorbs on the electrode. Wang and Lin (1988), however, have investigated an intermittent cleaning procedure for glassy carbon electrodes, which removes the electrode fouling and reduces the required overpotential. The technique, which has been applied for cyclic voltammetry, but could presumably be adopted for general use, involves the application of a cycling square wave between each voltammetric sweep. In the

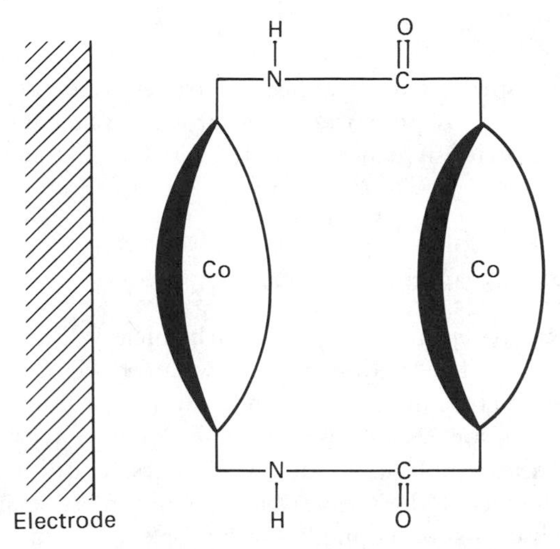

Fig. 5.19 Dicobalt cofacial porphyrin, adsorbed on an electrode surface as an electron promoter for oxygen reduction.

case of NADH the optimal treatment variables have been found to be a 10 Hz wave between ± 1.8 V for 15 s.

In general, however, various attempts have been made to modify the electrode in order to improve the electron transfer; medola blue (7-dimethylamino-1,2-benzophenoxazine) adsorbed on graphite, for example, reduces the overpotential for electron transfer, through the formation of a charge-transfer complex (Gorton *et al.*, 1984).

One of the major disadvantages of adsorption can be its instability. The anchoring of these reagents via a covalent bond has, therefore, also been extensively developed. The most usual approach is to use suitably activated metal or carbon surfaces, such as metal oxides or oxidized carbon. Oxidation of the latter produces a high density of –COOH groups to which the π-rich promoters can be attached, e.g.

$$|\!-\!C(=O)Cl + NH_2CH_2CH_2\!-\!C_6H_3(OH)\!-\!OH \longrightarrow |\!-\!C(=O)NHCH_2CH_2\!-\!C_6H_3(OH)\!-\!OH$$

or attachment to metal oxides could, for example, be achieved via the reactive silane,

$$M\!-\!-\!|\!-\!-OH + XSiX_2R \longrightarrow M\!-\!-\!|\!-\!-OSiX_2R + HX$$

Mediated Electron Transfer

A rather different approach to modified electrodes is to use a modifier that is electroactive at the redox potential of the protein. The catalytic system that has been discussed previously can be adapted for this mediated electron transfer:

$$Med_{(red)} \rightleftharpoons Med_{(ox)} + ne^- \qquad \text{(electrochemical)}$$
$$Med_{(ox)} + Bio_{(red)} \rightarrow Med_{(red)} + Bio_{(ox)} \qquad \text{(chemical)}$$

where $Bio_{(red)} \rightarrow Bio_{(ox)} + ne^-$ is slow electrochemically, Med = mediator and Bio = bio-redox system. The mediator is a low-molecular-weight redox couple with fast electrode kinetics, and which can also undergo homogeneous electron transfer with a solution species, in this case the bio-redox macromolecule. Among the many redox mediator molecules that have been tested, quinones, ferricyanide, organic dyes and ferrocene derivatives have seemed most successful here. The principal properties of a suitable mediator can be summarized:

(i) fast heterogeneous electron-transfer kinetics;
(ii) rapid homogeneous electron-transfer with the redox protein;
(iii) stable in the oxidized and reduced forms; and
(iv) no reaction with other solution species (particularly O_2).

Unlike the electron promoters already mentioned, these mediators do not need tc be adsorbed or otherwise attached to the electrode surface. Indeed homogeneous electron transfer between protein and mediator can take place at some distance from the electrode. However, for the purpose of a *biosensor*, a solution reagent that must be added prior to measurement is unsatisfactory and to be of any practical use the mediator must be suitable for immobilization with the enzyme on the electrode surface.

Various ferrocene derivatives have been identified to function either as solution mediators, or in the immobilized form. Ferrocene, functionalized, for example, by formation of the trichlorosilylferrocene derivatives will react with anodized Pt or Au to give an electrode showing the one-electron redox couple at ~0.45 V versus SCE, attributable to ferrocene. The preparation is moisture-sensitive and must be carried out under non-aqueous conditions, but the resultant electrode is reported to be extremely durable, with a shelf life exceeding 8 weeks.

Technically, however, covalent modification of the electrode surface is generally more difficult than entrapment via polymerization. The latter technique is therefore becoming increasingly popular. Polymer films often adhere well to the electrode providing either a mono- or multi-layer matrix which may be electroactive or electrochemically inert. Electroactive polymers fall into three categories:

(i) ion-exchange polymers
(ii) redox polymers
(iii) conducting polymers

ION-EXCHANGE POLYMERS

Ion-exchange polymer films become electroactive by doping with electroactive ions. Exchange of the ClO_4^- anion with $Fe(CN)_6^{3-}$ in the polyvinylpyridine film, for example, gives a modified electrode showing ferricyanide redox activity (Oyama and Anson, 1980a).

The fluoropolymer, Nafion, is a cation exchanger with hydrophilic and hydrophobic zones. A redox couple retained in the hydrophobic zones is electroactive there, but isolated from the electrode. It therefore requires a mobile redox cation to act as electron carrier (Oyama *et al.*, 1980a, b; Oyama and Anson, 1980b):

$$-(CF_2CF_2)_x(CFCF_2)_y-$$
$$|$$
$$O-(C_3F_6)-O-CF_2CF_2-SO_3-Na^+$$
(Nafion)

Introduction of a variety of redox species into the Nafion polymer by ion exchange gives a redox active polymer showing diffusion-controlled behaviour (i.e. $v^{1/2}$ proportional to peak current; see above, voltammetric techniques). However, the apparent diffusion coefficient (D_{app}) within the polymer will be influenced by the electron-exchange mechanism (Ruff *et al.*, 1971):

$$D_{app} = \frac{k\pi\delta^2 C}{4}$$

where D = diffusion coefficient, δ = distance for electron transfer, C = concentration of electroactive species and k = rate constant for electron transfer. The bulk diffusion process within the polymer could be described in terms of electron hopping through the polymer from reduced to oxidized sites or in terms of diffusion.

D_{app} for Nafion doped with $Ru(bpy)_3^{2+}$ (bpy = 2,2′-bipyridine) is reported to contain a large contribution from electron transfer (White *et al.*, 1982; Martin *et al.*, 1982) whereas $Cp_2Fe\text{-}TMA^+$ ([(trimethylammonio)methyl]ferrocene) does not. Apparently inconsistent with this finding, though, is the independence of D_{app} on concentration for $Ru(bpy)_3^{2+}$.

Generally the peak separation of the anodic and cathodic peaks for the redox couples in these polymers is quite large. This is normally associated with a heterogeneous rate constant lower than in aqueous solution. Martin *et al.* (1982) propose however that this is at least partially due to interactions between different electroactive sites.

However, regardless of the exact mechanism of electron transfer in the film, electroneutrality would be maintained by ion migration. Rate limitation by this process appears to be a function of ion size and film thickness, and only influences the value of D_{app} for thick films ($>20\ \mu m$). Ion doping of the film can however be exchanged in this migration process.

$Ru(bpy)_3^{2+}$ doped Nafion cycled to a steady state in $Ru(bpy)_3^{2+}$ free electrolyte and then doped briefly with $Os(bpy)_3^{2+}$ gives a mixed layer film (White *et al.*,

1982). Initially the film shows only a very small peak due to $Os(bpy)_3^{2+}$ and a significant increase in the $Ru(bpy)_3^{2+/3+}$ redox couple, attributed to electron mediation through the outer layer giving accelerated electron transport. The enhanced properties of such mixed films are, however, only short lived and as the two species equilibrate within the film with continued cycling, the $Os(bpy)_3^{2+}$ peak increases and the catalysis is reduced.

This type of redox film modification has been used to catalyse the reduction of oxygen (Faulkner, 1984) with cobalt(II) tetraphenyl-porphyrin acting as a redox catalyst for the oxygen reduction:

$$\text{Catalyst}_{(ox)} + ne^- \rightleftharpoons \text{Catalyst}_{(red)}$$
$$\text{Catalyst}_{(red)} + O_2 \rightarrow \text{Catalyst}_{(ox)} + OH^-$$

The redox catalyst was retained within the hydrophobic zones of the Nafion and the electron shuttling provided by a hexammine-ruthenium(III) cation.

Developing a polyelectrolyte film such as Nafion offers some considerable flexibility because of the range of electroactive counterions that can be exchanged into the film. However, since this ion exchange is an equilibrium reaction it can be slowly reversed in the absence of the ion in solution. In theory, therefore, the electroactivity of these films is more transient than when the redox couple is at a *fixed site* on the electrode surface.

REDOX POLYMERS

For the polymer system where the electroactive species is attached to the polymer backbone, it can be anticipated that leaching is likely to occur at a lesser rate than where the species is held by electrostatic interactions. Polyvinylferrocene typifies a polymer where the redox centre is an integral part of the polymer (Fig. 5.20). In this model the heterogeneous charge transfer rate constant is typically 10^2–10^3 times smaller than the solution species, vinyl ferrocene, but when compared with the diffusion coefficient (D), for the polymer matrix, the value $k^0/D^{1/2}$ is comparable with the solution parameters. Indeed in the latter case changes in the viscosity of the solution gave corresponding changes in the apparent value of k^0 (Leddy and Bard, 1985).

Covalent linkage of mediators to polystyrene, cellulose or other similar supports has also been recorded, and their catalytic activity retained—at least partially. However, the kinetics of electron transfer with the electrode are drastically reduced, compared directly with the solution reaction (Cenas *et al.*, 1984).

CONDUCTING POLYMERS

The conducting polymers are particularly attractive since many are easily formed by electrochemical polymerization. Polymers that can be grown in aqueous solution such as polypyrroles, polyphenoxides and polyanilines, would offer the possibility of formation in the presence of a redox enzyme. Indeed all these

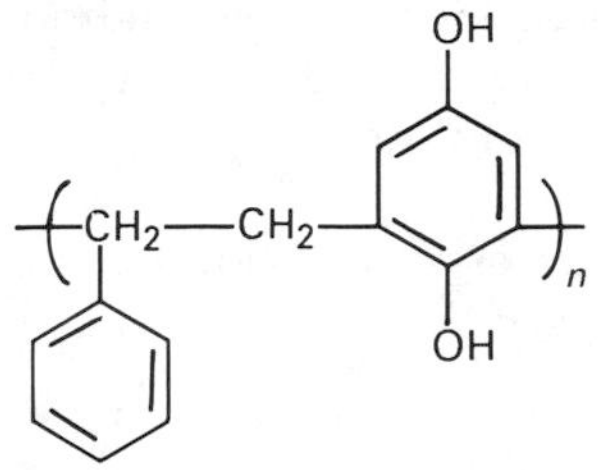

PSQ (sytrene-hydroquinone)

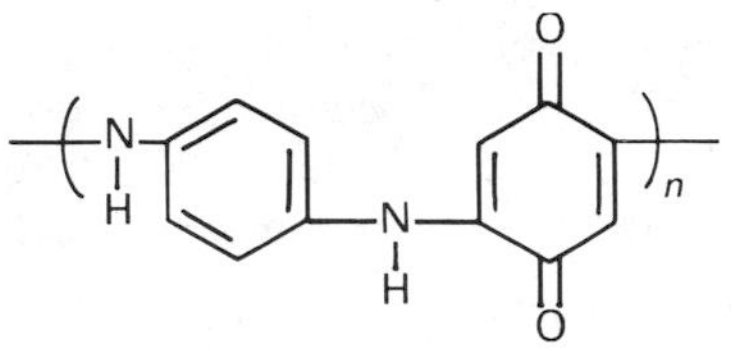

PFQ (p-phenylenediamine-benzoquinone)

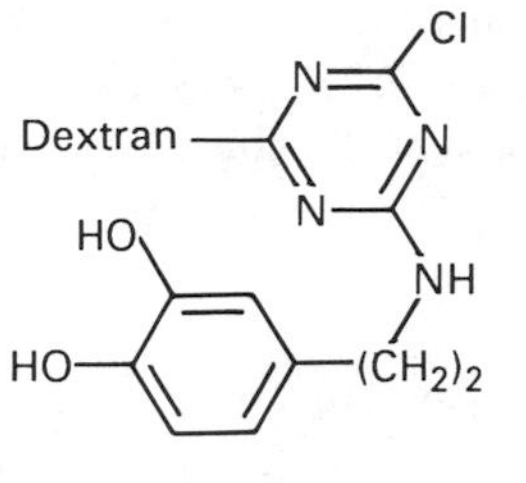

PDD (dopamine modified dextran)

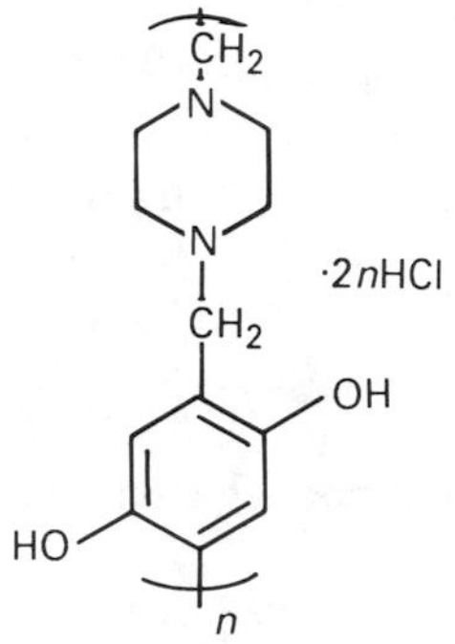

PPQ (hydroquinone-pyperazine)

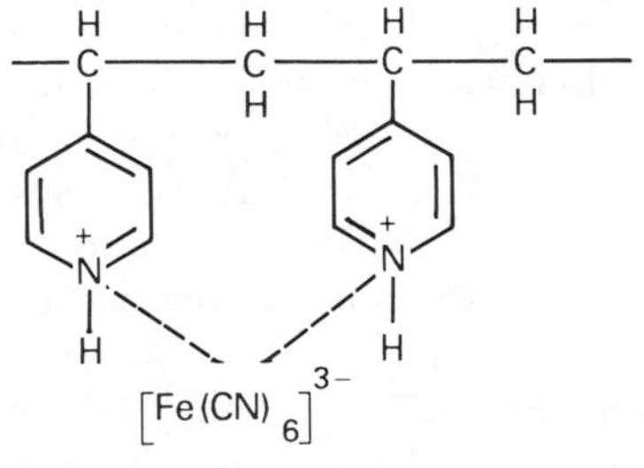

$PVPH^{+} - [Fe(CN)_6]^{3-}$

Fig. 5.20 Structures of some synthetic redox polymers.

polymers electropolymerized in the *presence* of glucose oxidase have been shown to entrap the enzyme in the polymer matrix (see Chapter 8).

The pyrrole polymer, for example, is formed primarily by α,α'-coupled pyrrole units resulting in a cationic structure. This film is inherently electroactive, the redox reaction involving oxidation of the delocalized π-system. The polymer can be reversibly cycled between oxidized (conducting) and neutral (insulating) states.

The conductivity of the polymer is very much dependent on the conditions of the polymerization. Aqueous polymerizations do not give such compact highly conducting polymers as non-aqueous formation, but do provide an ideal immobilization procedure for the biomolecule.

Efficient direct electron transfer to the redox centre of enzymes immobilized in these pyrrole polymers has not been anticipated since

(1) the redox potential of the film is too far from the redox potential of the protein, and anyway
(2) only the oxidized form of the film is conducting so electron transfer across the neutral film is not enhanced.

Further electroactivity can be induced in the film by the introduction of electroactive pendant groups or by entrapment of low-molecular-weight redox species. Entrapment of phthalocyanine, for example, gives an electroactive film, and other film modifications of particular benefit for biosensors will also be considered in Part II.

Micro-electrode Fabrication and Application

Amperometric sensors, whether they involve modified or unmodified electrode materials, require in the first instance the presence of the conducting material which constitutes the base electrode. Electrode fabrication has traditionally involved the embedding of a conducting wire of suitable material and desired diameter in an insulating collar (consider the Clark oxygen electrode). Modern technologies, however, allow precisely defined microstructures to be constructed as these base electrodes. The development of microelectrode devices has particular significance for *in vivo* sensor devices, but has also been a necessary feature of array electrodes for multi-analyte or multiple mono-analyte detection devices. There are also, of course, obvious cost benefits in the reduced material volumes required in their manufacture, and where a mass production method can be employed, in the manufacture itself.

Microelectrode array sensors have included irregular and regular arrays, based formerly on graphite and carbon fibre. Techniques other than microlithography generally produce a comparatively ill-defined geometry and tend to be limited to an active site $>0.1\ \mu m^2$. Lithographic techniques, however, allow a lower limit, and can be employed for the manufacture of electrodes in carbon, metals and semiconductor materials on various supports, such as silicon, glass, plastics and ceramics.

THIN-FILM FABRICATION

Photolithographic production of thin-film electrodes offers a high degree of flexibility in electrode geometry and excellent reproducibility. The exact procedure is dependent on the nature of the substrate material, the electrode material and, of course, the application. However, while some steps may vary, the general principle remains essentially the same; for example (Fig. 5.21), the procedure of Prohaska *et al.* (1985, 1986) is:

- A thin metal layer (Au ~0.1 μm) is evaporated onto a suitable substrate (glass)
- The metal is covered with a layer of photoresist (may be sensitive to visible, UV, X-ray photons or electron beam) and the surface exposed through a mask to give the pattern of the electrode(s) and contacts. Either positive or negative photoresist can be employed. The uncovered metal film is etched off leaving the selected electrode areas
- A thick layer of Si_3N_4 is deposited by plasma-enhanced vapour deposition to give an insulating layer
- A second covering of photoresist is laid, and the surface exposed to uncover the electrode area
- The electrode area is revealed by plasma etching

The exposed electrode surface may be electrochemically plated with other metals for different electrode applications.

With electrodes of this size, significant and undesirable signal interference could occur by cross talk between adjacent evaporated conducting paths and through thin 'insulating' layers, thus introducing a miniaturization limit. However, one of the main limitations of the system size is the reference electrode. In two-electrode systems this should be larger than the working electrode, while in three-electrode systems a third 'large' electrode must also be included.

With *in vivo* or implanted devices, the contact of the sensor with the tissue fluids quickly leads to a protein deposition on the surface which, with such a small working area, is chronically detrimental, whether it be at the working or the reference electrode.

A solution has been sought to this problem by positioning the electrode in an insulating layer chamber, which makes contact with the sample through a '*recording site hole*' (Fig. 5.22). The chamber can house either the working or reference half-cells or the complete electrode assembly, depending on the application, and might typically be 300–15 000 μm^2 with a depth of the order of 1.5 μm.

In essence this design is very similar to a classical membrane covered electrode, where the actual electrode is maintained within its own electrolyte microenvironment, and the sample enters the chamber through a selectively permeable membrane.

An oxygen electrode, therefore, was constructed with an Au working electrode positioned between two Ag/AgCl references in a chamber filled with electrolyte. As with the membrane-covered Clark-type oxygen electrode, the response time depends on the diffusion coefficient and the thickness of the electrolyte layer. In this case, the response time to a step change in concentration was of the order of 10 s, which is considerably longer than would be predicted by a one-dimensional model describing diffusion through a tube of rectangular cross-section obeying Fick's law (Prohaska *et al.*, 1987). However the chamber electrode model, where the 20 μm 'recording site holes' are not positioned immediately above the working

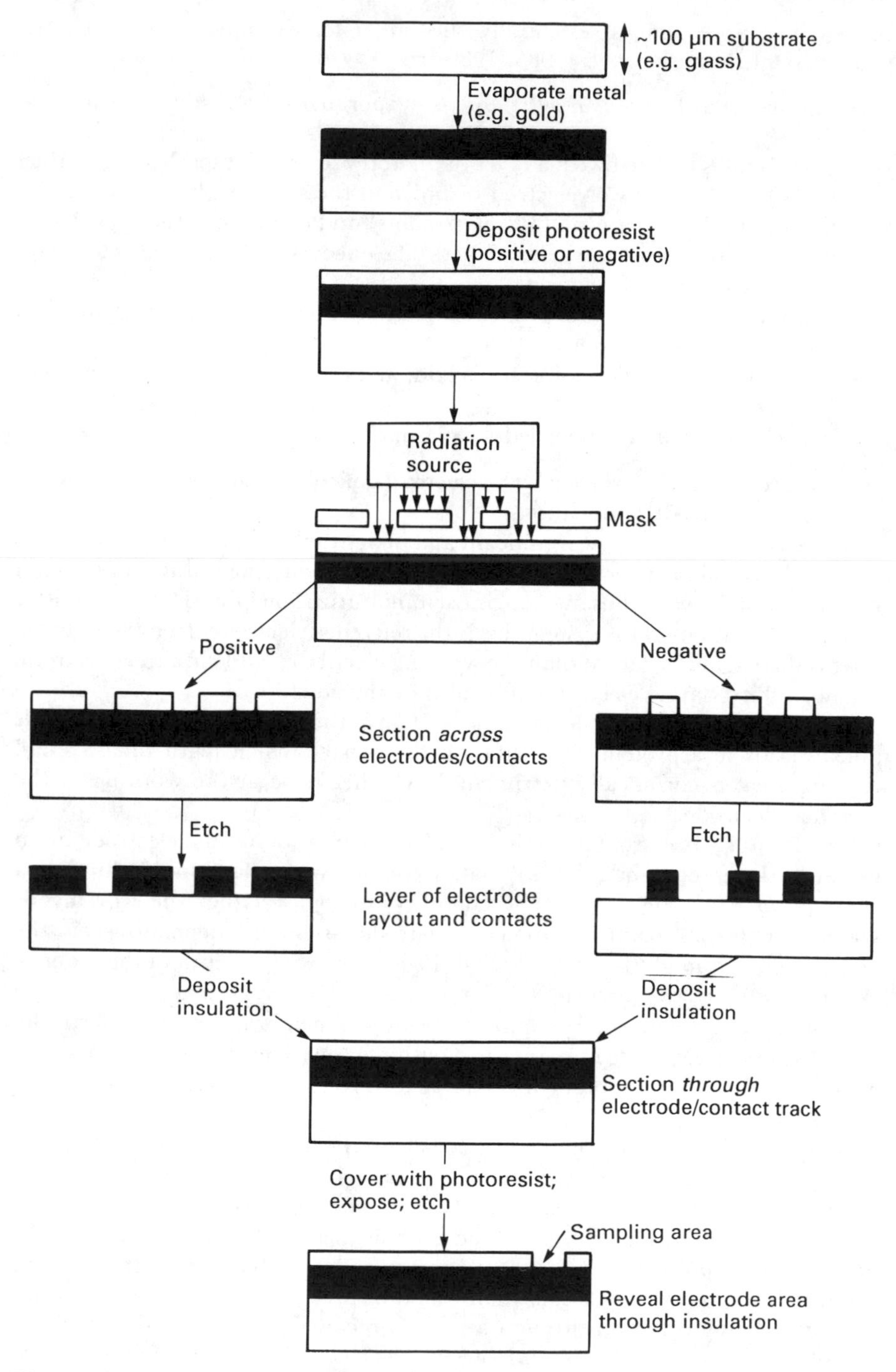

Fig. 5.21 Thin-film fabrication sequence.

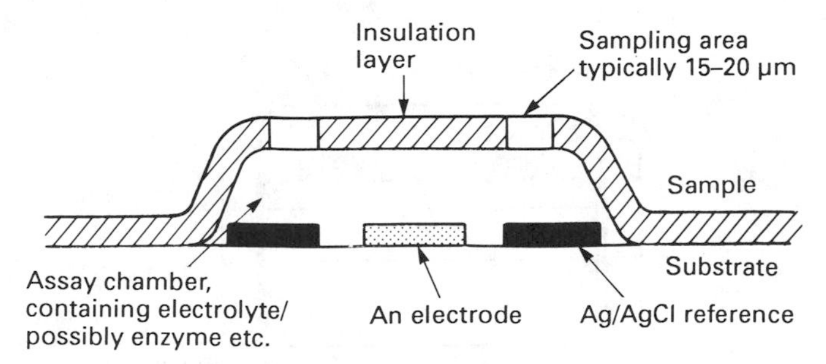

Fig. 5.22 The structure of the chamber electrode.

electrode, will have a longer diffusional path, thus deviating considerably from the one-dimensional model. In fact, deviations from classical electrochemical models can be quite considerable. For electrodes $<20\,\mu m^2$ non-linear diffusion is the predominant mode of mass transport, so that the current response would be expected to be different from the larger model.

Karube and Kubo (1988) have reported micro H_2O_2 and O_2 amperometric electrodes, developed using available integrated circuit technology. Gold electrodes 100 μm wide were deposited on a silicon nitride surface, partially insulated with a coating of Ta_2O_5. As with other similar geometries, the low currents generally involved in these devices and the close proximity of the electrodes allows the two-electrode configuration to be adopted.

H_2O_2 could be measured at the anode polarized at 1.1 V with respect to the counter-electrode, and oxygen measurement was performed by modification of the electrode to provide a 'Clark-type cell' with 0.1 M KOH aqueous electrolyte retained in a well at the electrode behind a Teflon membrane (Fig. 5.23).

Measurements from ultra-microelectrodes could be expected to present some considerable instrumental problems of sensitivity since currents less than femto-amperes are likely to be involved. Arrays of microelectrodes connected in parallel, however, yield higher currents than the sum of the individual isolated elements, when the spacing between the electrodes is small ($<50\,\mu m$), but they retain the characteristics of a single microelectrode. Arrays of 100 or more parallel straight-line electrodes (Thorman *et al.*, 1985) have been tested. The equivalent area on a disc electrode would have a radius 0.69–5 μm, but there is considerable current enhancement associated with the array design. Siu and Cobbold (1976) have reported a multicathode O_2 sensor based on this type of geometry.

Interdigitated arrays have also been produced. Sanderson and Anderson (1985) have investigated the filar electrode model of the two elements of the array at different potentials. The electrochemical oxidation product produced at one electrode diffuses across the space and is reduced at the second electrode. The opposing electrode limits the maximum diffusion layer thickness, thus amplifying the flux.

Chidsey *et al.* (1986) have shown good electron transport characteristics for redox polymer films formed over an interdigitated Pt array on a borosilicate

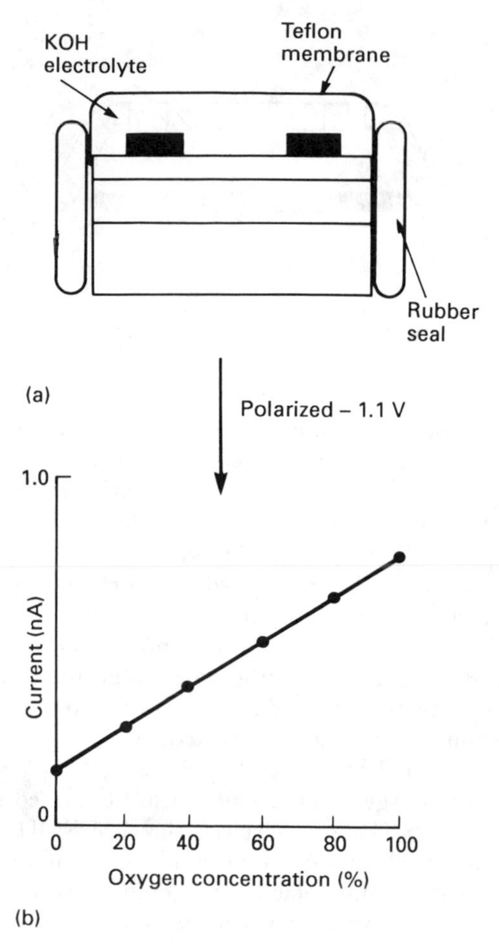

Fig. 5.23 The micro Clark-type cell (Karube and Kubo, 1988).

substrate, with a 2.5 μm interdigital space. These films, formed by electrochemical polymerization from solution of the monomer, include redox species whose application in modified electrodes has already featured in this chapter, namely poly[Os(bpy)$_2$(vpy)$_2$])ClO$_4$)$_2$ (bpy=2,2′-bipyridine; vpy=4-vinylpyridine), poly[Ru(bpy)$_2$(vpy)$_2$](ClO$_4$)$_2$ and the iron cyanol complex Prussian blue.

SOLID-STATE AND MOLECULAR TRANSISTORS

The development of miniature amperometric sensors with total integration between electrode sensing surface and integrated circuit involves a process which can be divided into two parts: fabrication, then mounting and encapsulation. The first part is the simultaneous production of multiple (several thousands) silicon wafer chips giving the required *p*-type and *n*-type regions by photolithographic techniques, and producing the silicon dioxide gate insulator by thermal oxidation of the silicon surface. Silicon nitride or further layers may be introduced by vapour deposition. Electrical contacts are made to the source, drain and substrate and the entire wafer broken into the individual chips on completion of the microfabrication process.

In view of the aforegoing discussion on modified electrodes and the considerable attraction of conducting polymers as immobilization matrices for biorecognition molecules, it is interesting to consider how in some respects many of the conducting polymers resemble solid-state transistors, and to contemplate how this property could be utilized in 'amperometric'-type assays. Polypyrrole, for example, is conducting in its oxidized form, but insulating in the reduced form. The conductivity between the two forms varies by a factor of more than 10^{10}.

Kittlesen *et al.* (1984) have electrochemically deposited polypyrrole across a 1.4 μm spaced array of 3 μm gold microsensors, deposited on a silica substrate in a manner as described above. The application of a potential difference across the polypyrrole linked array, where one end was held at the negative insulating potential and the other end at the positive conducting potential, induced a potential profile as represented in Fig. 5.24. A large potential drop occurs across

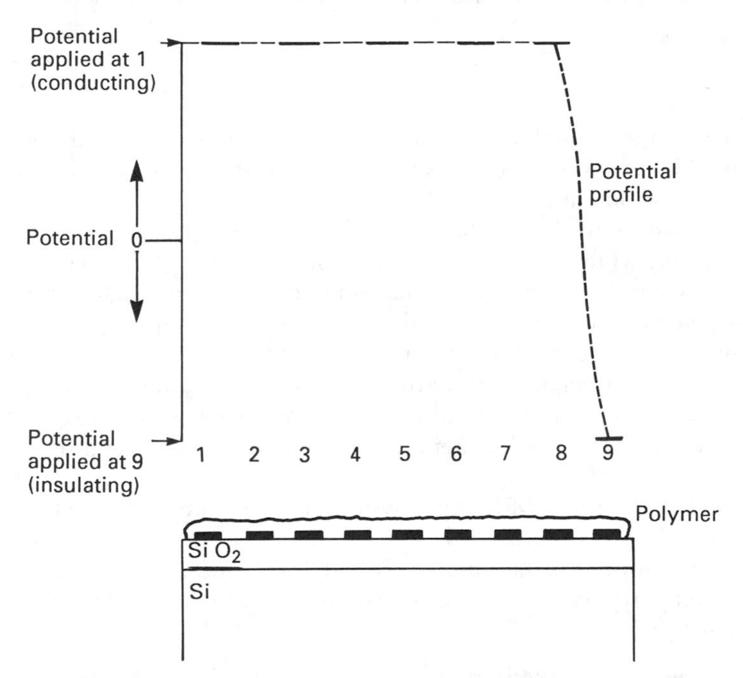

Fig. 5.24 The diode behaviour of conducting films such as polypyrrole. The potential profile across a nine-electrode array covered with polymer film, as a result of the application of a potential difference between electrodes 1 and 9. Electrode 1 is held at a potential where the polymer is conducting and electrode 9 at a potential where it is insulating.

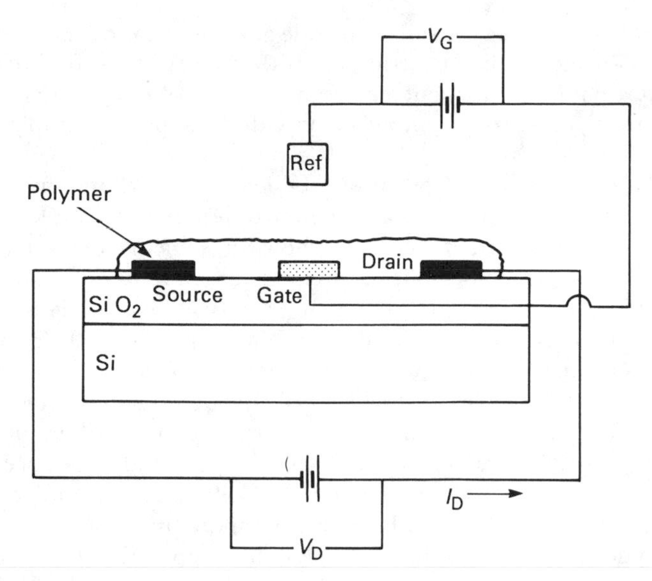

Fig. 5.25 Construction of a FET-type system employing conducting polymer.

the narrow region immediately between the negative electrode and the adjacent electrode while the other electrodes remain unaffected at a potential close to the positive applied potential.

This contrasts with the constantly varying potential profile across a so-called redox polymer, which exhibits *maximum* conductivity at [Ox] = [Red]. In this instance a linear change in the concentration of redox centres between the two controlled electrodes is seen, as predicted by the Nernst equation.

The polypyrrole configuration simulates diode-type behaviour, since if one electrode is set at a negative insulating potential (V_{set}) and the other electrode varied (V_{app}), a threshold potential will be reached below which no current flows between the two electrodes. Further increases in V_{app} will cause a linear increase in current.

Constructed as part of a FET-type system, a device can be devised mimicking the fundamental characteristics of a solid-state transistor. An array of three Au microelectrodes provides the gate, source and drain contacts (Fig. 5.25) (White *et al.*, 1984) and polypyrrole is deposited over the whole area. For a given V_G, I_D is a function of V_D:

V_G negative (insulating)	I_D small for $V_D < 0.5$ V
$V_G > -0.2$ V (conducting)	I_D larger
$V_G > V_T$ (redox potential)	I_D proportional to V_D

Discussions of such systems would seem perhaps more relevant to the previous chapter, but it is the potential dependent change in the conduction (or *charge transfer*) properties of such films that has been exploited for example, in a poly-(3 methylthiopene)-based device. The 'threshold' potential of this polymer is suitable for switching the film 'on and off' by the presence of the oxidizing and reducing species O_2 and H_2 respectively (Thackeray and Wrighton, 1986). Similarly Paul *et al.* (1985) have described a polyaniline film which can be turned 'on and off' by redox reagents.

Other electroactive materials with transistor-type characteristics would show different threshold potentials, and so may offer the possibility for integration of the redox systems of the enzymes and the mediators directly into the 'transistor', rather than utilizing a conventional amperometric electrode. Indeed any of the electroactive species classically targeted in amperometric devices could be converted if a suitable polymer could be devised.

References

Albery, W.J., Brooks, W.N., Gibson, S.P. and Hahn, C.E.W. (1978). *J. Appl. Physiol.* **45**, p. 637.

Albery, W.J., Eddowes, M.J., Hill, H.A.O. and Hillman, A.R. (1981a). *J. Am. Chem. Soc.* **103**, p. 3904.

Albery, W.J., Hahn, C.E.W. and Brooks, W.N. (1981b). *Br. J. Anaesth.* **53**, p. 447.

Amos, L.J., Duggal, A., Mirsky, E.J., Ragonesi, P., Bocarsly, A.B. and Fitzgerald-Bocarsly, P.A. (1988). *Anal. Chem.* **60**, pp. 245–249.

Bourdillon, C., Bourgeous, J. and Thomas, D. (1979). *J. Am. Chem. Soc.* **102**, p. 4231.

Brabec, V. and Dryhurst, G. (1978). *J. Electroanal. Chem.* **89**, p. 161.

Brdička, R. (1933a). *Collect. Czech. Chem. Commun.* **5**, p. 112.

Brdička, R. (1933b). *Collect. Czech. Chem. Commun.* **5**, p. 148.

Brooks, W.N. (1980). Electroanalytical Techniques for Anaesthetic Gases. D. Phil. Thesis, Oxford.

Brooks, W.N., Hahn, C.E.W., Foëx, P., Maynard, P. and Albery, W.J. (1980). *Br. J. Anaesth.* **52**, p. 715.

Cecil, R. and Weitzman, P.G.J. (1964). *Biochem. J.* **93**, p. 1.

Cenas, N.K., Pocius, A.K. and Kulys, J.J. (1984). *J. Electroanal. Chem.* **173**, p. 583.

Chidsey, C.E., Feldman, B.J., Lundgren, C. and Murray, R.W. (1986). *Anal. Chem.* **58**, p. 601.

Clark, Jr., L.C. (1956). *Trans. Am. Soc. Art. Int. Org.* **2**, p. 41.

Collman, J.P., Denisevich, P., Konai, Y., Morrocco, M., Koval, C. and Anson, F.C. (1980). *J. Am. Chem. Soc.* **102**, p. 6027.

Eddowes, M.J. and Hill, H.A.O. (1979). *J. Am. Chem. Soc.* **101**, p. 4461.

Faulkner, L.R. (1984). *Chem. Eng. News*, Feb. 28.

Gorton, L., Torstensson, A., Jaegfeldt, H. and Johansson, G. (1984). *J. Electroanal. Chem.* **161**, p. 103.

Hall, E.A.H. (1986). *Neonatal Physiological Measurements*, Ed. Rolfe, P., p. 212. London, Butterworths.

Hall, E.A.H., Conhill, E.J. and Hahn, C.E.W. (1988). *J. Biomed. Eng.* **10**, p. 319.

Hurrell, H.C. and Abruña, H.D. (1988). *Anal. Chem.* **60**, pp. 254–258.

Janik, B. and Elving, P.J. (1968). *Chem. Rev.*, **68**, p. 2951.

Leddy, J. and Bard, A.J. (1985). *J. Electroanal. Chem.* **189**, pp. 203–219.

Kamin, R. and Wilson, G. (1980). *Anal. Chem.* **52**, p. 1198.

Karube, I. and Kubo, I. (1988). *Analytical Uses of Biological Compounds for Detection, Medical and Industrial Uses*. NATO ASI Series, Reidel Publishing Company, pp. 207–217.
Kittlesen, G.P., White, H.S. and Wrighton, M.S. (1984). *J. Am. Chem. Soc.* **106**, p. 7389.
Mancy, K.H., Okun, D.A. and Reilley, C.N. (1962). *J. Electroanal. Chem.* **4**, p. 65.
Martin, C.R., Rubinstein, I. and Bard, A.J. (1982). *J. Am. Chem. Soc.* **104**, pp. 4817–4824.
Nadjo, L., Savèant, J.M. and Su, K.B. (1985). *J. Electroanal. Chem.* **196**, p. 23.
Nicholson, R.S. and Shain, I. (1964). *Anal. Chem.* **30**, p. 706.
Oyama, N. and Anson, F.C. (1980a). *J. Electrochem. Soc.* **127**, p. 247.
Oyama, N. and Anson, F.C. (1980b). *J. Electrochem. Soc.* **127**, p. 640.
Oyama, N., Sata, K. and Anson, F.C. (1980a). *J. Electroanal. Chem.* **115**, p. 149.
Oyama, N., Shimomura, T., Shigehara, K. and Anson, F.C. (1980b). *J. Electroanal. Chem.* **112**, p. 271.
Paul, E.W., Ricco, A.J. and Wrighton, M.S. (1985). *J. Phys. Chem.* **89**, p. 1441.
Prohaska, O.J. (1985). *Transducers '85*, pp. 402–405.
Prohaska, O.J., Olcaytug, F., Pfunder, P. and Dragaun, M. (1986). *IEEE Trans. Biomed. Eng. BME* **33**(3), pp. 223–229.
Prohaska, O.J., Kohl, F., Goiser, P., Olcaytug, F., Urban, G., Jachimowicz, A., Pirker, K., Chu, W., Patil, M., La Manna, J. and Vollmer, P. (1987). *Transducers '87*, p. 187.
Ruff, I., Friedrich, V.J., Demeter, K. and Csillag, K. (1971). *J. Phys. Chem.* **75**, p. 3303.
Sanderson, D.G. and Anderson, L.B. (1985). *Anal. Chem.* **57**(12), pp. 2388–2393.
Savèant, J.M. and Su, K.B. (1984). *J. Electroanal. Chem.* **171**, p. 341.
Savèant, J.M. and Su, K.B. (1985). *J. Electroanal. Chem.* **196**, p. 1.
Severinghaus, J.W., Weiskopf, R.B., Nishimura, M. and Bradley, A.F. (1971). *J. Appl. Physiol.* **31**, p. 640.
Stankovich, M.T. and Bard, A.J. (1977). *J. Electroanal. Chem.* **85**, p. 173.
Siu, W. and Cobbold, R.S.C. (1976). *Med. Biol. Eng.* **14**, p. 109.
Thackeray, J.W. and Wrighton, M.S. (1986). *J. Phys. Chem.* **90**, p. 6674.
Thorman, W., van den Bosch, P. and Bond, A.M. (1985). *Anal. Chem.* **57**, p. 2764.
Wang, J. and Lin, M.S. (1988). *Anal. Chem.* **60**, pp. 499–502.
White, H.S., Leddy, J. and Bard, A.J. (1982). *J. Am. Chem. Soc.* **104**, pp. 4811–4817.
White, H.S., Kittlesen, G.P. and Wrighton, M.S. (1984). *J. Am. Chem. Soc.* **106**, p. 7389.
Wise, K.D., Smart, R.B. and Mancy, K.H. (1980). *Anal. Chim. Acta* **116**, p. 297.

Chapter 6

Photometric Assay Techniques

Energy Transitions

The Quantum Theory requires that when the internal energy of a molecule is raised by the absorption of a quantum of electromagnetic radiation, then the energy change (ΔE) is given by

$$\Delta E = h\nu = hc/\lambda$$

where h is Planck's constant, ν is the frequency, λ is the wavelength, and c is the velocity of electromagnetic radiation in a vacuum, and corresponds exactly to the difference between two energy levels of the molecule.

If the internal energy is considered as the sum of electronic, vibrational and rotational energy

$$E_{int} = E_{elec} + E_{vib} + E_{rot}$$

and transition between two energy levels may be accompanied by changes in vibrational and rotational energy. Depending on the nature of the absorption, energy over the entire electromagnetic spectrum might be required. In practice, absorption is concentrated in certain regions (Fig. 6.1).

Ultra-violet and Visible Absorption Spectra

The most generally employed absorption studies involving biomolecules are those composing of electronic energy level transitions. For organic molecules most of

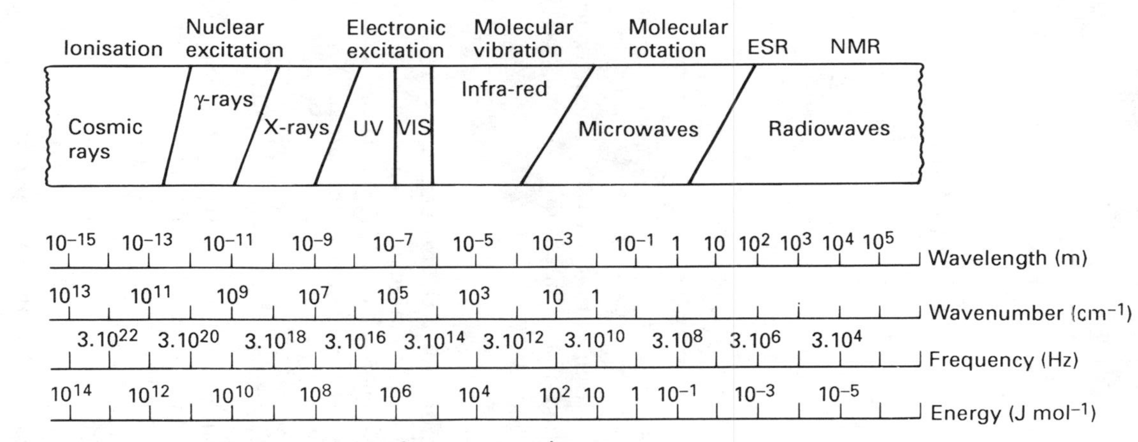

Fig. 6.1 The regions of the electromagnetic spectrum.

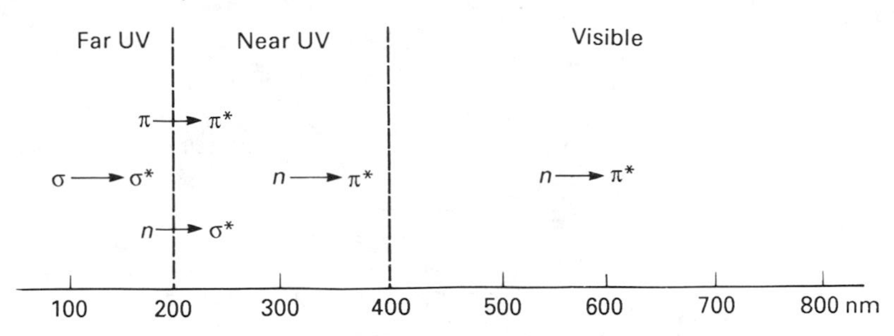

Fig. 6.2 The regions of the electronic spectrum and the type of transition that occurs in each.

these occur within the ultra-violet and visible region. The most important modes of excitation are:

- promotion of electrons from *bonding* to *antibonding*: $(\sigma \rightarrow \sigma^*)$ and $(\pi \rightarrow \pi^*)$
- promotion of electrons from *non-bonding* to *antibonding*: $(n \rightarrow \sigma^*)$ and $(n \rightarrow \pi^*)$.

Excitation of electrons from the σ level requires the highest energy (Fig. 6.2), and thus the shortest wavelength, so that saturated compounds that contain only σ electrons absorb in the far UV < 190 nm.

Electronic absorption spectra alone rarely provide a complete structural solution, and while the shape and position of the absorption bands is a useful diagnostic tool the major application to quantitative assay comes from utilization of the *Beer–Lambert law*, which expresses the relationship between absorption and concentration:

$$\log(I_0/I) = \varepsilon cl$$

where I is the intensity of transmitted light, I_0 is the intensity of the incident light, ε is the molar extinction coefficient, c is the concentration and l is the path length. It follows from this relationship that molecules that show an electronic absorption spectrum can be estimated in solution, as can the path of a reaction involving a change in the spectral properties between reactants and products.

Fluorescence and Phosphorescence

After absorption of light and formation of an electronically excited species, radiative deactivation may take place via fluorescence or phosphorescence (Fig. 6.3). In most cases absorption of a quantum leads to an excitation from a singlet ground state to a singlet excited state, and immediately following this absorption there is a rapid radiationless energy transfer: internally to the lowest excited singlet state or else via intersystem crossing to the lowest triplet state. The excited molecule primarily formed has a typical lifetime of 10^{-8} s and may re-emit a quantum of either the same or a different frequency; this emission is either *fluorescence* or *phosphorescence*.

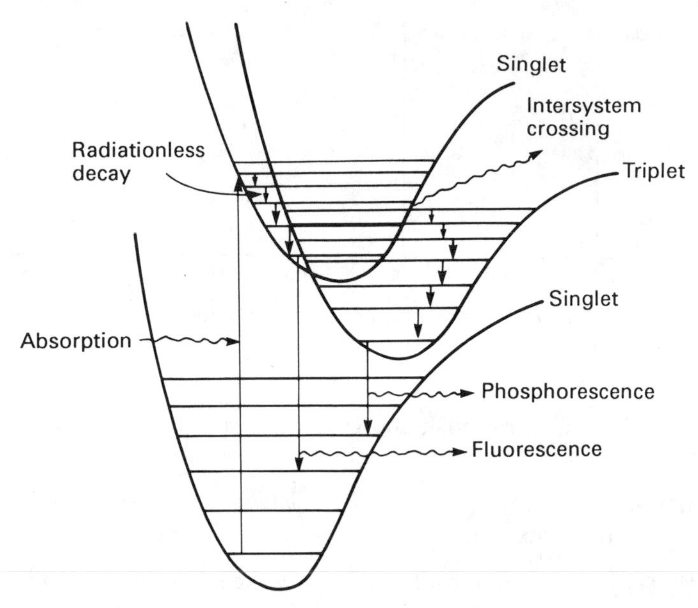

Fig. 6.3 The sequence of steps leading to phosphorescence and fluorescence.

During the lifetime of the excited molecule it may be able to participate in a chemical reaction, thus transferring energy to another solution species and returning to a ground state. In using the emission spectrum as an analytical tool, the possibility of incidental quenching must be considered or else quantified and utilized.

The kinetics of quenching of fluorescence can be described by the competition between two parallel processes:

$$\text{Fluorescence } M^* \xrightarrow{k_f} M + h\nu$$

$$\text{Quenching } M^* + Q \xrightarrow{k_q} M + Q^*$$

giving a deactivation rate

$$\frac{-d[M^*]}{dt} = k_f[M^*] + k_q[M^*][Q]$$

and for an initial absorption of intensity I_o and fluorescence intensity I_f the fluorescence yield will be:

$$\frac{I_f}{I_o} = \frac{k_f[M^*]}{k_f[M^*] + k_q[M^*][Q]} = \frac{1}{1 + (k_q/k_f)[Q]}$$

Bioluminescence and Chemiluminescence

A number of chemical transformations that emit light (*luminesce*) are known. Luminescence emitted from biological objects can be divided into two types:

(1) A specific enzyme linked strong bioluminescence. This involves the participation of such compounds as ATP, FAD, NAD, O_2, H_2O and Ca^{2+}.
(2) Non-specific, low-level luminescence, both spontaneous or externally induced.

The production of the excited state and its subsequent emission of light can be described by:

$$A+B\xrightarrow{k}P^{*}\xrightarrow{k_1}P+h\nu$$

If the rate-limiting step is described by

$$\frac{d[P^{*}]}{dt}=k[A][B]$$

then in bioluminescent assays the light intensity I is measured as a function of the analyte concentration (A, where B is in excess),

$$I(t)=\Phi\frac{d[A_t]}{dt}$$

where Φ is the quantum yield of bioluminescence and $[A_t]$ the concentration of A at time t.

Since $I=f(t)$ the time variable is important in the analysis, and various sets of parameters can be chosen for the assay (Fig. 6.4).

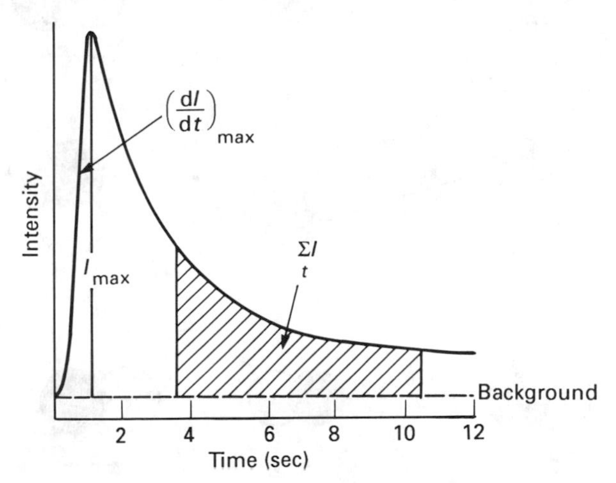

Fig. 6.4 Luminescent intensity as a function of time, illustrating various possible signal capture 'windows'.

Infra-red Transitions

Transitions between vibrational energy levels in the molecule correspond to absorption in the infra-red region. The allowed energy levels are given for a simple harmonic oscillator model from a solution of the *Schrödinger wave equation*,

$$\varepsilon_v = (v + \tfrac{1}{2})\hbar w$$

or to the first approximation for the anharmonic molecular model,

$$\varepsilon_v = (v + \tfrac{1}{2})\hbar w - (v + \tfrac{1}{2})^2 \hbar w_e x_e$$

where w_e = oscillation frequency and x_e = anharmonic constant. In order that a molecule can interact with the incident radiation it must experience some change in momentum to 'accommodate' the momentum of the incoming energy. Vibrational changes will only be infra-red active if the vibration involves a change in the dipole moment of the molecule. Thus, in the linear triatomic carbon dioxide molecule, a symmetrical stretching vibration does not induce a change in dipole and is not infra-red active, whereas asymmetrical stretching *is* infra-red active (Fig. 6.5).

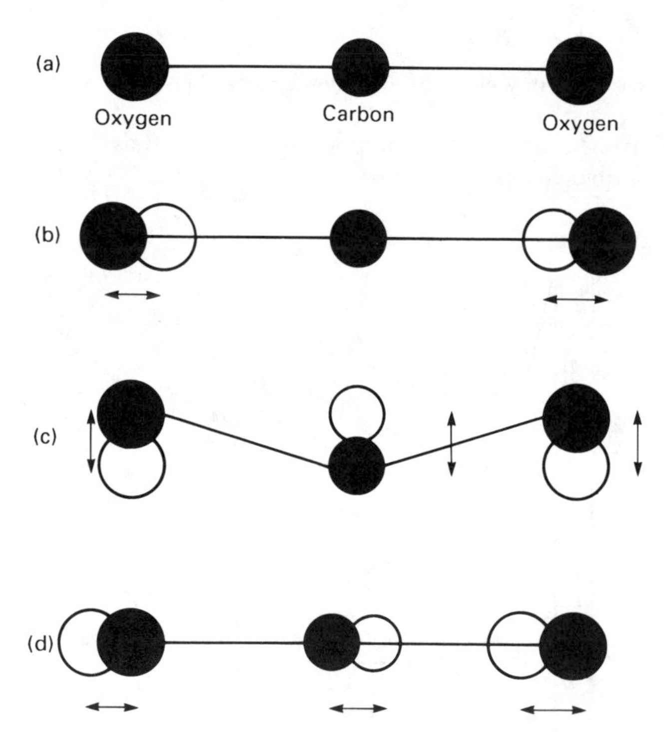

Fig. 6.5 (a) Vibration modes of carbon dioxide. (b) Symmetric stretching mode. (c) Bending mode. (d) Asymmetric stretching mode.

For a non-linear molecule containing n atoms there are $(3n-6)$ theoretical fundamental vibration modes. As discussed above, not all these modes will be infra-red active, but nevertheless it follows that the infra-red spectrum of a large molecule will be very complex. Inherent in this complexity is the fact that it represents a unique fingerprint of the bonds in the molecule and their environment, so that any one vibration could theoretically be employed to follow a reaction where ambient environmental factors can be eliminated.

Light Scattering

Static light-scattering techniques are a familiar method for particle-size-related bioassays at the microorganism, cellular and sub-cellular levels. When collimated plane-polarized light illuminates a particle, the incident field induces a fluctuating dipole and light is scattered perpendicular to the dipole axis, in directions dependent on the size of the particle relative to the wavelength.

When the wavelength (λ) is large compared with the particle volume (V), the scattered light is emitted equally in all directions with an intensity (*Rayleigh scatter*):

$$I \propto V^2/\lambda^4$$

but with the particle size comparable with or larger than the incident wavelength, then interference between the scattered light at different points on the particle reduces the amount of light scattered back towards the source and produces 'lobes' of high and areas of zero intensity (Fig. 6.6).

For dynamic measurements, where the particles are moving under Brownian motion, the intensity of light collected at a fixed angle will fluctuate according to that motion, since it will be dependent on the time taken for the particle to diffuse a distance comparable with a wavelength. It can therefore be correlated with particle size.

The scattered light, measured at a fixed angle, has a phase shift Φ, and a scattering vector,

$$K = \frac{4\pi n}{\lambda} \sin\left(\frac{\theta}{2}\right)$$

where n = refractive index of the suspension phase and θ = scattering angle. The intensity of the detected light fluctuates on a time scale comparable with the time required for a particle to diffuse a distance sufficient to induce a phase shift $\Phi = \pi$. This distance is inversely proportional to K.

Raman Scattering

Complementary to infra-red is Raman spectroscopy since here too a transition between vibrational energy levels is induced. Unlike infra-red, however, a monochromatic excitation frequency is chosen, usually in the ultra-violet-visible

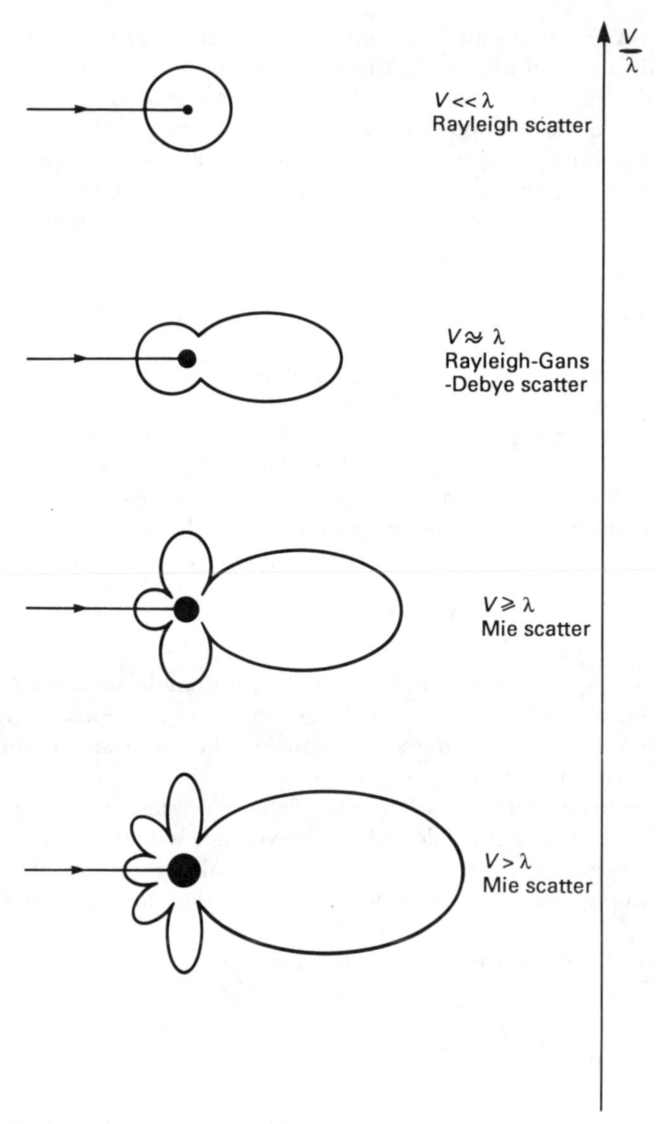

Fig. 6.6 Relative light scatter perpendicular to the dipole axis in all directions, as a function of particle volume (V).

region and the scattered light recorded perpendicular to the incident beam. As already mentioned above, most of the scattered light is of the same frequency as the incident frequency but there is a small proportion that is scattered and detected at new frequencies, either side of this frequency. These changes in frequency which correspond to the $(3n-6)$ fundamental vibration modes are termed *Raman scattering*. In order for a mode to be Raman active there must be a change in *polarizability* involved in the vibration. For an induced dipole (μ),

$$\mu = \alpha E$$

where α = polarizability. At an incident frequency ν, the electric field (E) varies according to

$$E = E_o \sin 2\pi \nu t$$

so that

$$\mu = \alpha E_o \sin 2\pi \nu t$$

and such an oscillating dipole will scatter light at the incident frequency (ν)—*Rayleigh scattering*.

If the molecule undergoes a change in polarizability then for a vibration oscillation ν_{vib},

$$\alpha = \alpha_o + \beta \sin 2\pi \nu_{vib} t$$

where β = rate of change of polarizability with vibration, so that

$$\mu = (\alpha_o + \beta \sin 2\pi \nu_{vib} t) E_o \sin 2\pi \nu t$$
$$\mu = \alpha_o E_o \sin 2\pi \nu t + E_o \beta / 2 \{\cos 2\pi (\nu - \nu_{vib}) t - \cos 2\pi (\nu + \nu_{vib}) t\}$$

showing that when $\beta \neq 0$ light will also be scattered with frequency components ($\nu \pm \nu_{vib}$) (Raman scattering).

This change in polarizability may be considered in terms of a change in the size or shape of the electron cloud surrounding the molecule, so that in contrast to the infra-red spectrum for the linear triatomic molecule cited above, the symmetrical stretching mode will be Raman active, while the asymmetrical stretching vibration will be Raman inactive. Indeed, for a symmetrical linear molecule, or one with a centre of symmetry, a vibration will not appear in both the infra-red and Raman spectra.

Application of Ultra-violet-visible Spectra

PROTEINS

As one might expect from their σ-bond character, the majority of proteins exhibit similar absorption spectra with a wavelength maximum around 278 nm. Exceptions to this are those containing the unsaturated amino acid residues like tryptophan, tyrosine and phenylalanine, and those containing disulphide bridges. It follows therefore that only significant changes in these latter π-electron residues is likely to cause an appreciable difference in the absorption spectrum. The method does not therefore suggest itself as a useful sensitive transducer of polypeptide interactions.

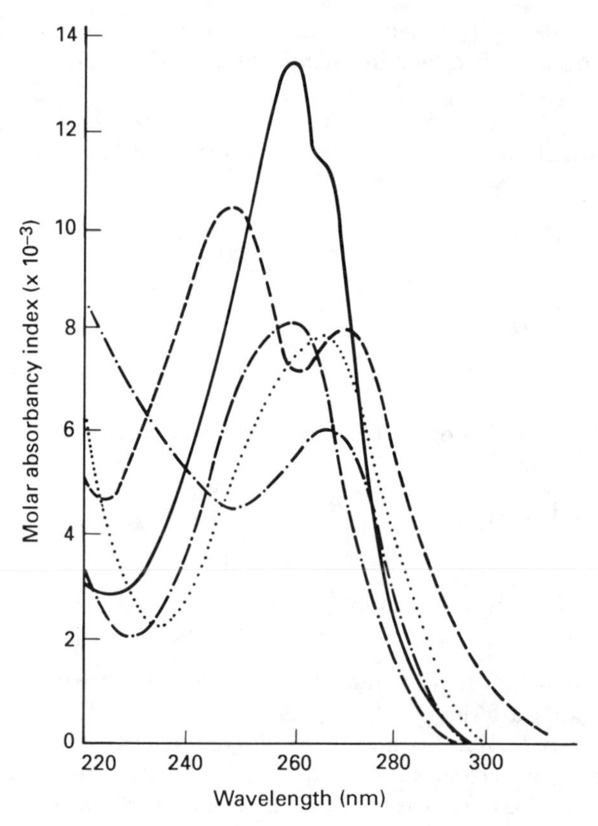

Fig. 6.7 Ultra-violet-visible absorption spectra for the common DNA bases: (——) adenine; (– – – –) guanine; (– – · – – ·) uracil; (· · · · ·) thymine; (— · — ·) cytosine.

POLYNUCLEOTIDES

Similar conclusions might be drawn about the direct assay of polynucleotides. The absorption spectra of the individual bases is distinct and intense (Fig. 6.7; Table 6.1), and as will be shown later in this section, this property has been successfully employed in enzyme-linked assays involving a coenzyme nucleotide.

In a polymeric nucleotide, each monomeric unit represents a chromophore with a strong ultra-violet absorption. Those containing all four bases show a broad absorption maximum between 256 and 265 nm and a second maximum in the far ultra-violet at around 195 nm. The absorbance is less than that expected from the sum of the individual bases, due to coupling between adjacent bases (*hyperchromism*). It follows therefore that single-stranded polynucleotides show less hyperchromism than double-stranded polynucleotides and that quantitative analysis is likely to be complicated by this phenomenon. One could imagine, however, a DNA probe where a short length of single-stranded DNA which has been

Table 6.1 Molar absorbencies for the common DNA bases

Base	*Molar absorbency index* ($M^{-1}\,cm^{-1}$) *at 260 nm*
Adenine	13.4×10^3
Guanine	7.2×10^3
Cytosine	5.55×10^3
Uracil	8.2×10^3
Thymine	7.4×10^3

identified as the complementary 'master' sequence for a virus could be used as a template to 'recognize' that virus. In the absence of interfering bases and polynucleotides the 'reaction' could be followed by monitoring the ultra-violet spectrum. However, although the 'recognition' system is feasible, it can be calculated from the expected molar extinction coefficients (Table 6.1) that the minimum detection level is likely to be $> 10^{10}$ molecules/μl, that is, greater than desirable for any viral assay.

Indicator-linked Bioassay

In Chapter 3 the use of coloured indicators to monitor pH was mentioned as an alternative to the pH electrode. It is the use of such indicators (or labels) with specific spectral properties that has formed the basis of the traditional bioassay techniques. The pH indicators are obvious candidates for analysis in the ultra-violet–visible region of the spectrum, since their protonated and non-protonated forms normally have very distinct absorption spectra in this region (e.g. methyl red, see Fig. 6.8). By linking such an indicator to the reaction involving a change in pH, it can therefore be analysed in this way.

Many other colorimetric-type assays have been devised where the indicator dye reacts with a functional group on the target species. When treated with copper sulphate in alkaline solution, for example, compounds containing the peptide group react giving complexes with the copper(II) ion, e.g.

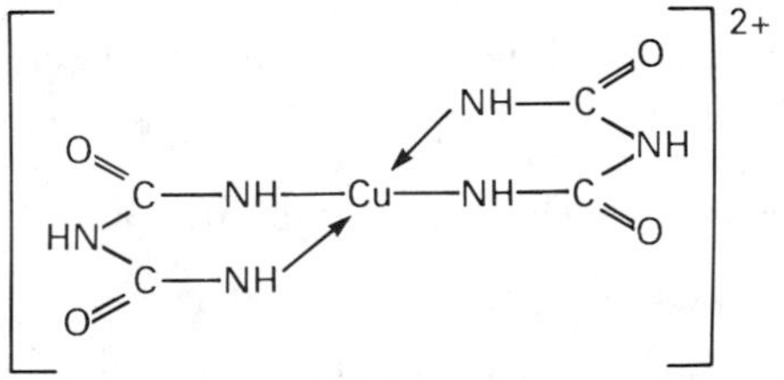

which are purple in colour and can be estimated by reading at 540 nm (Weichselbaum, 1946). This, the *Biuret method*, gives a value for total protein.

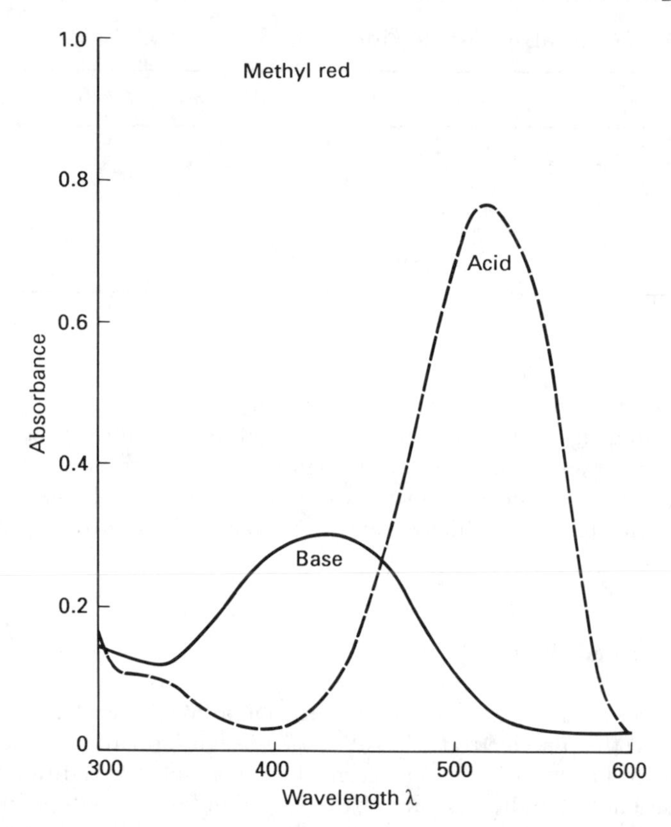

Fig. 6.8 The ultra-violet-visible absorption spectrum of methyl red in its acid and base forms.

Glucose may be assayed through a condensation reaction between the aldehyde group on glucose and a primary aromatic amine, *o*-toluidine. The resulting compound develops a stable blue-green colour that can be estimated at 630 nm (Dubowski, 1962).

Both these methods suffer from lack of specificity. The latter reaction, for example, will select all aldohexoses and not just glucose, so that in the presence of galactose it will give a false-positive reading indicating glucose. Specificity can be induced by capitalizing on the substrate-recognition properties of enzymes and linking the assay to an enzyme-catalysed reaction. For example, glucose can be selected in the presence of other sugars by the use of glucose oxidase (GOD):

$$\text{glucose} + O_2 + H_2O \xrightarrow{GOD} \text{gluconic acid} + H_2O_2$$

Estimation of the hydrogen peroxide produced in this glucose selective reaction gives an indirect assay of the glucose concentration. This can be done by reaction with 4-aminoantipyrine in the presence of phenol and peroxidase and measurement of the coloured compound that is formed at 500 nm (Trinder, 1969).

Reactions involving enzymes requiring a non-peptide coenzyme or prosthetic group can however often be monitored by taking advantage of the optical properties of the coenzyme.

Free NADH, for example, shows an absorption maximum at 340 nm (Fig. 6.9), with a strong fluorescence at 400 nm, both of which change in magnitude and position on binding to protein and/or substrate. This alteration in spectrophotometric properties is frequently employed in assays of nicotinamide nucleotide-dependent enzymes. Lactate dehydrogenase (LDH), for example, is determined from the reaction,

$$\text{pyruvate} + \text{NADH} + \text{H}^+ \xrightarrow{LDH} \text{L-lactate} + \text{NAD}^+$$

by measuring the NADH oxidation rate (Wroblewski and LaDue, 1955; Deutsche Gesellschaft für Klinische Chemie, 1972). Similarly glutamate–pyruvate transaminase (GPT),

$$\alpha\text{-ketoglutarate} + \text{L-alanine} \xrightarrow{GPT} \text{L-glutamate} + \text{pyruvate}$$

can be related to the same auxiliary reaction catalysed by lactate dehydrogenase (Wroblewski and LaDue, 1956).

In the case of the glutamate oxaloacetate transaminase (GOT) catalysed reaction,

$$\alpha\text{-ketoglutarate} + \text{L-aspartate} \xrightarrow{GOT} \text{L-glutamate} + \text{oxaloacetate}$$

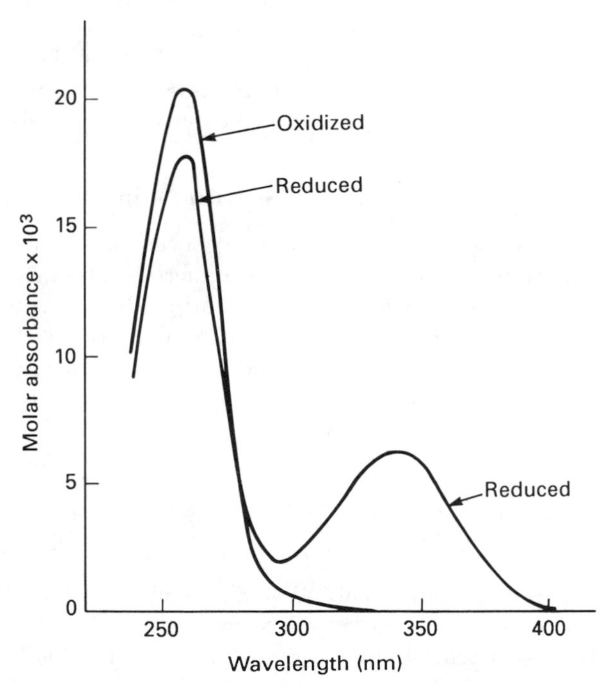

Fig. 6.9 Absorption spectra of NAD in oxidized and reduced forms.

enzymic activity can be determined by measuring the NADH oxidation rate in the secondary reaction catalysed by the malate dehydrogenase (MDH) (Deutsche Gesellschaft für Klinische Chemie, 1972):

$$\text{oxaloacetate} + \text{NADH} + \text{H}^+ \xrightarrow{MDH} \text{L-malate} + \text{NAD}^+$$

Since there are over 250 NADH/NAD^+-dependent enzymes, applications of such a system are immense.

LUMINESCENT ASSAY

Photometric bioassay can also be linked to labels for emission spectroscopy. A particularly attractive alternative would be the luminescent agents, since here no excitation source is required.

Bioluminescence is a light-emission phenomenon primarily associated perhaps with the firefly. Various biochemical sequences can be responsible for the production of light. In its simplest form the mechanism can be considered to be associated with the *luciferase* enzyme-catalysed reaction producing an excited state, which then decays with the emission of light of characteristic wavelength. In many cases it is the enzyme-catalysed oxidation of a heterocyclic organic molecule (a *luciferin*) to produce a singlet excited state that initiates the spectral emission:

$$\text{luciferin} \xrightarrow{\textit{luciferase},\ O_2,\ H_2O} \text{oxyluciferin}^* \rightarrow \text{oxyluciferin} + h\nu\ (\lambda = 562\text{ nm})$$

The luciferins are very varied in structure although many show common features (Table 6.2), and some have been successfully synthesized. Several of the luciferase-catalysed reactions involve cofactors such as ATP, FMN or NADH, so a route into other metabolic reactions involving these cofactors is suggested. Firefly luciferase, for example, has been successfully linked to the detection of analytes in enzyme sequences involving ATP.

$$\text{ATP} + \text{luciferin} + O_2 \xrightarrow{\textit{luciferase}} \text{AMP} + \text{PP} + \text{oxyluciferin} + CO_2 + H_2O + h\nu$$

Because of its great sensitivity (sub-femtomole concentrations are claimed) this method has been extended to the analysis of numerous substances of analytical interest. Among these the determination of creatine kinase has been produced commercially as a kit for the diagnosis of myocardial infarction (Lundin, 1982). Increased total creatine kinase levels are found in serum following several types of muscle damage. This level is particularly useful in the diagnosis of myocardial infarction. The principle of the assay is based on the monitoring of the ATP formed in the creatine kinase reaction:

$$\text{AMP} + \text{creatine phosphate} \xrightarrow{\textit{creatine kinase}} \text{ATP} + \text{creatine}$$
$$\text{ATP} + \text{luciferin}_{red} + O_2 \xrightarrow{\textit{luciferase}} \text{AMP} + \text{PP} + \text{luciferin}_{ox} + CO_2 + h\nu$$

Bacterial luciferases do not appear to involve a luciferin in their light-emitting mechanism, but an excited state complex with the reduced flavin cofactor FMNH. The process is believed to proceed along the following lines,

$$FMNH_2 + O_2 + \text{RCHO} \xrightarrow{\textit{luciferase}} \text{FMN} + \text{RCOOH} + H_2O + h\nu\ (\lambda = 478\text{–}505\text{ nm})$$

Table 6.2 Examples of luciferin structures

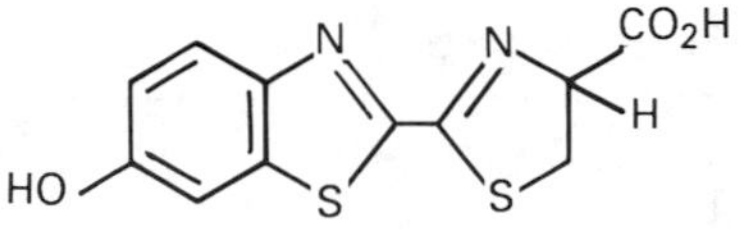

Firefly (Coleoptera)

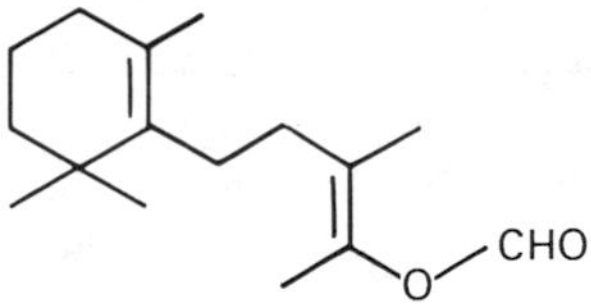

Fresh water limpet
(Latia neritoides)

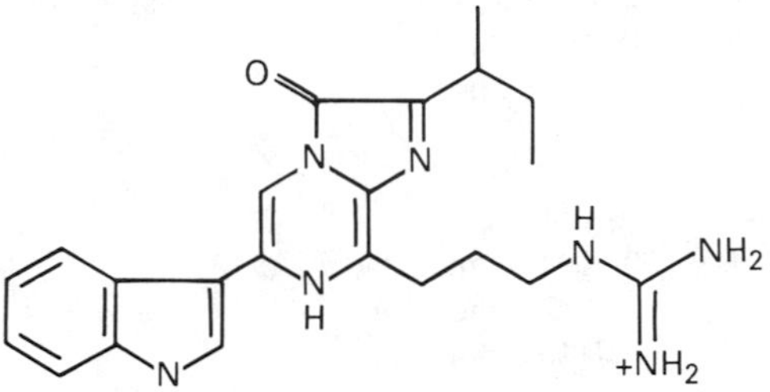

Cypridina hilgendorfii

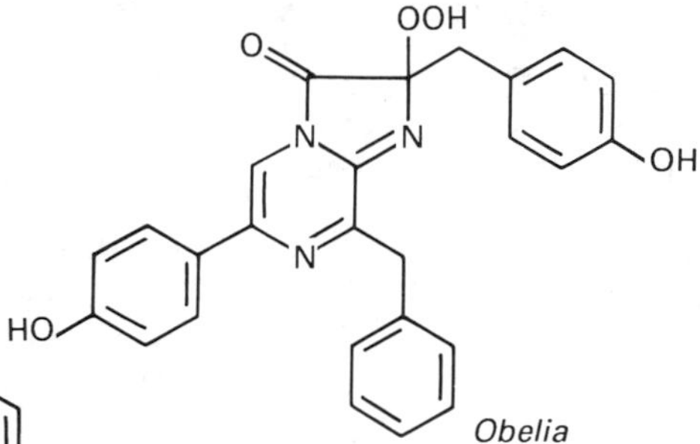

Obelia

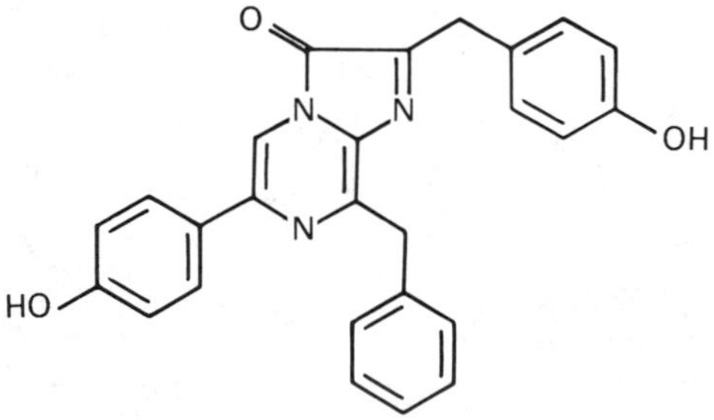

Renilla reniformis

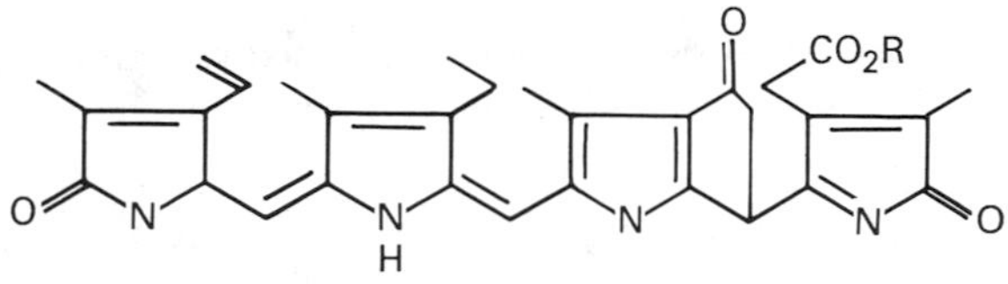

Pyrocystis lunula

Table 6.3 NADPH-coupled assays based on bacterial bioluminescence *in vitro*

Analyte	*Sensitivity or range*
Substrates	
Acetylpyridine–NADH	0.1–1.4 pmol
Aldehydes	0.1–100 pmol
Ammonia	0.05–0.6 mol
Androsterone	0.8–1000 pmol
Ethyl alcohol	0.003–0.012%
FAD	0.01–100 pmol
Fatty acids C_{14}–C_{18}	—
FMN	1 fmol
$FMNH_2$	1 fmol
Glucose	0.15–1.5 nmol
Glucose-6-phosphate	2–1000 pmol
Glycerol	0.3–15 pmol
Glycerol-1(3)-phosphate	—
Glycogen	—
3-Hydroxybutyrate	
D-Lactate	0.5–15 nM
Malate	0.2–2 pmol
Methotrexate	0.5–2 pmol
NAD^+	0.2–2.4 pmol
NADH	1 fmol to 0.1 μmol
$NADP^+$	15 fmol
NADPH	0.5–1000 pmol
Oxaloacetate	—
Pyruvate	—
Testosterone	0.8–1000 pmol
TNT	10 amol (10^{-18} M)
Enzymes	
Alcohol dehydrogenase	0.01–10 pmol
ATP : NMN adenyltransferase	5 nmol to 1 μmol

so that each of the substrates can be analysed via this route. However, the majority of analytical applications involve coupling the oxidation of pyridine nucleotides to the production of $FMNH_2$:

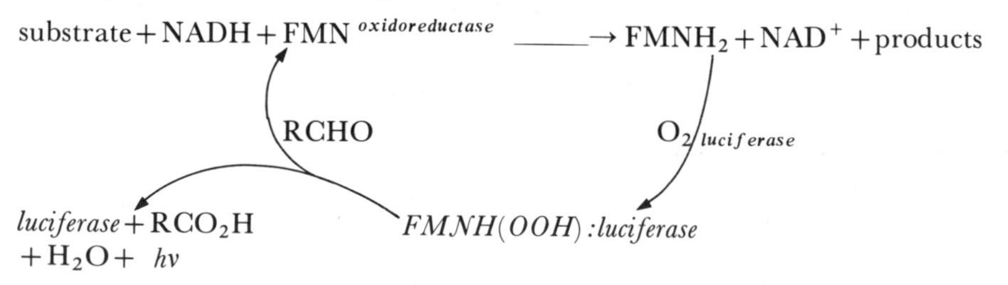

This luciferase is more readily isolated than its firefly counterpart, and a number of NADH-linked assays have been devised (Table 6.3) with detection limits as low as 10^{-15} mol NADH and linearity over five orders of magnitude being reported. Ethanol for example, may be determined according to the following scheme (Stanley, 1978):

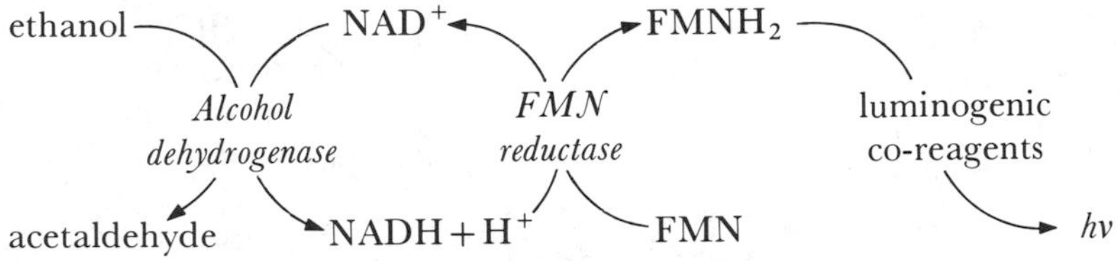

ethanol → acetaldehyde (*Alcohol dehydrogenase*); NAD^+ → $NADH + H^+$

$NADH + H^+$ → NAD^+ (*FMN reductase*); FMN → $FMNH_2$

$FMNH_2$ + luminogenic co-reagents → *hv*

The use of the bacterial bioluminescent system for immunoassay can be demonstrated in the detection of attomole (10^{-18} M) levels of TNT (Wannlund *et al.*, 1982). The assay is based on a competitive binding between TNT–antibody and sample TNT or glucose-6-phosphate dehydrogenase–TNT conjugate (Fig. 6.10). The bound enzyme-labelled TNT is assayed via the NAD/FMN/luciferase route:

$$\text{G-6-P} + \text{NAD}^+ \xrightarrow{\textit{G-6-PDH-TNT}} \text{glucono-6}'\text{-lactone-6-P} + \text{NADH} + \text{H}^+$$

$$\text{NADH} + \text{H}^+ + \text{FMN} \xrightarrow{\textit{oxidoreductase}} \text{NAD}^+ + \text{FMNH}_2 \xrightarrow{\textit{luminogenic reagents}} hv$$

where G-6-P is glucose-6-phosphate and G-6-PDH the corresponding enzyme.

CHEMILUMINESCENT ASSAY

Chemiluminescent indicators can be used in much the same way as other optical labels. Chemiluminescence in solution is almost entirely confined to an oxidative process usually involving peroxide and/or oxygen. Except for the cyclic hydrazides and acridinium salts, the quantum yields in aqueous solutions are low due to quenching by other molecules. The most widely used chemiluminescence agents are therefore the cyclic hydrazides derived from luminol:

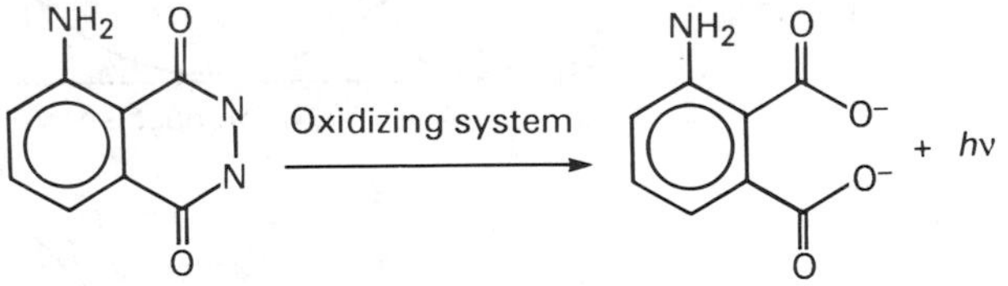

Luminol $\xrightarrow{\text{Oxidizing system}}$ 3-aminophthalate + *hv*

(5-amino-2, 3-dihydrophthalazine-1, 4-dione)

A variety of oxidizing systems have been used but the most frequently employed system is hydrogen peroxide, which requires a catalyst such as the enzyme peroxidase. Since any of the reagents involved in the chemiluminescence reaction may be assayed, a wide range of analytes can be targeted, linked primarily to peroxide or peroxidase. Luminol itself can of course also be used as a label in

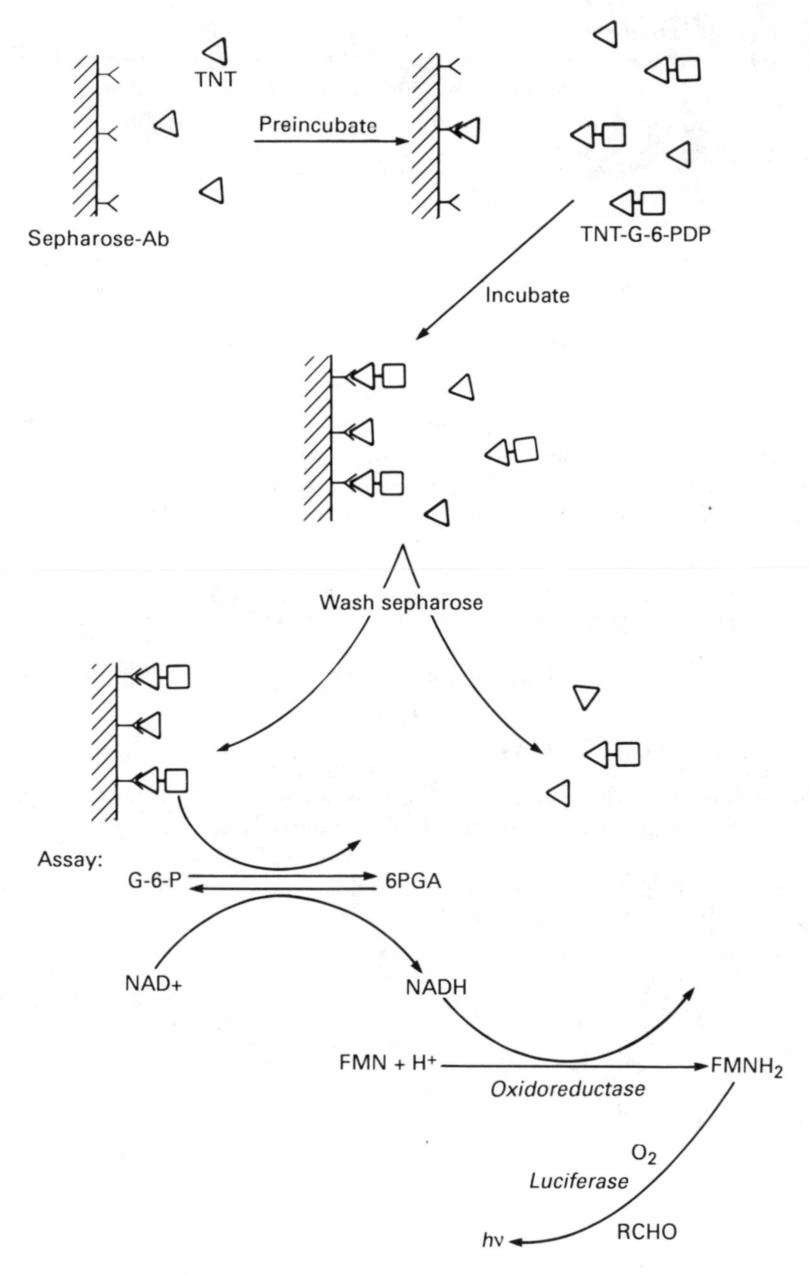

Fig. 6.10 The procedure for the amplified bioluminescent immunoassay of TNT. Sample TNT is incubated with immobilized antibody then enzyme-labelled TNT added and incubated. Bound enzyme label is activated with substrate (glucose-6-phosphate) and NAD^+, and assayed with the bacterial luminescent system.

Table 6.4 Chemiluminescent immunoassay examples

Analyte	*Label*	*Chemiluminescent reactants*
Human IgG	Luminol	Hydrogen peroxide–hemin
Testosterone	Luminol derivative	Hydrogen peroxide–Cu(II)
Thyroxine	Luminol derivative	Microperoxidase
Biotin	Isoluminol	Hydrogen peroxide–lactoperoxides or potassium superoxide
Hepatitis B surface antigen	Isoluminol derivative	Microperoxidase–peroxide
Rabbit IgG	Isoluminol	Microperoxidase–peroxide
Oestriol-16α-glucuronide	Isoluminol	Microperoxidase–peroxide
Cortisol	Isoluminol	Microperoxidase–peroxide

immunoassay (Table 6.4). The assay for biotin is based on the chemiluminescent efficiency enhancement which is noted on forming the conjugate between the isoluminol and protein. Normally the quantum yield for chemiluminescence is typically as low as 1%, and this apparent arrest of quenching on binding protein is a desirable effect. In the model biotin–avidin immunoassay, the signal increased by an order of magnitude, thus obviating the need for a separation and clean-up step (Schroeder *et al.*, 1976).

However, as described before a better sensitivity is often obtained indirectly with an enzyme label and in view of the peroxidase involvement in the chemiluminescence, this is an ideal enzyme label for chemiluminescent detection.

The low quantum yields of chemiluminescent measurements is a potential limitation to their employment. Techniques to improve this yield have therefore been investigated. The major loss is due to quenching by solution species, so that if this loss can be intercepted then improved yields should be noted. In a development by Campbell and Patel (1983), a fluorescent acceptor was used to transfer the chemiluminescent energy. The technique (Fig. 6.11) involved a

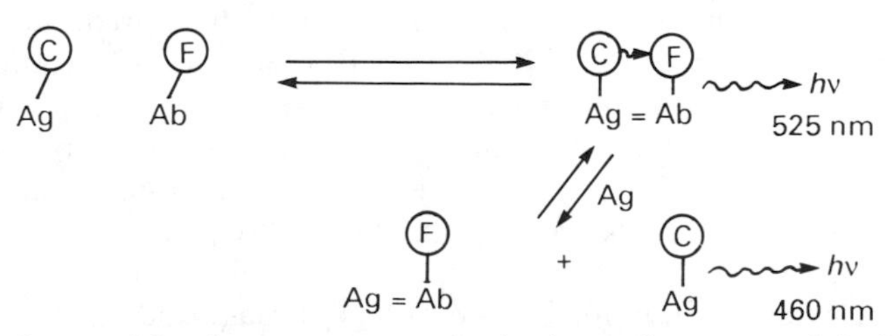

Fig. 6.11 Competitive immunoassay employing a fluorescent-labelled antibody and chemiluminescent-labelled antigen, where the fluophore acts as an acceptor of the chemiluminescent energy in the antibody–antigen doubly labelled complex.

competitive immunoassay between a fluorescent-labelled antibody and sample antigen or chemiluminescent-labelled antigen. In the antibody–antigen complex, energy transfer could occur between labels and detection performed at the fluorescence wavelength. Not only does this technique improve the quantum yield, but also allows both bound and unbound labelled antigen to be distinguished and measured separately.

FLUORESCENCE

Fluorescent indicators can also be employed in their own right. Many biological molecules show an inherent fluorescence, associated with the amino acid derivatives of tryptophan or tyrosine, nucleic acids or other metabolites such as carbohydrates and porphyrins. The reduced form of the nicotinamide cofactor NADH also shows a fluorescent emission on excitation at ≈ 350 nm as does the oxidized flavin nucleotide on excitation at ≈ 450 nm. Exploitation of this inherent property for bioassay is however not straightforward since the emission is weak and often difficult to distinguish above a background signal. Several examples have however been developed.

The fluorescence of chlorophyll depends on the quantity of photosynthetic biomass and its physiological state. Damage to the photosynthetic system is immediately identified by a decrease in the fluorescence with a long time constant (delayed fluorescence) and an increase in the immediate fluorescence. By the introduction of the time dimension into the assay, photosynthetic biomass can be estimated under various environmental conditions. The effect of environmental pollution has been studied *in vivo* in leaves and trees (Schimazaki and Sagahara, 1980; Critchley and Smillic, 1981; Voss *et al.*, 1984) and the development of the technique considered for the analysis of environmental damage.

In general, however, fluorescent labels are more widely used. As with the other indicators already discussed in this chapter, these can be linked to enzymic and immunoassays. Antibodies or antigens can be labelled directly (Fig. 6.12a) with fluorescent indicators such as fluorescein isothiocyanate, but a more sensitive technique employs an enzyme labelled antigen whose substrate produces a fluorogenic product (Fig. 6.12b) (Hösli *et al.*, 1978). Using 4-methyl-umbelliferyl-β-D-galactopyranoside, a fluorogenic substrate for β-galactosidase, extremely small quantities of antigen or antibody have been measured. Each enzyme molecule typically produces 10^4–10^5 fluorescent molecules per minute.

In these examples too sensitivity can be increased by separation on the time axis. The use of such labels as the europium chelates with decay times in the microsecond range allows a 'sampling window' technique to be employed following a nanosecond pulse at the excitation frequency. This technique separates the shorter decay times associated with background fluorescence and can improve the detection limits by two orders of magnitude.

One of the limitations of fluorescent indicators can also be that of quenching by solution species. This limitation can be turned to some use if the measurement is made via a quenching agent (Ullman *et al.*, 1976). This is analogous to the energy transfer between the chemiluminescent marker and the fluorescent marker

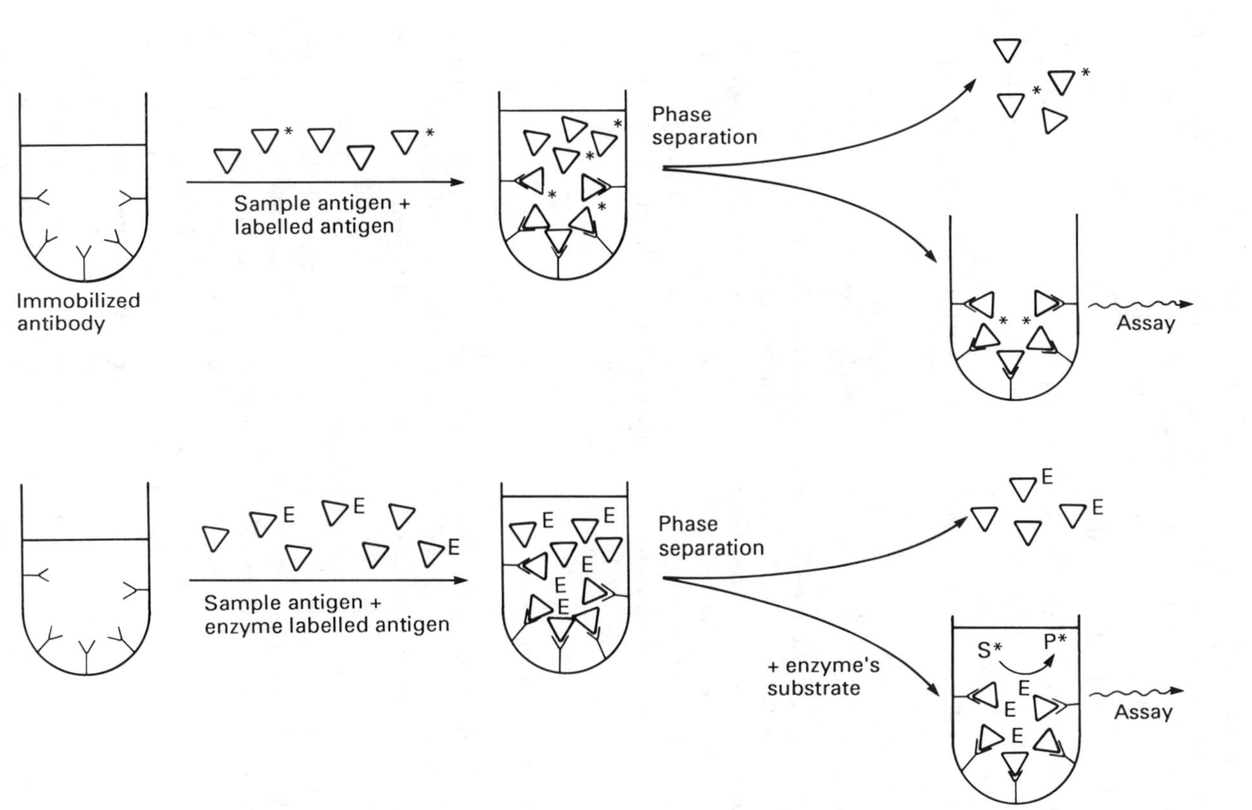

Fig. 6.12 Immunoassay via fluorescent labels: (a) employing fluorescent-labelled antigen and (b) enzyme-labelled antigen and fluorescent substrate.

already described, and is based on the increase in efficiency of energy transfer when the fluorescer and the quencher are brought into close proximity by virtue of the antigen–antibody binding. For example, rhodamine (Rh), with absorption spectrum overlapping the fluorescence spectrum of fluorescein (Fl), will act as a fluorescence quencher. In the immuno-reaction

$$\text{Ab–Fl} + \text{Ag–Rh} \rightleftharpoons \text{Fl–Ab}:\text{Ag–Rh}$$

the free labelled antibody (Ab–Fl) emits a fluorescent spectrum. This is quenched in the Ab:Ag complex, where the rhodamine is sufficiently close to allow energy transfer from the fluorescein.

A competitive immunoassay can therefore be devised based on this technique, for example:

$$\underset{(\textit{fluoresces})}{\text{Ab–Fl}} \xrightarrow{\text{sample Ag+ Ag–Rh}} \underset{(\textit{fluoresces})}{\text{Ag}:\text{Ab–Fl}} + \underset{(\textit{quenched})}{\text{Rh–Ag}:\text{Ab–Fl}}$$

In principle, such assays as these show the fundamental requirements for the bio-unit of the biosensor, and each of the systems described could now be reconsidered and designed with the necessary reagents immobilized to an optical transducer. While many of these established assay methods may finally be unsuitable for adaptation into the reagentless biosensor, they provide a good starting block for further development. Their possible integration with a transducer and indeed the design of an optical transducer will be considered later.

Vibrational Spectroscopy

In the past, vibrational spectroscopy techniques have only found limited application in the quantitative assay of biomolecules but identification of vibrational mode excitation as the physico-chemical event to be transduced in an optical biosensor has possible applications.

Polypeptides exhibit characteristic vibrational spectra that are significantly influenced by the secondary structure of the protein. One could envisage, therefore, applications in antibody–antigen interactions and enzyme–substrate interactions where some conformational change is involved. The absorptions of particular interest in the infra-red are the N–H stretching vibration ($\sim 3290\ \text{cm}^{-1}$); $>\text{C}=\text{O}$ stretch ($\sim 1650\ \text{cm}^{-1}$); and N–H deformation ($\sim 1535\ \text{cm}^{-1}$). As a result of vibrational interactions between adjacent peptide groups, these bands may be shifted or split. They were assigned more definitely for polypeptides in various conformations with the aid of infra-red dichroism (Miyazawa and Blout, 1961). Since maximum absorption by the molecule occurs when the transition moment of the vibration is *parallel* with the electric vector of light, then for a molecule in a fixed orientation, irradiation by plane-polarized light will give enhancement of stretching modes lying parallel to the plane of the incident light and complete abolition of those lying perpendicular to the plane of incident light.

The idealized biosensor (see Fig. 1.3) requires that the biospecific recognition molecule be immobilized at the transduction surface of the biosensor, i.e. with the orientation defined by the method of immobilization and ideally fixed. Application of such an orientation-dependent analysis would therefore be relevant.

Similar considerations apply to polynucleotides where vibration bands due to $>C=O$, $>C=N$ and $>C=C<$ are of particular interest.

Considerable improvements in signal processing and the introduction of such techniques as Fourier transform infra-red and multichannel Raman spectroscopy have introduced an analytical capability to vibrational spectroscopy that was hitherto impossible. Methodology is also now being developed that allows atmospheric pollution to be monitored by a mobile system using passive Fourier transform infra-red interferograms (Small *et al.*, 1988). By the use of suitable digital filters with vibration frequency characteristics of target species, the instrument can be tuned to specific pollutants, for example SF_6. Other sensing applications are also reported. Aromatic hydrocarbons have been identified and estimated with a detection limit of 20 ppb (20 ppt considered possible) and a linear response to the Raman intensity over three orders of magnitude (Campion and Woodruff, 1987), while the reaction of cytochrome oxidase with oxygen, followed by time-resolved Raman spectroscopy was able to detect intermediates in the reaction and provide data concerning the mechanism (Babcock *et al.*, 1984). Such experiments reveal the potential of vibrational spectroscopy to monitor binding events between biological molecules. This would be highly specific and particularly relevant to antibody–antigen complex formation, where direct assay of the binding event without the use of labels is most desirable.

The Optical Transducer

The initiation, translation and interpretation of optical events for these bioassays has traditionally involved a suitable photometric light source, irradiating a sample solution, which has undergone sufficient pretreatment so as to prevent undesirable effects from scatter, absorption or other interfering optical side effects. While these analyses are proven methods, they are bioassays rather than biosensors, with the source and detector not in intimate contact with the assay reagents. In order to make the transition to a biosensor device it is necessary to identify a transduction element suitable for optical processes.

SIMPLE WAVEGUIDES

The production of optical fibres, initially predominately for communication purposes, represents little more than a decade of intensive development, and has been facilitated by an increasing availability of optically clear polymeric materials for a wide range of wavelengths. These fibres, or more generally optical waveguides as a whole, provide an optical transduction system possessing the properties required of a biosensor.

Transmission of light through these materials is as a guided wave, where the wave is propagated by total internal reflection. At the interface between two

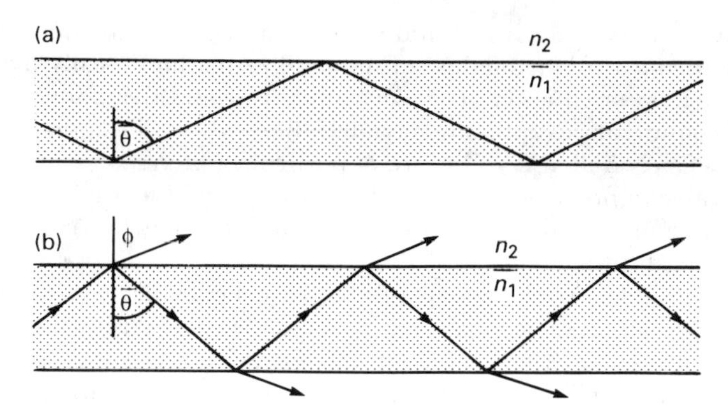

Fig. 6.13 Ray trajectories in a step profile wave guide for (a) bound rays and (b) refracting rays.

optically transparent media of different refractive indexes the light will be refracted or reflected according to the angle of incidence at the interface and the refractive indexes of the two phases.

A planar dielectric step-profile waveguide can be simply modelled by a ray diagram as shown in Fig. 6.13. The path of the ray constructs zigzag trajectories where the ray is (a) totally reflected at every reflection, in which case it is a bound ray (Fig. 6.13a), or (b) partly reflected and refracted (Fig. 6.13b). Since the incident, reflected and refracted rays are coplanar then:

$$\frac{\sin\theta}{\sin\phi}=\frac{n_2}{n_1}=n \qquad \text{(\textit{Snell's law})}$$

When $\phi=\pi/2$ the refracted ray will lie along the interface, and it follows that the power transmission coefficient will be zero and,

$$\sin\theta=\frac{n_2}{n_1}$$

This angle, termed the *critical angle* (θ_c) is required for total internal reflection of the incident ray.

$$\text{For } \sin\theta \geqslant \frac{n_2}{n_1} \rightarrow \text{total internal reflection}$$

$$\text{For } \sin\theta < \frac{n_2}{n_1} \rightarrow \text{refraction} + \text{reflection}$$

The plane harmonic wave can also be described by three vectors **k**, **E**, **H**. The electric (**E**) and magnetic (**H**) fields are perpendicular to the direction of propagation (**k** is the propagation or wave vector). For the case of unpolarized light incident on a plane boundary separating two different optical media, the incident, reflected and transmitted waves have the forms:

Incident wave: $\mathbf{E}=\mathbf{E}_i e^{-i\omega t}\exp\{i\mathbf{k}n_1(-z\cos\theta+x\sin\theta)\}$
Reflected wave: $\mathbf{E}=\mathbf{E}_r e^{-i\omega t}\exp\{i\mathbf{k}n_1(z\cos\theta+x\sin\theta)\}$
Transmitted wave: $\mathbf{E}=\mathbf{E}_t e^{-i\omega t}\exp\{i\mathbf{k}n_1(-z\cos\phi+x\sin\phi)\}$

where $\omega=2\pi\nu=$ angular frequency and $\mathbf{k}=2\pi/\lambda=$ angular wavenumber.

The electric vector of the incident wave may be parallel to the plane of the boundary, in which case it is called *transverse electric* or TE polarization, or else the magnetic vector is parallel to the boundary plane, in which case it is called *transverse magnetic* or TM polarization. The general case for the directions of the electric and magnetic vectors of the incident, reflected and transmitted waves is shown in Fig. 6.14 for these two cases. The boundary conditions for the interface require that the tangential components of the electric and magnetic fields be continuous across the boundary,

For TE polarization: $\mathbf{E}_i+\mathbf{E}_r=\mathbf{E}_t$
For TM polarization: $\mathbf{H}_i-\mathbf{H}_r=\mathbf{H}_t$

The amplitude ratio for reflection in the TE and TM cases respectively will be:

$$r_s=\frac{\cos\theta-n\cos\phi}{\cos\theta+n\cos\phi}$$

$$r_p=\frac{-n\cos\theta+\cos\phi}{n\cos\theta+\cos\phi}$$

Eliminating ϕ by Snell's law gives

$$r_s=\frac{\cos\theta-\sqrt{(n^2-\sin^2\theta)}}{\cos\theta+\sqrt{(n^2-\sin^2\theta)}}$$

and (*Fresnel's equations*)

$$r_p=\frac{-n^2\cos\theta+\sqrt{(n^2-\sin^2\theta)}}{n^2\cos\theta+\sqrt{(n^2-\sin^2\theta)}}$$

The *reflectance* is defined as the fraction of the incident light energy that is reflected. Since the energy is proportional to the absolute square of the field amplitude,

$$R_s=|r_s|^2$$
$$R_p=|r_p|^2$$

EXTERNAL REFLECTION

In the case of *external reflection* where $(n_2/n_1)=n>1$ (Fig. 6.15a) these amplitude ratios are real for all values of θ. For the TM case the reflection is zero when $\theta=\tan^{-1}n$, i.e.

$$\sin\theta=\cos\phi$$
$$\theta+\phi=\pi/2$$

so that at this point the reflected light is rendered linearly polarized, with the electric vector transverse to the plane of incidence and the transmitted light is

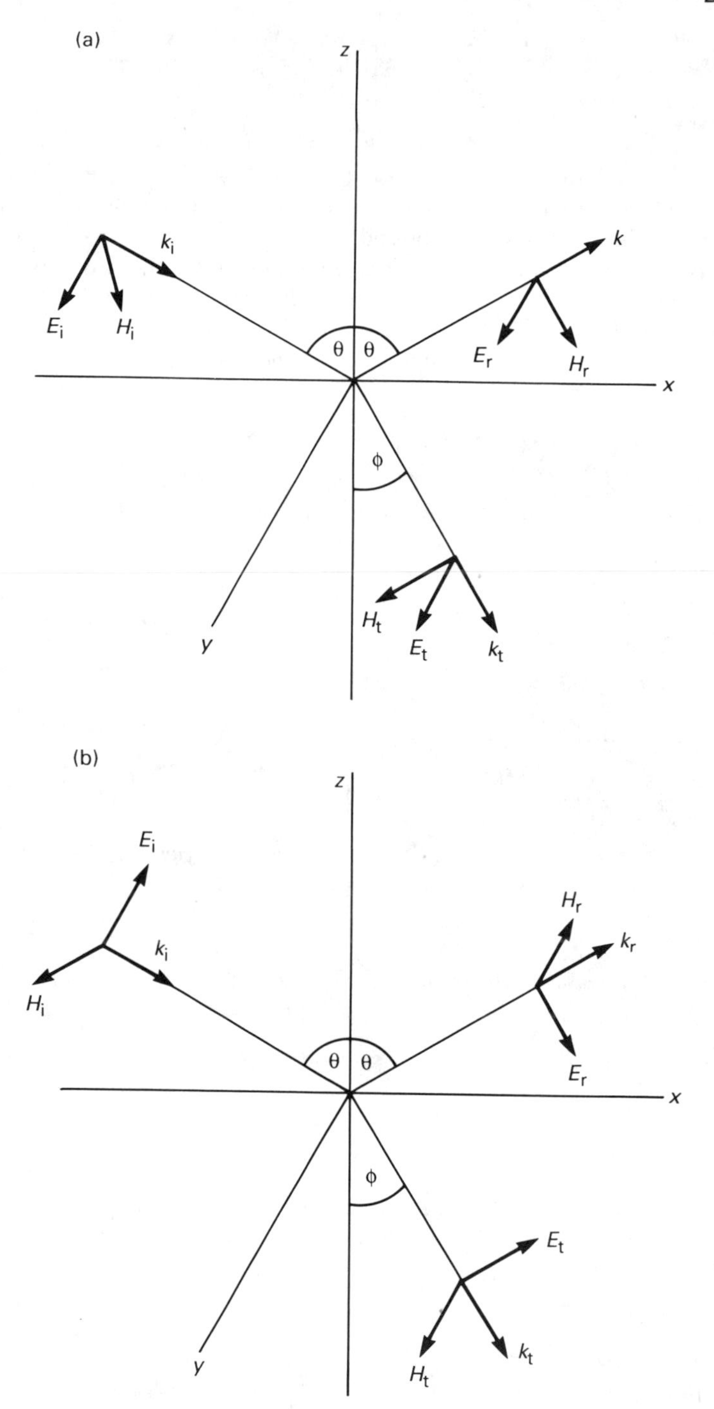

Fig. 6.14 Wave vectors and associated fields for (a) transverse electric (TE) and (b) transverse magnetic (TM) polarization.

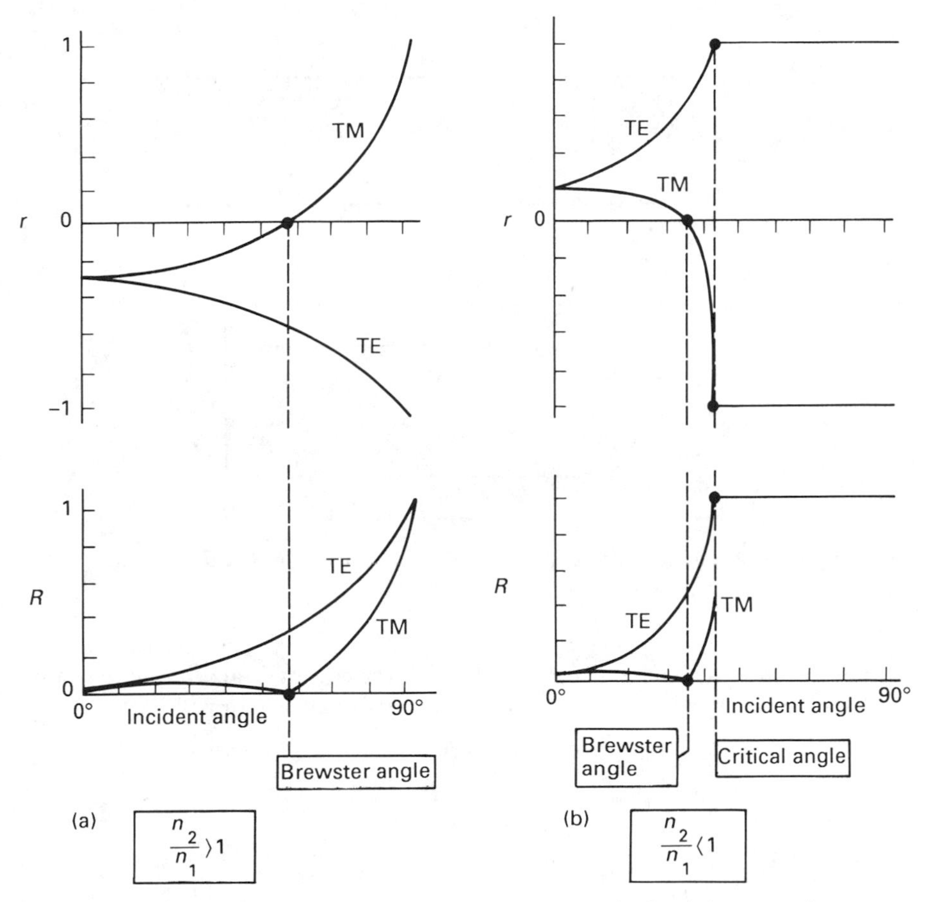

Fig. 6.15 Reflectance as a function of incident angle for (a) external reflection and (b) internal reflection.

partially polarized. The angle required to achieve this is known as the *Brewster angle*.

INTERNAL REFLECTION

In the case of *internal reflection* where $n<1$ (Fig. 6.15b), the amplitude ratios have real solutions for $\sin\theta<n$, i.e. for θ less than the critical angle. However, where the angle of incidence exceeds the critical angle, the ratios become complex:

$$r_s=\frac{\cos\theta-i\sqrt{(\sin^2\theta-n^2)}}{\cos\theta+i\sqrt{(\sin^2\theta-n^2)}}$$

$$r_p=\frac{-n^2\cos\theta+i\sqrt{(\sin^2\theta-n^2)}}{n^2\cos\theta+i\sqrt{(\sin^2\theta-n^2)}}$$

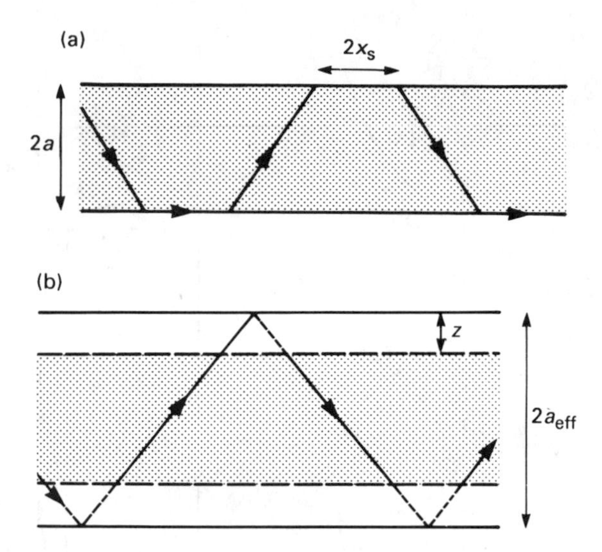

Fig. 6.16 The lateral shift $2x_s$ (a), of the ray path is equivalent to an effective core width of $2a_{\text{eff}}$ (b).

In this instance the reflectance $R=1$, showing the case of total internal reflection, and the phase change between incident and reflected waves will be:

$$TE: \quad \Phi=\frac{2\tan^{-1}(\sin^2\theta-n_2^2/n_1^2)^{1/2}}{\cos\theta}$$

$$TM: \quad \Phi=\frac{2\tan^{-1}(\sin^2\theta-n_2^2/n_1^2)^{1/2}}{(n_2^2/n_1^2)\cos\theta}$$

Under these conditions, the angle of transmission, ϕ, is complex and given by

$$\cos\phi=(1-\sin^2\phi)^{1/2}=\frac{i(n_1^2\sin^2\theta-n_2^2)^{1/2}}{n_2}$$

so that the *transmitted wave* is an *evanescent wave* normal to the interface, i.e.

$$E=E_t\exp\{i\omega t+ikn_1(\sin\theta x+kn_1(\sin^2\theta-n_2^2/n_1^2)^{1/2}z\}$$

where only the z term is real, so that the field is a decaying wave in the z direction with a penetration depth,

$$d_p=\frac{-1}{kn_1\sqrt{(\sin^2\theta-n_2^2/n_1^2)}}$$

PHASE CHANGE

In the initial model of total internal reflection, the ray analysis approximation has $\lambda=0$, thus ignoring phase changes and assuming that power only flows along the geometric optics trajectories. It would be more accurate to consider the incident beam as a collection of individual rays, each with their own slightly different angle of incidence, giving a value for Φ expanded for all the values of θ. This can be interpreted in terms of a shift x_s in the position of the reflected beam. In a simple ray model, this would be represented by an effective wave guide width which is greater than the true width, and accommodates the lateral shift x_s by the penetration depth, d_p, as shown in Fig. 6.16.

$$d_p = \frac{-1}{kn_1\sqrt{\{\sin^2\theta - (n_2/n_1)^2\}}}$$

$$= \frac{-\lambda}{2\pi n_1\sqrt{\{sin^2\theta - (n_2/n_1)^2\}}}$$

and

$$x_s = \frac{d_p}{\cot\theta}$$

$$= \frac{-1}{2kn_1\cos\theta} \times \frac{2\sin\theta}{\sqrt{\{\sin^2\theta - (n_2/n_1)^2\}}}$$

$$= \frac{1}{2kn_1\cos\theta} \times \frac{d\Phi}{d\theta}$$

where Φ is the phase change as has been defined earlier.

The continuum implied by this model would not be self consistent, so that in fact only discrete ray directions are allowed, which can be described by imposing the boundary conditions that the total transverse phase shift occurring in one complete ray period must be an integer multiple of 2π (Tien, 1971). This restricts the allowed values of θ.

In the simple single ray hypothetical model, with zero electric field at the interface and transverse phase change $4kn_1a\cos\theta$, the allowed directions would be satisfied by:

$$4kn_1a\cos\theta = 2m\pi$$
$$\Rightarrow \quad 2a\cos\theta = m\lambda/2$$

where m is an integer termed a mode number.

When $\sin\theta = n_2/n_1$, a maximum value of m is obtained:

$$m_{\max} = \left(\frac{4a}{\lambda}\right)\left[1 - \left(\frac{n_2}{n_1}\right)^2\right]^{1/2}$$

If $m_{\max} < 2$ then the guide will be single moded; for $m_{\max} \geqslant 2$ it will be multimoded. This approximates the relationship between core width, wavelength and propagation in the x-direction. The standing waves of the electric field across the guide which describe the first few modes are shown in Fig. 6.17.

In practice, however, the allowed values of θ must be corrected to account for phase changes associated with reflection at the boundaries.

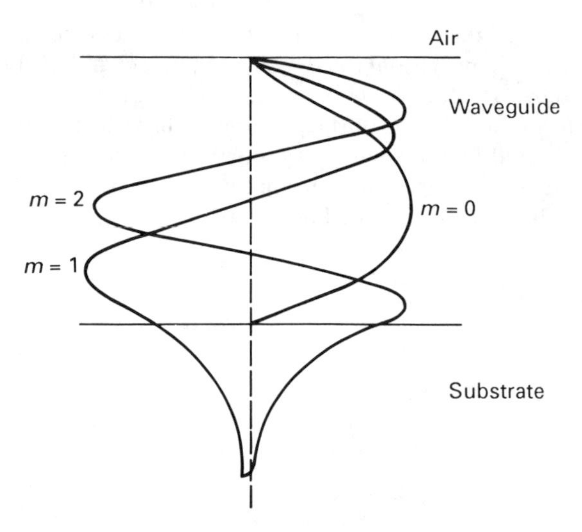

Fig. 6.17 The standing waves established for the lowest modes, showing the light amplitude variation across the wave guide.

FIBRES

In considering an optical fibre then the preceeding treatment of planar wave guides is applicable, with the introduction of a third dimension. A ray may cross the fibre axis between each reflection (meridional ray) (Fig. 6.18a) or else it may take a helical path, never crossing the fibre axis (skew ray) (Fig. 6.18b). In the former case, the behaviour of the ray corresponds to that in a planar wave guide, but in the latter case, the point of reflection is on a cylindrical surface, so that in addition to β, the angle with the axial direction, the trajectory must be described by α, the angle between the ray path and a tangent to the interface at the point of reflection (Fig. 6.19). It can be seen from this figure that for skew rays $\theta+\beta\neq\pi/2$, so that an angle of incidence described by θ and β with $\theta>\theta_c$ *and* $\beta<\beta_c$, would not be accounted for in the treatment of a planar wave guide. Rays falling into this category are termed *tunnelling rays*.

BENDS

If a straight section of planar waveguide leads into a significantly curved section, then all rays that are bound on the straight section will be refracting or tunnelling on the bend. This is analogous to the tunnelling model for a fibre. On the bend, the ray path follows straight lines between reflections (Fig. 6.20):

(a) alternately between inner and outer surfaces; and
(b) only from the outer interface (whispering gallery ray).

If $\theta<\theta_c$ the ray refracts at the outer surface, but if $\theta>\theta_c$ and $\beta<\beta_c$ it tunnels. At the inner surface if $\theta'<\theta_c$ and $\beta'>\beta_c$ the ray is transmitted, but since the interface is convex $\beta'>\theta_c$ gives total internal reflection.

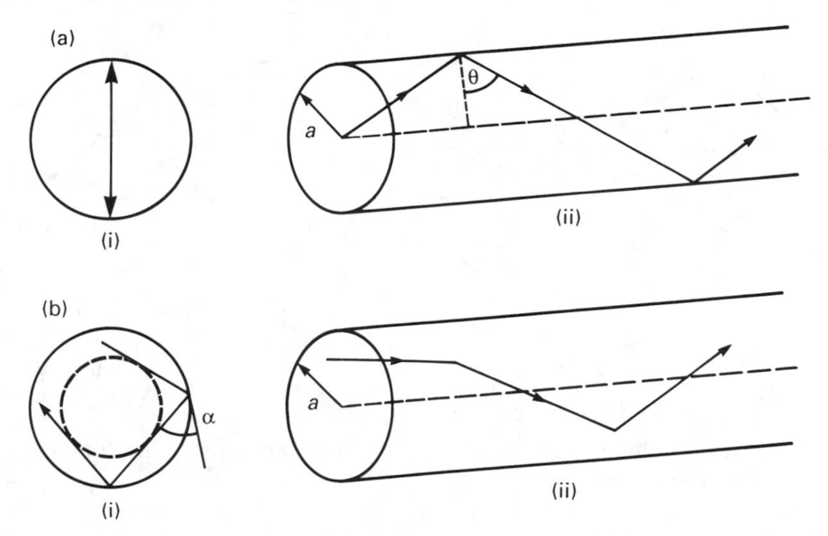

Fig. 6.18 Ray paths within the core of a fibre showing (a) the meridional ray and (b) the helical path of the skew ray.

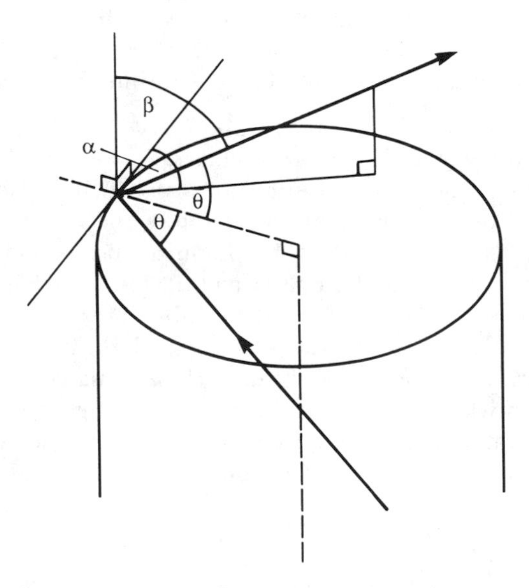

Fig. 6.19 Description of a ray incident on the interface of a step profile fibre by angle α, β and θ.

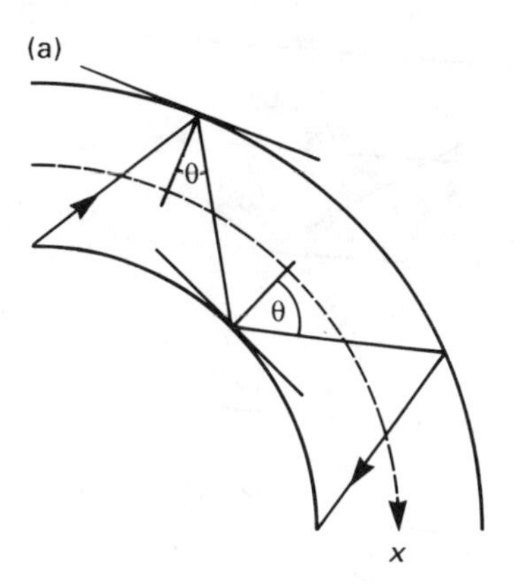

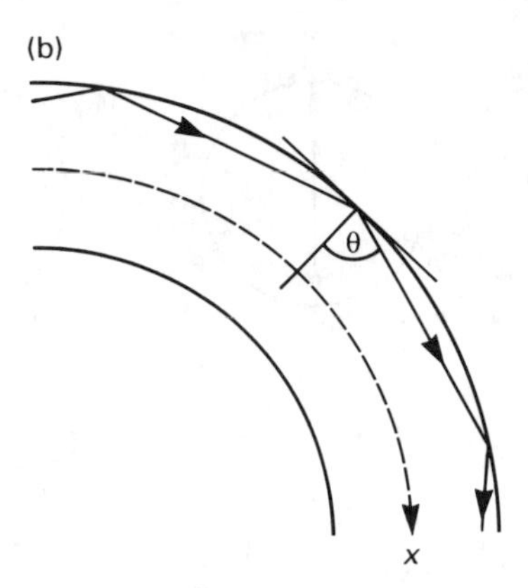

Fig. 6.20 Ray paths on a bend. (a) Reflected between inner and outer surfaces; (b) 'whispering gallery' rays.

Waveguides in Sensors

Manipulation of the various parameters concerned with propagation of the wave along a waveguide, has been primarily directed towards optical telecommunications. There are however an infinite number of ways in which guided light may be modulated by an environmental parameter, for sensing and signal processing applications. Waveguide sensors could be divided into two basic classes.

(1) *Extrinsic*: The waveguide (usually a fibre) serves as a light source and light collector and the modulation process takes place externally from the fibre, usually by way of an attenuation process which is modulated by the measurand. For example, light absorption by the sample in an in-line cell modulator, will be described by Lambert–Beer's law and its effect determined by measurand intensity and the layer thickness of the sample. The light is carried to and from the cell by fibres.

(2) *Intrinsic*: The measurand interacts directly with the light in the waveguide. Phase, polarization and intensity may all be modulated within the guide by a measurand which lies within the penetration depth for the evanescent field in the rarer medium adjacent to the guide. Since the electric field amplitude which is transmitted at a point of total internal reflection can be written as,

$$E = E_t \exp(-z/d_p)$$

the attenuation of the signal will be greatest immediately at the interface. Within a distance equivalent to about 20% of the wavelength, the field has dropped below 70% of its amplitude at the interface.

Sensitivity can be enhanced by interaction with the sample over an extended path length. In a simple ray model approximation, the number of reflections ($\mathcal{N}$) (and thus sample interactions) is a function of length (L), guide thickness ($2a$) and θ,

$$\mathcal{N}=\frac{L}{2a}\cot\theta$$

and it follows that where d_e is the effective thickness (i.e. the thickness of film that would be required to give the same absorption in a transmission experiment), then the reflectivity (R) at each reflection will be:

$$R=1-\alpha d_e$$

where α is the absorption coefficient and

$$d_e=\frac{n_2 E_t^2 \mathrm{d_p}}{2n_1\cos\theta}$$

The evanescent wave can therefore be employed to excite or monitor the spectral characteristics of any molecule in the field for a particular working wavelength.

Device Construction

Unlike electrochemical sensors, optical sensors do not require a reference sensor. However, the response characteristics are generally considerably improved by comparing the analyte signal with a 'reference blank'. There are several ways of achieving this:

(1) Source intensity fluctuations can be monitored directly at the analytical wavelength and adjustments made in the signal recorded.
(2) Source can be split between sample and reference 'cell' and a relative measurement made.
(3) Detection can be made at an analytical and reference wavelength, so that variations due to scatter and source fluctuations can be eliminated by the ratio of the two signals.

The two latter techniques are the most practical solutions for sensor development. The actual construction and material of the waveguide sensor will depend on which of the two basic classes of device, defined above, are to be employed, and the wavelength that will be required for source and detection.

Plastic waveguides, for example, are suitable for wavelengths above about 450 nm and glass above 350 nm. Below this wavelength fused silica, and for infra-red measurements germanium crystal guides, are usually employed, thus considerably escalating the basic cost of the optical transducer.

For measurements made in the evanescent field, a portion of the interface must be exposed to the sample (Fig. 6.21a), but configuration of the sensor in an extrinsic device can follow various designs. Figure 6.21(b), for example, shows an equivalent configuration to that generally found in a spectrophotometer with light passing straight through the sample and being collected at the other side.

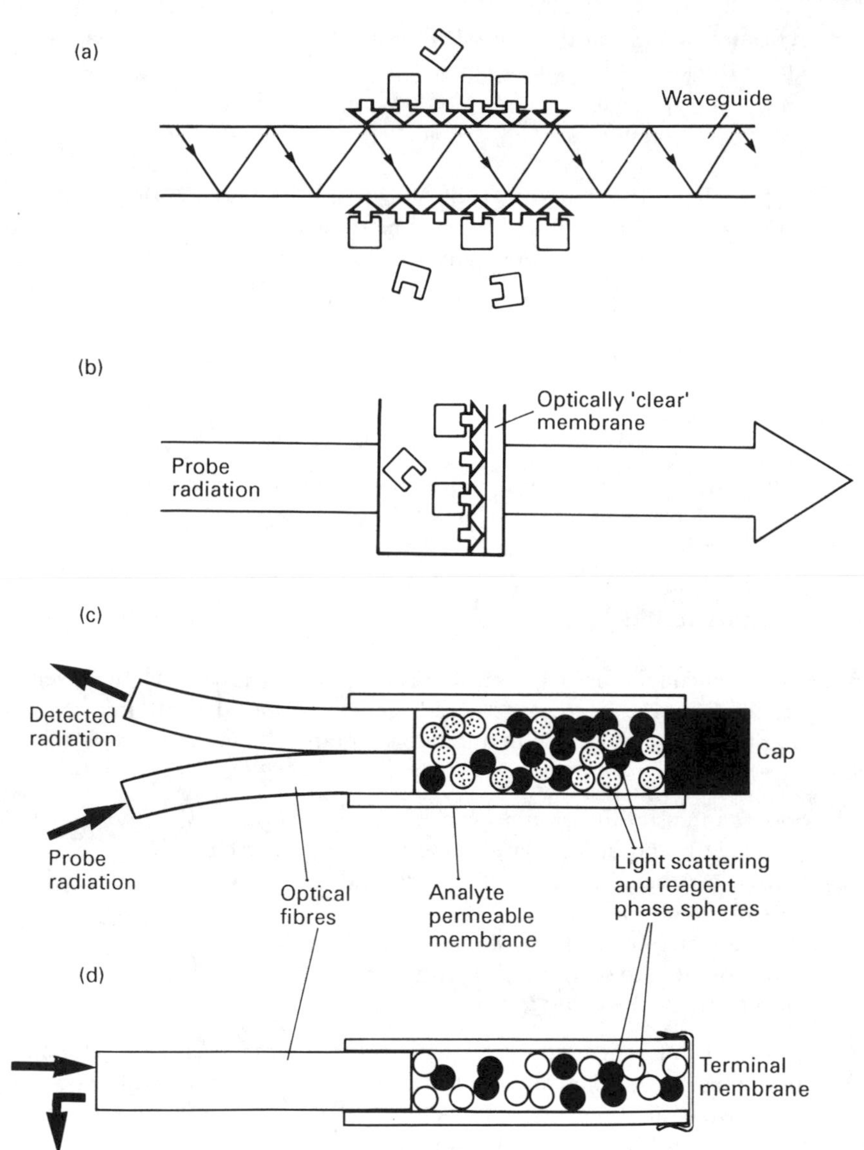

Fig. 6.21 Optical sensor configurations. (a) Evanescent field monitoring; (b) solid-state optical sensor; (c) light-scattering probe with common source and detection wavelength; (d) light-scattering probe with source and detection distinguished by wavelength.

Figures 6.21(c) and (d) employ light-scattering spheres in the device, in order to scatter a proportion of the light back in the source direction. In (c) a bifurcated waveguide is used for source and detector, but when probe and detection wavelengths can be distinguished (as in fluorescence) the same fibre serves to transmit light both to and from the reagent phase.

The immobilized reagent phase that is required in the construction of any optical biosensor is not always an easily defined medium. The majority of optical chemical sensors already reported have employed an indicator. Various indicator interactions can be modified for this model:

Analyte will react with indicator according to the equilibrium:

$$A + Ind \rightleftharpoons AInd$$

with equilibrium constant,

$$K = \frac{[AInd]}{[A][Ind]}$$

Where total indicator concentration, C_{ind}, in the immobilized phase is

$$C_{ind} = [Ind] + [AInd]$$

the free indicator will thus be

$$[Ind] = \frac{C_{ind}}{(1 + K[A])}$$

and the combined indicator

$$[AInd] = \frac{K[A]C_{ind}}{(1 + K[A])}$$

It follows therefore that no simple linear relationship exists between signal due to free indicator and analyte [A] or between combined indicator and analyte, and that both measurements are dependent on C_{ind}. It was suggested earlier that if two wavelengths were used, making the measurement with respect to a reference blank, then an improved signal quality could be obtained. In this instance the signal due to [AInd] referenced to [Ind]:

$$\frac{[AInd]}{[Ind]} = K[A]$$

also gives an assay independent of C_{ind}.

The first optical chemical sensors have been of the extrinsic class, where the optical waveguide was employed to transport light to and from a discrete modulation unit. Like their solution-based counterparts, these assay devices rely on matching a particular optical indicator to the target analyte. The initial focus of these sensors has been towards the monitoring of parameters already established by the electrochemical sensors commonly in use, namely pH, O_2, CO_2, NH_3 and ion concentration.

pH Optical Sensors

Optical pH sensors have employed a variety of different indicators. Their attraction is the prospect of an improvement in stability over the pH electrode, but while competing in this area the range of an optical pH sensor is usually only a maximum of 1 to 2 pH units around the end point of the particular indicator employed, thus limiting their use as a general broad range pH probe. For a specific application within a known pH range, however, they have proved successful.

PHENOL-RED pH PROBE

Peterson *et al.* (1980) have described a pH probe based on the indicator phenol red, for the range pH 7.0–7.4. Polyacrylamide-coated polystyrene microspheres were prepared containing the dye covalently attached to the vinyl chain. These indicator microspheres were packed with 1 μm polystyrene spheres in H^+-permeable cellulose tubing to form the pH modulator (Fig. 6.21, design c). Fibres made the optical connections with the cell and the light scattered back towards the source direction was detected after multiple scattering by the microspheres. The detector made a two wavelength measurement, comparing the transmittance at 558 nm (where the basic form of phenol red absorbs) to a reference value at 600 nm (where neither base nor acidic forms absorb). The pK_a of the bound dye was 7.57, compared with 7.9 for the soluble dye allowing pH measurements to be made in the 7.4–7.0 range to 0.01 pH units. In blood samples and *in vivo* this probe compared favourably with the pH electrode.

Grattan *et al.* (1987/88) have shown that low cost source and detection components can be used in conjunction with the phenol red indicator to give a device capable of estimating pH. The source radiation was provided by an ultra-bright green LED ($\lambda \sim 565$ nm) referenced to an infra-red LED at 810 nm, and detection for both wavelengths was at a Si p-i-n diode. The integration of such low-cost miniature non-specialist components into the optical probe makes the transition to a portable device and the realization of a sensor becomes viable.

FLUORESCENT pH PROBE

The simplest type of fluorescent pH sensor would involve a single wavelength measurement, for example fluoresceinamine covalently bound to cellulose shows an increase in fluorescence with pH (Saari and Seitz, 1982). Fluorescein is a proteolytic acid with p$K_{a1}=2.2$, p$K_{a2}=4.4$ and p$K_{a3}=6.7$. The absorption and fluorescence spectra at different pH values are related to the different forms (Fig. 6.22). At high pH the dianion is the major species, with absorption maximum around 491 nm and fluorescence, monitored at around 530 nm, has a high quantum yield (0.9). The monoanion on the other hand is not excited in this region, so that two regions can be identified in the pH-related response. Between pH 2.8 and 5.5, the neutral molecule and monoanion exist, while at pH 5.5 to 7.0 the monoanion and dianion are the major species.

Fluorescein isothiocyanate (FITC) has similar spectral properties to those of

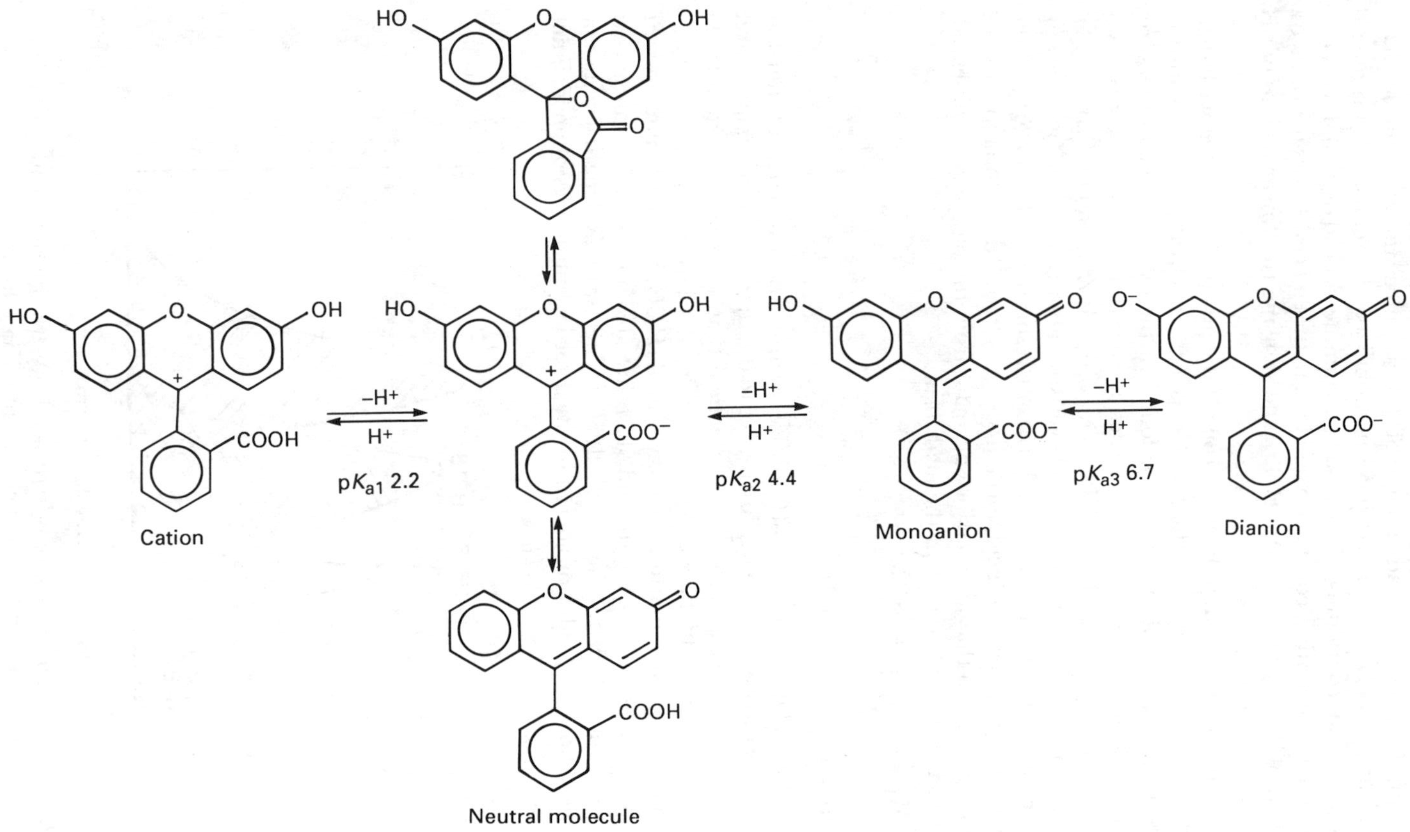

Cation

$-H^+$

H^+

pK_{a1} 2.2

pK_{a2} 4.4

Monoanion

pK_{a3} 6.7

Dianion

Neutral molecule

Fig. 6.22 Proteolytic forms of fluorescein.

fluorescein, but with the advantage of the isothiocyanate group for easy immobilization. Fuh *et al.* (1987) have immobilized the indicator to silanized porous glass beads containing a long chain alkylamine. A bead of similar diameter to a multimode fused silica core fibre, was glued to the end of the fibre to produce a pH sensor, with a dynamic range pH 3–7 and showing two linear regions of pH 2.8–5.0 and 5.5–7.0.

Quenching of fluorescent indicators is one of the inherent difficulties with their use and in the immobilized state this problem is increased by the close proximity of adjacent molecules. A solution to this is found in the two wavelength measurements made with fluorescent indicators such as the trisodium salt of 8-hydroxy-1,3,6-trisulphonic acid (HPTS) (Wolfbeis *et al.*, 1983; Zhujun and Seitz, 1984). The acid and base forms have characteristic absorption spectra, showing excitation wavelengths of 405 nm and 470 nm respectively (Fig. 6.23). In the excited state, however, the acid form undergoes rapid deprotonation to the excited base state, so that an emission spectrum from the base form is observed, even when the ground state is in the acid form:

$$\begin{array}{ccc} \mathrm{HHPTS}^{*} & \rightarrow & \mathrm{HPTS}^{-*} \\ \uparrow & & \upharpoonleft\downharpoonright \\ \mathrm{HHPTS} & \rightleftharpoons & \mathrm{HPTS}^{-} + \mathrm{H}^{+} \end{array}$$

pH is therefore related to the ratio of fluorescence intensities excited at 405 nm and 470 nm. This ratio measurement is insensitive to fluctuations due to temperature, quenching or ionic strength so is more practical than the previous example.

Several of the coumarins have also been considered to be useful in this way. Both HPTS and 7-hydroxycoumarin-3-carboxylic acid (HCC) have been immobilized to glass membranes, mounted at the end of a bifurcated optical fibre to produce a remote pH probe (Offenbacher *et al.*, 1986). The two-wavelength excitation method was, however, only considered justified for circumstances of poor signal

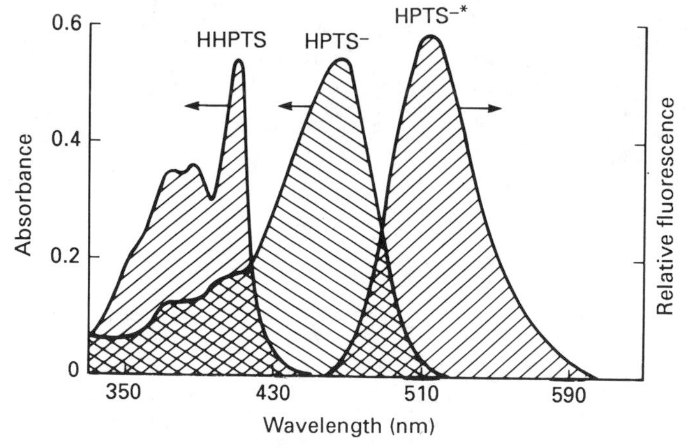

Fig. 6.23 Absorption spectra of 8-hydroxy-1,3,6-trisulphonic acid (HPTS) in acid and base forms and fluorescent spectrum of base form.

intensity, since the additional complexity of control and signal handling would be unnecessary for general routine sampling. In fact HCC with excitation wavelengths of 405 nm and 360 nm would also require more expensive optical ware suitable for the near ultra-violet if measurement was made at 360 nm.

The HPTS sensor in particular reports good stability and an accuracy of ± 0.01 pH unit in the range 6.4–7.5. Performance in blood is however only achieved if the sensing area is protected with an H^+-permeable membrane such as cellulose. This presumably excludes proteins from the surface, but the result is achieved at the expense of response time.

The fluorescent indicator 4-methylumbelliferone has also been employed in pH sensors. Various approaches have been made (Opitz and Lübbers, 1983; Lübbers and Opitz, 1983), but this indicator requires excitation in the ultra-violet region of the spectrum (318 nm) so that its employment is not encouraged.

EFFECT OF IONIC STRENGTH ON pH MEASUREMENT

One of the frequent problems with the use of indicators to monitor pH is the effect of ionic strength on the pK_a and the measurement. Lübbers and Opitz (1983) showed the ionic strength (Ion) could be estimated from the behaviour of two fluorescent indicators with different charges (z), since:

$$pK_a^{\mathrm{Ion}} = \frac{pK_a + 0.512(z_B^2 - z_{HB}^2)\mathrm{Ion}^{1/2}}{(1 + 1.6\mathrm{Ion}^{1/2})}$$

Wolfbeis and Offenbacher (1986) have shown that in practice, for an immobilized indicator, this function of ionic strength is dependent on the microenvironment at the immobilization surface. HCC was immobilized at a porous glass surface (a) leaving residual charged $-NH_3^+$ groups at the surface and (b) where the charged groups have been reacted to give $-NH-$acetyl. In the former case it was estimated that each indicator molecule was surrounded by 2–3 $-NH_3^+$ groups, thus simulating the condition of high ionic strength and rendering the surface immune to bulk changes in ionic strength.

In contact with the same sample these two sensors displayed different estimates of pH: the former independent of ionic strength and the latter dependent on ionic strength. The difference between these two readings, which is equivalent to the difference in pK_a values, can be correlated with the ionic strength of the sample from the above equation. This arrangement thus allows simultaneous estimation of pH and ionic strength.

Ion-selective Probes

Inherent quenching of fluorescence by solution species is a common limitation to fluorescent assays. For many analytes, however, it provides a means of linking the target species to a reversible fluorescent indicator.

The kinetics of quenching of the excited-state fluorescent indicator are in competition with the kinetics of the natural fluorescence decay mechanism. The rate of quenching will thus be proportional to the lifetime of the excited state and the collision rate between quencher and fluorescent species. The latter in turn will be proportional to the concentration of quencher. The relationship is given by the *Stern–Volmer equation*:

$$\frac{I_o}{I} = 1 + K\text{pO}_2$$

where I and I_o are the relative fluorescence intensities in the presence and absence of oxygen, pO_2 is the partial pressure of oxygen and K is the overall quenching constant, including the rate constants for all modes of decay. This equation is equivalent to that derived earlier (see above) for immobilized indicators

$$\frac{C_{\text{ind}}}{[\text{Ind}]} = 1 + K[\text{A}]$$

assuming that fluorescence intensity is proportional to unquenched reagent.

Urbano *et al.* (1984) designed a fluorescence quenching sensor targeted to halides, using acridinium- and quinolinium-based fluorescent reagents, covalently bound to a glass support via a carbodiimide linkage; the former indicator showed the higher sensitivity. Quenching efficiency of either fluorophore is $I^- > Br^- > Cl^-$, with the detection limit for I^- with the acridinium dye of 0.15 mM and for Cl^- of 10 mM.

Although the probe does not respond to nitrate, phosphate, sulphate or perchlorate, its specificity is not good; various anions such as cyanide, cyanate and isothiocyanate interfere, the latter with similar efficiency to that observed with iodide. Except for well-characterized applications, where this multiple species response is not critical, this does not offer a viable probe which can be widely exploited.

The use of crown compounds in the development of ion-selective membranes has allowed ion-selective electrodes to be designed with a selectivity for specific target ionic species (Chapter 3). The combination of an ionophore with a chromogenic reaction permits the same estimation to be performed spectrophotometrically.

Zhujun *et al.* (1986) have assembled the necessary components to create a Na^+-selective optical probe. The ionophore (Ion) was immobilized on light-scattering silica powder at the end of a bifurcated fibre, and the assembly dipped in a 'reagent solution' contained by a dialysis membrane. The optical label which is present in the reagent solution is the anionic fluorophor, Fluor(8-anilino-1-naphthalene sulphonic acid). This forms ion pairs with the immobilized ionophore–Na^+ complex, so that fluorescence intensity from this complex serves as a measure of alkali metal ion concentration. Fluorescence of excess label in the reagent phase is suppressed by the formation of ion pairs with the cationic polyelectrolyte (Poly), copper(II)–polyethyleneimine quenching polymer:

$$\underset{(quenched)}{\text{Ion} + \text{Poly–Fluor}_{(\text{soln})}} \overset{Na^+}{\rightleftharpoons} \underset{(\text{fluoresces})}{\text{Ion–Na}^+\text{–Fluor} + \text{Poly}_{(\text{soln})}}$$

The assay is therefore based on a competition for the fluorophore between the immobilized ionophore–Na^+ complex and the solution quenching polymer, and will be determined by the respective affinity constants for the two ion-pair complexes. (Note the similarity in the principle here and that involved for the bioaffinity assays discussed in Chapters 8 and 9.)

A more integrated chromogenic ionophore has been proposed by Alder *et al.* (1987), who synthesized a photoactivated crown compound selective for K^+, by reacting 2-hydroxy-1,3-xylyl-18-crown-5 with the diazonium salt derived from 4-nitroaniline and immobilizing the modified ionophore at the end of an optical fibre. The detection limit for K^+ was 0.5 mM, which would be well within the requirements for clinical applications, if the host ionophore could be improved to give a better k_{K^+/Na^+} selectivity constant. Since the serum sodium levels are approximately 30 times greater than potassium, then a considerable K^+ selectivity is required to give an accurate determination. For this ionophore, where the selectivity for K^+ is ≈ 6 times greater than for Na^+, the Na^+ 'background' signal would be ≈ 4 times the K^+ signal!

Even with $k_{ab} \ll 1$, therefore, the relevance of this value can only be assessed together with information concerning the relative concentrations of 'a' and 'b' in the particular monitoring environment.

Gas-sensing Optical Probes

CARBON DIOXIDE PROBES

Optical carbon dioxide sensors are based on the same basic principle as found in the carbon dioxide electrode (see Chapter 3), namely an inner pH-sensitive element in contact with a HCO_3^- solution, surrounded by a CO_2-permeable membrane. The concentration of CO_2 dissolved in the bicarbonate layer determines the pH of the layer. Since the pH-sensing element is not in direct contact with the sample, these probes can be less susceptible to fluctuations in sampling environment. In theory, a device could be constructed employing any of the optical pH probes with a suitable pH range, but as with membrane-covered sensors in general, the final performance will be dependent on optimization of the membrane characteristics, in order to achieve maximum selectivity for CO_2 and minimum response time.

AMMONIA-SENSING PROBES

Similar criteria are required for an ammonia probe to those already established for CO_2, except that here a higher pH can be anticipated. A suitable indicator has been found to be *p*-nitrophenol. The non-protonated form of this dye absorbs at 404 nm and an optical sensor has been constructed which responds to ammonia over a large range (Arnold and Ostler, 1986). The experimental assembly of this device employed two fibres (source and detector) mounted in contact with one

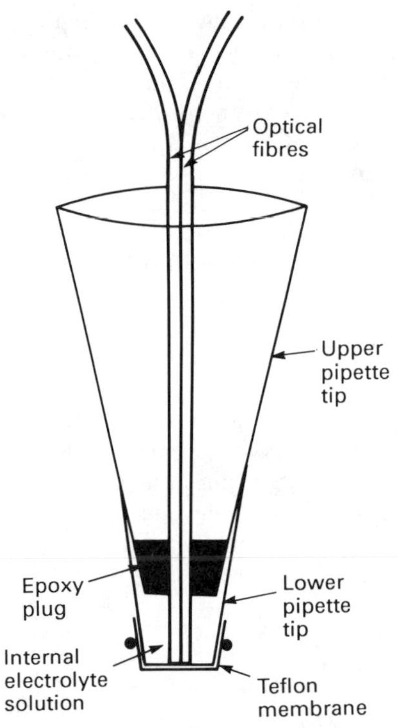

Fig. 6.24 Scheme of an optical fibre ammonia gas sensor.

another in an ammonium chloride solution containing *p*-nitrophenol (Fig. 6.24). The light is returned to the detector due to reflection of the source radiation at the Teflon membrane beneath the two fibres, so the efficiency of capture is dependent on correct cell geometry.

Both in this example and the CO_2 probe concept cited above, the indicator does not respond directly with the analyte, but the response is due to the dissociation equilibrium, in this case

$$NH_3 + H^+ \rightleftharpoons NH_4^+; \; K_a^{am} = \frac{[NH_4^+]}{[NH_3][H^+]}$$

causing a change in pH in the internal solution which is reflected by the indicator.

The earlier treatment for the reaction between analyte (A) and indicator (Ind) (see above) can therefore be extended by substitution of the 'analyte' (A) ($[H^+]$ here) from the above equilibrium:

$$K_a^{Ind} = \frac{K_a^{am}[NH_3][HInd]}{[NH_4^+][Ind^-]}$$

$$= \frac{K_a^{am}(C_{Ind} - [Ind^-])[NH_3]}{[Ind](C_{am} - [NH_3])}$$

this gives 'free' indicator $[\mathrm{Ind}^-]$ as:

$$[\mathrm{Ind}^-]=\frac{K_a^{am}[\mathrm{NH_3}]C_{\mathrm{Ind}}}{(K_a^{\mathrm{Ind}}C_{am}-K_a^{\mathrm{Ind}}[\mathrm{NH_3}]+K_a^{am}[\mathrm{NH_3}])}$$

When $[\mathrm{NH_3}] \ll C_{am}$ and $K_a^{\mathrm{Ind}}[\mathrm{NH_3}] < K_a^{\mathrm{Ind}}C_{am}$ then this reduces to

$$[\mathrm{Ind}^-]=\frac{K_a^{am}C_{\mathrm{Ind}}[\mathrm{NH_3}]}{K_a^{am}C_{\mathrm{Ind}}}$$

so that a linear relationship exists between the partial pressure of ammonia in the sample and $[\mathrm{Ind}^-]$, which as before, shows a strong dependence on indicator concentration.

A rather different approach has been proposed by Giuliani *et al.* (1983). A capillary tube, acting as a waveguide (Fig. 6.25), was spray coated with an oxazine perchlorate dye. This dye responds to ammonia, rapidly changing colour from blue to red. Irradiated at 560 nm by a LED and using a photodiode detector, the absorbance of the dye can be related to ammonia concentration with a lower limit in the region of 60 ppm. This is an example of an intrinsic optical sensor, the measurand (in this case the dye) interacting directly with the light propagating through the waveguide in the evanescent field.

OXYGEN PROBES

Oxygen appears to be a particularly efficient fluorescent quencher of many aromatics, providing a technique which has been used for the detection of oxygen in the millitorr pressure range. Obviously, the fluorescent indicator must have sufficiently long a lifetime in the excited state to allow interaction with oxygen. A selection of different reagents have been investigated; one such example (Peterson *et al.*, 1984) is based in design on the pH probe already described by the same

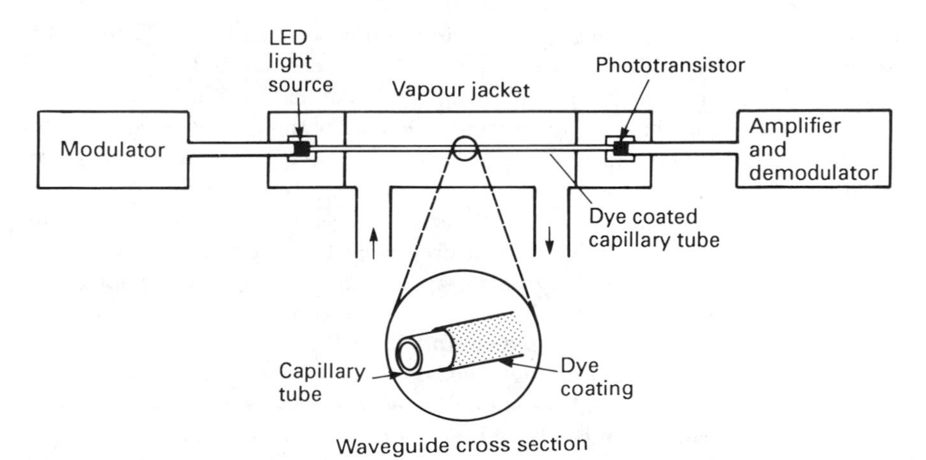

Fig. 6.25 Modified capillary tube optical wave guide sensor for ammonia.

group, but with perylene dibutyrate as fluorescence indicator, adsorbed on the polymeric support. This dye is excited at 468 nm and fluorescence detected at 514 nm. The I_o/I ratio was therefore taken as $I_{<500\ nm}/I_{>500\ nm}$ (i.e. blue intensity/green intensity), and the probe responded to oxygen in the 0–150 torr range.

Pyrenebutyric acid has also been used (Lübbers and Opitz, 1975; Wolfbeis *et al.*, 1984) but this requires excitation in the ultra-violet (342 nm), so that low-cost optical fibres cannot be employed. This indicator has, however, been investigated as a fluorescence layer for use at elevated temperatures (300–500 K), suitable for many industrial applications (Opitz *et al.*, 1988). An indicator-treated silicone membrane was coated on the end of a quartz waveguide to give a system which showed comparable behaviour at room temperature and 500 K. Since K, the constant in the Stern–Volmer equation, can be expressed as $K=\alpha\tau k$ (where α is the solubility of O_2, τ is the lifetime of the excited state and k is the collision quenching constant), then this term will show temperature dependence. The following relationship (Opitz *et al.*, 1988) has been derived to correct for working temperature:

$$K_T=\frac{K_o(T_{crit}+T)}{(T_{crit}-T)}$$

If this correction holds then at temperatures significantly less than C_{crit}, readings could, in principle, be corrected for relatively large deviations in operation temperatures. In this example C_{crit} was found experimentally to be 700 K and operation feasible up to 500 K.

An alternative approach proposed by Lippitsch *et al.* (1988) is to base the assay on the measurement of fluorescence lifetime, rather than intensity, so that the Stern–Volmer type equality becomes

$$\frac{\tau_o}{\tau}=1+K\mathrm{pO}_2$$

where τ is the fluorescent lifetime. The ruthenium complex, tris(2,2′-dipyridyl) ruthenium(II) dichloride hydrate produces a relatively long-lived fluorescence (>700 ns in methanol; >1 μs in ethanol) when excited at 460 nm, with $\lambda_{max}=610$ nm. Oxygen acts as an efficient quencher of this fluorescence. With the indicator adsorbed on kieselgel and included in a silicone membrane, the fluorescence lifetime in the absence of O_2 was measured to be 205 ns. This membrane was used in a blue LED source O_2 probe and the O_2 response compared for intensity and lifetime measurements. The latter method showed an extended linear range, but of greater significance perhaps is the long-term stability. Since the decay time is independent of fluorophore concentration, the probe is not affected by drift caused by indicator leaching and bleaching.

It would also be possible to design sensors based on analytical reactions that consume the reagent, although such devices would obviously only have a finite lifetime. This type of system has been investigated by Freeman and Seitz (1981) who have used the chemiluminescent reagent 1,1′3,3′-tetra ethyl-$\Delta^{2,2'}$bi (imidazolidine):

```
      CH3          CH3                                 CH3
       |            |                                   |
      CH2          CH2                                 CH2
       |            |                                   |
  CH2—N            N——CH2                              CH2—N
   |     \        /     |                               |    \
   |      C=C          |   +O2 ———→ 2                  |     O + hv
   |     /    \         |                               |    /
  CH2—N        N——CH2                                  CH2—N
       |            |                                   |
      CH2          CH2                                 CH2
       |            |                                   |
      CH3          CH3                                 CH3
```

Oxygen diffuses through a membrane from a sample flowing past the detection cell. This chemiluminescent oxygen probe is described as being more sensitive than a standard Clark oxygen electrode, but its long-term employment is inhibited by the changing response due to reagent consumption.

SULPHUR DIOXIDE PROBES

Oxygen is an efficient quencher for many fluorescent indicators. The quantum yield of fluorescence for the polynuclear hydrocarbons, e.g. benzo(b)-fluoranthene, is considerably reduced in the presence of oxygen. This particular fluorophore is also quenched by the gas SO_2 and in 'oxygen-free' assay environments, e.g. in exhaust gases, it can be employed in the estimation of SO_2, with a detection limit of 84 ppm SO_2 (Wolfbeis and Sharma, 1988). A probe may be constructed by adding the indicator to silicone pre-polymer and forming a 'fluorescent membrane' which may be mounted in an optical configuration, as already discussed. The probe will respond to SO_2, O_2, halogenated hydrocarbons but not to CO_2, CO or CH_4.

Attenuated Total Reflection (ATR) Infra-red

Intrinsic sensor devices have so far only been considered working in the ultra-violet-visible region of the spectrum. Although waveguides suitable for the propagation of light in the infra-red region are expensive compared with the plastic materials suitable to the longer wavelengths, the use of total internal reflection to excite the vibration modes of surface immobilized species has become an attractive method for the study of thin biological films, since it allows the acquisition of geometric information.

An early investigation by Fromherz *et al.* (1972), for example, indicated conformation changes associated with the interaction between haemoglobin and arachidonic acid, while Briggs *et al.* (1986) have been concerned with the interaction of the receptor signal sequence of the *E. coli* λ-phage with a lipid phase and Brauner *et al.* (1987) have examined the action of a lytic peptide and shown large conformational changes which can be monitored by vibration frequencies associated with CH_2 symmetric stretching modes.

Such studies have not been directed towards the analytical applications of ATR–infra-red as a diagnostic tool, but they indicate a sensitivity and a selectivity for a specific vibration frequency which is in contrast to the use of ultra-violet-visible spectroscopy. Even though the computational capability of Fournier transform–infra-red (FT–IR) does not at present necessarily suggest easy portability, the development of the infra-red interferogram, described earlier for the direct analysis of atmospheric pollutants, has indicated the potential applicability. Indeed, the acquisition of complete spectrums as demanded in these ATR–infra-red studies makes considerably greater instrumental demands than would be required in a narrow frequency band biosensor device. The ATR–infra-red technique already demonstrates many of the essential characteristics required of a biosensor, i.e. the integration of the signal-producing element with the optical transducer, so that its further exploitation is obvious.

Ellipsometry

As was seen in Fig. 6.15, the general behaviour of the reflectance of an incident ray differs for the transverse magnetic (TM) and transverse electric (TE) modes. For the TE mode, reflectance increases monotonically from its value at normal incidence, whereas for TM polarization a minimum is observed in the reflectance.

An incident ray of mixed TE and TM polarization will therefore be elliptically polarized on reflection. The intensity and the polarization of the reflected ray can give accurate information concerning the thickness and refractive index of layers deposited at the interface where reflection is occurring.

These features would appear particularly suitable for exploitation in immunological measurements. The original work in this area is now more than four decades old (Rothen, 1947), and its application to *in situ* measurement of the formation of biological films at reflecting interfaces has been used since the 1950s (Trurnit, 1953). Various immunological models have been investigated, generally with the antigen immobilized at the interface and the antibody or other non-specific proteins reacting with this surface from solution.

The antigens have been adsorbed onto a variety of different reflecting supports, and it is clear from the various data that the support is an important variable in the interpretation of the results. Giaever (1974) reports that bovine serum antigen (BSA) can be adsorbed onto nickel-coated glass slides to a limiting thickness of 300 nm, and this layer increased by 500 nm on binding anti-BSA. On gold films however (Azzam *et al.*, 1977), only 60 nm of BSA could be adsorbed, and this increased by 70 nm in the presence of anti-BSA, while a chromium interface allowed a 220 nm thick layer of albumin to bind anti-albumin up to 2700 nm (Cuypers *et al.*, 1978). The orientation of adsorption, and the degree of denaturing of the protein on adsorption are obviously critical in making determinations based only on the absolute layer thickness. The specificity of the measurement is however indicated; Poste and Moss (1972) showed that on an antigen-modified surface, the specific antibody caused an increase in the layer thickness which was at least twice that of a non-specific protein.

Rothen *et al.* (1969) suggested that by perfecting the immobilization of the bio-recognition component at the interface, sensitive immunoassay could be achieved. In an assay for human growth hormone, he claimed a minimum detection limit of 0.2 ng/ml. A significant limitation, however, in the conversion of this technique for biosensor exploitation, has been the complexity and expense of the equipment generally employed in order to make a full ellipsometric measurement. However, with the advent of laser diode light sources and photodiode detectors, miniature systems are now becoming available.

Downs *et al.* (1985) have designed a 'common path' interferometer with 0.1 nm resolution, giving similar diagnostic information to the classic ellipsometric experiment. The interferometer utilizes a birefringent lens and a microscope objective to separate the focal planes of the two polarizations. The target surface is placed in one of the focal planes and light from the other polarization is out of focus and reflected from a larger area. After reflection and recombination, the phase of one component is determined by the surface level of the focused spot, referenced to the defocused beam.

These and related optical surface-diagnostic techniques are readily applicable to the biosensor problem. Their evolution and exploitation alongside the supporting instrumentation advances will identify the most suitable biosensor transducer technique.

Light-scattering Analyses

The dynamic light-scattering technique of photon correlation spectroscopy (PCS), which analyses fluctuations in light scattered from populations of macromolecules under Brownian motion, is a non-invasive realtime measurement, which allows the accurate determination of diffusion coefficient and with globular proteins, molecular weight. Carr *et al.* (1988) have demonstrated the application of PCS to flowing systems in the determination of eluants from column chromatography. In a comparison with ultra-violet detection of the elution of protein mixtures, which gives no indication of the nature of the eluant, PCS could be used to determine the hydrodynamic diameter of each eluant, and it is proposed that this should be combined with the intensity measurements to give an indication of eluant concentration.

Other applications of this technique are also suggested: for example, *protein interactions* where a change in molecular size can be followed (e.g. antibody–antigen reaction). However, its general adoption as a routine analytical procedure has not been encouraged since PCS has traditionally required HeNe or argon–ion gas lasers and bulky photon correlation equipment. Brown (1988), however, has now reported the use of solid-state visible laser diodes, monomode optical fibres and avalanche photodiodes as replacements for these non-portable components, and now the construction of hand-held, lithium battery-operated devices. These low-cost, small-size (yet high-quality) units, already proved in eluant detection, further promote the use of PCS, or indeed any optical method employing these components, as a diagnostic tool.

Biosensor Construction

In theory, any of the bioassays discussed earlier in this chapter would be suitable for conversion into a wave guide based device. The possibilities need not as has been suggested, be restricted to the ultra-violet and visible regions of the spectrum and the use of secondary photometric labels, but can be extended to vibrational absorptions, light-scattering phenomena and surface plasmon effects which may be more directly related to a biorecognition molecule immobilized within the evanescent field, interacting with the target analyte. The use of optical excitation techniques for the study of species immobilized at an interface has now been employed for many applications, both biological and non-biological. In view of the requirements of a *biosensor*, all such interface analysis methods could be relevant to their future, if not present, development. Optical excitation of an interface and its monitoring could represent, therefore, the realization of the optical biosensor concept, and its discussion is thus pertinent to Part II of this book, where it will be discussed further.

References

Alder, J.F., Ashworth, D.C., Narayanaswamy, R., Moss, R.E. and Sutherland, I.O. (1987). *Analyst* **112**, p. 1191.
Arnold, M.A. and Ostler, T.J. (1986). *Anal. Chem.* **58**, p. 1137.
Azzam, R.M.A., Rigby, P.G. and Krueger, J.A. (1977). *Phys. Med. Biol.* **22**, p. 422.
Babcock, G.T., Jean, J.M., Johnston, C.N., Palmer, G., and Woodruff, W.H. (1984). *J. Am. Chem. Soc.* **106**, p. 8305.
Bravner, A., Boeufgra, J.M., Jacobson, S.H., Kaijser, B., Kaueniu, G., Svenson, S.B. and Wretlind, B. (1987). *J. Gen. Microbiol.* **133**, pp. 2825–2834.
Briggs, M.S., Cornell, D.G., Dluhy, R.A. and Gierasch, L.M. (1986). *Science* **233**, pp. 206–208.
Brown, R.G.W. (1988). *Photon Correlation Techniques and Applications*, Eds Abbiss, J.B. and Smart, A.E., American Institute of Physics.
Campbell, A.K. and Patel, A. (1983). *Biochem. J.* **216**, p. 185.
Carr, R.J.G., Rarity, J.G., Stansfield, A.G., Brown, R.G.W., Clark, D.J. and Atkinson, T. (1988). *Anal. Biochem.*
Campion, A. and Woodruff, W.H. (1987). *Anal. Chem.* **59**(22), p. 1305A.
Critchley, Ch. and Smillic, R.M. (1981). *Austral. J. Plant Physiol.* **8**, p. 133.
Cuypers, P.A., Hermens, W.T. and Hemker, H.C. (1978). *Anal. Biochem.* **84**, p. 56.
Deutsche Gesellschaft für Klinische Chemie (1972). *Z. Klin. Chem. Clin. Biochem.* **10**(4).
Downs, M.J., McGivern, W.H. and Ferguson, H.J. (1985). *Precision Engineering* **7**(4), p. 211.
Dubowski, K.M. (1962). *Clin. Chem.* **8**, p. 215.

Freeman, T.M. and Seitz, W.R. (1981). *Anal. Chem.* **53**, p. 98.
Fromherz, P., Peters, J., Muldner, H.G., Offing, W. (1972). *Biochem. Biophys. Acta* **274**(2), p. 644.
Fuh, S.M-R., Burgess, L.W., Hirschfeld, T. and Christian, G.D. (1987). *Analyst* **122**, pp. 1159–1163.
Giaever, I. (1974). *Bull. Am. Phys. Soc.* **19**, p. 564.
Giuliani, J.F., Wohltjen, H. and Javis, N.L. (1983). *Optics Letters* **8**(1), p. 54.
Grattan, K.T.V., Mouaziz, Z. and Palmer, A.W. (1987/88). *Biosensors* **3**(1), p. 17.
Hösli, P., Avrameas, S., Ullmann, A., Vogt, E. and Rodrigot, M. (1978). *Clin. Chem.* **24**, p. 1325.
Lippitsch, M.E., Pusterhofer, J., Leiner, M.J.P. and Wolfbeis, O.S. (1988). *Anal. Chim. Acta* **205**, p. 1.
Lübbers, D.W. and Opitz, N. (1975). *Z. Naturforsch. (C) Biosci.* **30C**, p. 532.
Lübbers, D.W. and Opitz, N. (1983). *Proc. Int. Meeting on Chem. Sensors Japan*, p. 609. Amsterdam, Elsevier.
Lundin, A. (1982). *Luminescent Assays: Perspectives in Endocrinology and Clinical Chemistry*, Eds Serio, M. and Pazzagli, M. **1**, p. 29. New York, Raven.
Miyazawa, T. and Blout, E.R. (1961). *J. Am. Chem. Soc.* **83**, p. 712.
Offenbacher, H., Wolfbeis, O.S. and Fuerlinger, E. (1986). *Sensors and Actuators* **9**, p. 73.
Opitz, N. and Lübbers, D.W. (1983). *Sensors and Actuators* **4**, p. 473.
Opitz, N., Graf, H-J. and Lübbers, D.W. (1988). *Sensors and Actuators* **13**, p. 159.
Peterson, J.I., Goldstein, S.R., Fitzgerald, R.V. and Buckhold, D.K. (1980). *Anal. Chem.* **52**, p. 864.
Peterson, J.L., Fitzgerald, R.V. and Buckhold, D.K. (1984). *Anal. Chem.* **56**, p. 62.
Poste, G. and Moss, C. (1972). *The Study of Surface Reactions in Biological Systems by Ellipsometry*, p. 206. New York, Pergamon Press.
Rothen, A. (1947). *J. Biol. Chem.* **168**, p. 75.
Rothen, A., Mathot, C. and Thiele, E.H. (1969). *Experientia* **25**, p. 420.
Saari, L.A. and Seitz, W.R. (1982). *Anal. Chem.* **54**, p. 821.
Schimazaki, K. and Sagahara, K. (1980). *Plant Cell Physiol.* **21**, p. 125.
Schroeder, H.R., Vogelhut, P.O., Carrico, R.J., Boguslaski, R.C. and Buckler, R.T. (1976). *Anal. Chem.* **48**, p. 1933.
Small, G.W., Kroutil, R.T., Ditillo, J.T. and Loerop, W.R. (1988). *Anal. Chem.* **60**, pp. 264–269.
Stanley, P.E. (1978). *Liquid Scintillation Counting*, Eds Johnson, P. and Crook, M. 5, p. 79. London, Heyden.
Tien, P.K. (1971). *Appl. Opt.* **10**, p. 2395.
Trinder, P. (1969). *J. Clin. Pathol.* **22**, p. 246.
Trurnit, H.J. (1953). *Arch. Biochem. Biophys.* **48**, p. 176.
Ullman, E.F., Schwarzberg, M. and Rubenstein, K. (1976). *J. Biol. Chem.* **251**, p. 4172.
Urbano, E., Offenbacher, H. and Wolfbeis, O.S. (1984). *Anal. Chem.* **56**, p. 427.
Voss, M., Renger, G., Kötter, C. and Gräber, P. (1984). *Weed Sci.* **32**, p. 675.
Wannlund, J., Egghart, H. and DeLuca, M. (1982). *Luminescent Assays: Perspectives in Endocrinology and Clinical Chemistry*, Eds Serio, M. and Pazzagli, M. **1**, p. 125. New York, Raven.
Weichselbaum, T.E. (1946). *Am. J. Clin. Pathol.* **7**, p. 40.
Wolfbeis, O.S. and Offenbacher, H. (1986). *Sensors and Actuators* **9**, p. 85.
Wolfbeis, O.S. and Sharma, A. (1988). *Anal. Chim. Acta* **208**, p. 55.
Wolfbeis, O.S., Fuerlinger, E., Kroneis, H. and Marsoner, H. (1983). *Fres. Z. Anal. Chem.* **314**, p. 119.

Wolfbeis, O.S., Offenbacher, H., Kroneis, H. and Marsoner, H. (1984). *Mikrochim. Acta*, p. 153.
Wroblewski, F. and LaDue, J.S. (1955). *Proc. Soc. Exper. Biol. Med.* **90**, p. 210.
Wroblewski, F. and LaDue, J.S. (1956). *Proc. Soc. Exper. Biol. Med.* **91**, p. 569.
Zhujun, Z. and Seitz, W.R. (1984). *Anal. Chim. Acta* **160**, p. 47.
Zhujun, Z., Mullin, J.L. and Seitz, W.R. (1986). *Anal. Chim. Acta* **184**, p. 251.

PART 2

The Biosensor Genus: Its Friends and Relations

Chapter 7

Cell-based Biosensors

Development of the Electrochemical Biosensor Model

Recognition systems so far considered for employment in sensor devices have been single molecules or molecular complexes. This chapter will deal with the use of the larger recognition systems found in whole cells, tissues and whole organisms, while later chapters will concentrate more on the use of the isolated cell components. The major contribution to cell-based sensors has been made by electrochemical devices. Microbial-based sensors exhibit certain economic advantages in that the cells are cheaper to isolate than the individual cell components. Cell activity is also frequently less susceptible to incidental fluctuations in environmental conditions than for the isolated enzymes, and this can contribute to an extended lifetime of the microbial sensor. Compared with enzyme-based devices however the cell based system typically has a longer response time, probably associated with the additional transport step through the cell wall.

Where the active biorecognition compound is an enzyme, the construction of an electrochemical biosensor, whether it be using whole cells or the cellular components, can be described by a general model. In the simplest case, the enzyme kinetics follow the sequence:

$$\mathrm{S+Enz} \underset{k_{-1}}{\overset{k_1}{\rightleftharpoons}} \mathrm{SEnz} \underset{k_{-2}}{\overset{k_2}{\rightleftharpoons}} \mathrm{PEnz} \underset{k_{-3}}{\overset{k_3}{\rightleftharpoons}} \mathrm{P+Enz}$$

With the enzyme immobilized in a layer next to the electrode (Fig. 7.1a), mass transport for S and P across the membrane–solution interface and through the membrane layer will be described by the partition coefficient and the membrane diffusion coefficient, respectively. In this assay, S can be estimated via measure-

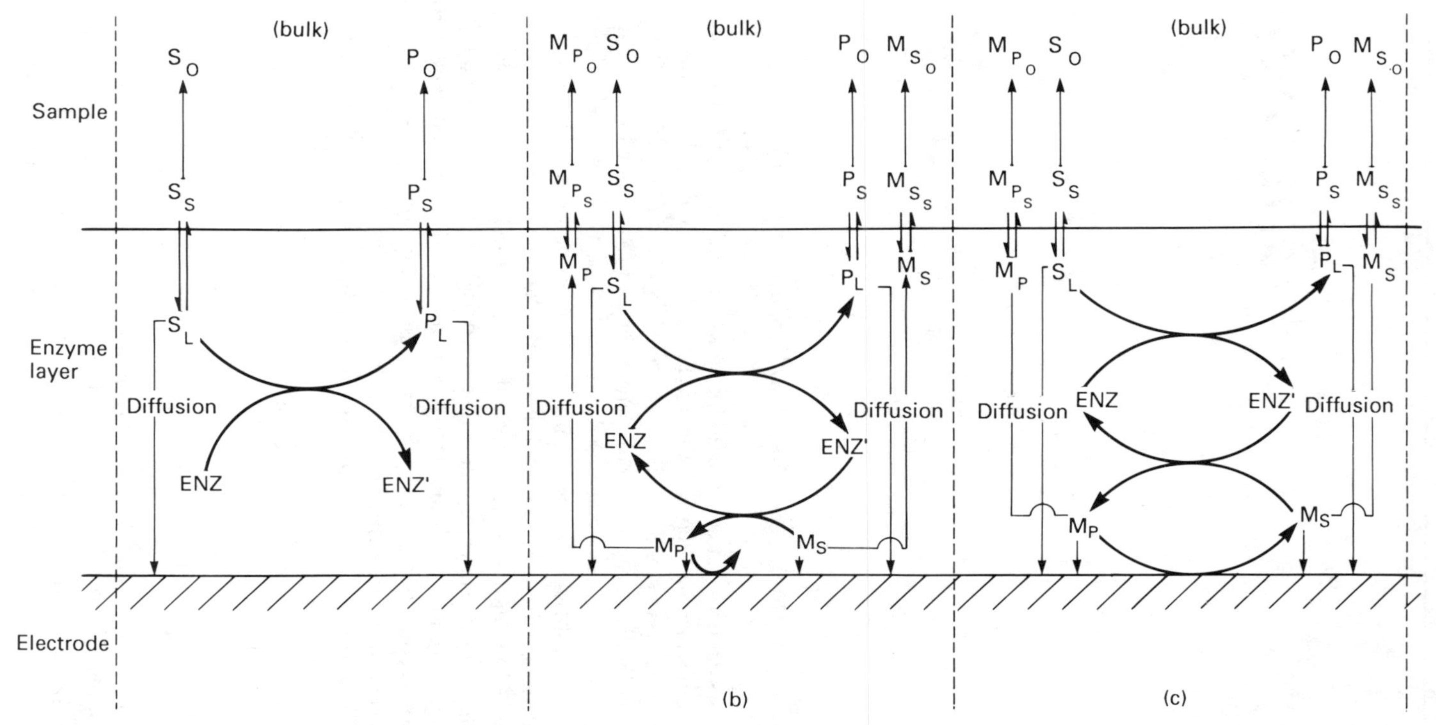

Fig. 7.1 Transport of species through the enzyme layer for various general models of enzyme electrodes.

ment of P, possibly at a potentiometric electrode, in which case simple diffusion of P back into the bulk solution is mainly responsible for its depletion.

When a redox enzyme is involved, then conversion of S to P causes an equivalent change in the redox state of the enzyme, $Enz \rightarrow Enz'$. Various natural electron-transfer mediators (M) exist to convert Enz′ back to Enz, $M_S \rightarrow M_P$ (Fig. 7.1b),

$$S + Enz \rightleftharpoons P + Enz'$$
$$M_S + Enz' \rightleftharpoons M_P + Enz$$

and here the mass transfer of M_S and M_P must also be considered. The assay of S can now be related to P, M_S or M_P. If in the latter instance M_S or M_P are detected amperometrically, then they are consumed by the estimation, so that this and diffusion into bulk are responsible for their depletion.

For example, where the enzyme employed is an oxidase utilizing O_2 as mediator, the electron transfer results in

$$\begin{array}{ccc} M_S \rightarrow M_P & \equiv & O_2 \rightarrow H_2O_2 \\ Enz \rightarrow Enz & & Enz_{(red)} \rightarrow Enz_{(ox)} \end{array}$$

so that the assay can be made via peroxide determination at an electrode,

$$H_2O_2 - 2e^- \rightarrow 2H^+ + O_2$$

thus consuming peroxide in the measurement.

'Artificial' mediators used to regenerate the enzyme (Fig. 7.1c) should ideally be immobilized with the enzyme in the layer next to the electrode. If this is the case, and *neither* M_S *nor* M_P can diffuse into the bulk, then depletion of M_P is caused by the electron-transfer reaction with the electrode in which M_S is regenerated:

$$S + Enz \rightleftharpoons P + Enz'$$
$$M_S + Enz' \rightleftharpoons M_P + Enz$$
$$M_P \rightleftharpoons M_S \text{ (at the electrode)}$$

The kinetics of all the various equilibria involved in the particular sensor construction under discussion will contribute to the response of the signal that is to be measured.

As with all types of biosensors the major problems concerned with cell-based systems are those of (i) immobilization of the biological component, (ii) identification of a physico-chemical parameter that can be monitored and will reflect the concentration of analyte, and (iii) selectivity and sensitivity of the measurement for the chosen analyte.

Even the cell reaction *itself* has been monitored via mediators and thus will be discussed later, but the vast majority of cell sensors are based on a secondary measurement of some low-molecular-weight species via potentiometric or amperometric electrodes.

Table 7.1 Examples of amperometric cell-based sensors

Analyte	*Immobilized biolayer*
Ethanol	*Acetobacter xylinum*
	Trichosporon brassicae
Acetic acid	*Trichosporon brassicae*
Methanol	Unidentified bacteria
Glucose	*Pseudomonas fluorescens*
Total sugars	*Brevibacterium lactofermentum*
Vitamin B_1	*Lactobacillus fermenti*
Nitrogen dioxide	Nitrifying bacteria, *Nitrobacter* sp.
Ammonia	*Nitrosomonas europaea*/*Nitrobacter* sp. (nitrifying bacteria)
Methane	*Methylomonas flagellata*
Mutagens	*Bacillus subtilis* rec$^-$
	Salmonella typhimurium TA 100
BOD	*Clostridium butyricum*
	Trichosporon cutaneum
Peroxide	Bovine liver
Dopamine	Banana pulp
Tyrosine	Sugar-beet
Cholesterol	*Nocardia erythropolis*

Amperometric Cell-based Sensors

MICROBIAL BIOSENSORS

Ethanol

The first cell-based sensor (Table 7.1) was reported in 1975 by Divies for the determination of ethanol. The sensor employed a bacterium requiring ethanol as substrate and was able to follow the concentration of ethanol by monitoring respiratory activity, i.e. the bacterium *Acetobacter xylinum* was immobilized on an amperometric oxygen electrode (see Chapter 5), and the consumption of oxygen by the bacteria related to the concentration of ethanol:

$$\text{ethanol} + O_2 \rightarrow \text{acetic acid} + H_2O$$

A similar sensor for ethanol has been commercialized in Japan, employing the bacterium *Trichosporon brassicae* (Hikuma *et al.*, 1979a, b; Karube *et al.*, 1980). The microorganism was immobilized between two gas permeable membranes in front of an oxygen electrode (Fig. 7.2). Any interference from the use of non-volatile nutrients such as glucose or phosphate, was prevented by the incorporation of the external gas-permeable membrane.

Tuning for Acetic Acid

In yeast fermentation broths (pH$>$6) the sensor showed no response to other volatiles such as acetic acid, propionic acid, formic acid or methanol. Working at a sample pH well below the pK_a for acetic acid (pH 4.75, 30°C) however, the equilibrium

$$CH_3COOH + H_2O \rightleftharpoons CH_3COO^- + H_3O^+$$

lies well to the left and the membrane, while being impermeable to acetate ions, allows the assay of acetic acid in the range 5–72 mg l^{-1}.

Under these conditions this sensor was also sensitive to propionic acid, *n*-butyric acid and of course ethanol, but employed in a fermentation broth for glutamic acid production these acids are present in concentrations too low to effect the measurement. In fact application of this measurement to yeast fermentation is particularly relevant, the optimization of acetic acid concentration being critical since when the microorganisms are grown on acetic acid as the carbon source, an excess causes inhibition of growth.

In both these sensors analyte concentration was related to a decrease in the dissolved oxygen in the electrolyte layer next to the electrode (Fig. 7.3), and thus to a decrease in the electrode current due to oxygen. With steady-state measurements, >10 min can be required to obtain a reading, but with the application of non-steady state techniques and curve analysis this can be considerably shortened.

An important feature is revealed by these examples concerning the design of a biosensor relevant for a particular application. A major requirement of a successful biosensor is specificity. Both these sensors are now reported to be commercialized in Japan and yet in common with most cell-based sensors they are not as specific as isolated enzymes. They are however *specific for the target analyte under the operating conditions for which they were designed*. The introduction of the additional variable of *operating conditions* is essential to assessing potential biosensor performance.

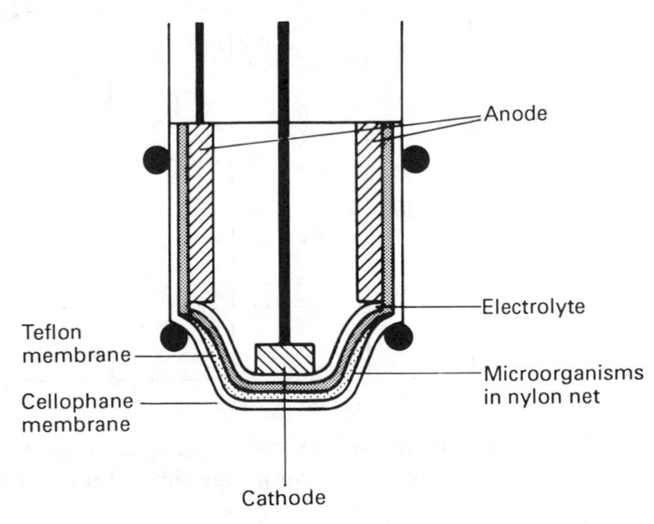

Fig. 7.2 Microorganism-based sensor: assay via O_2 electrode.

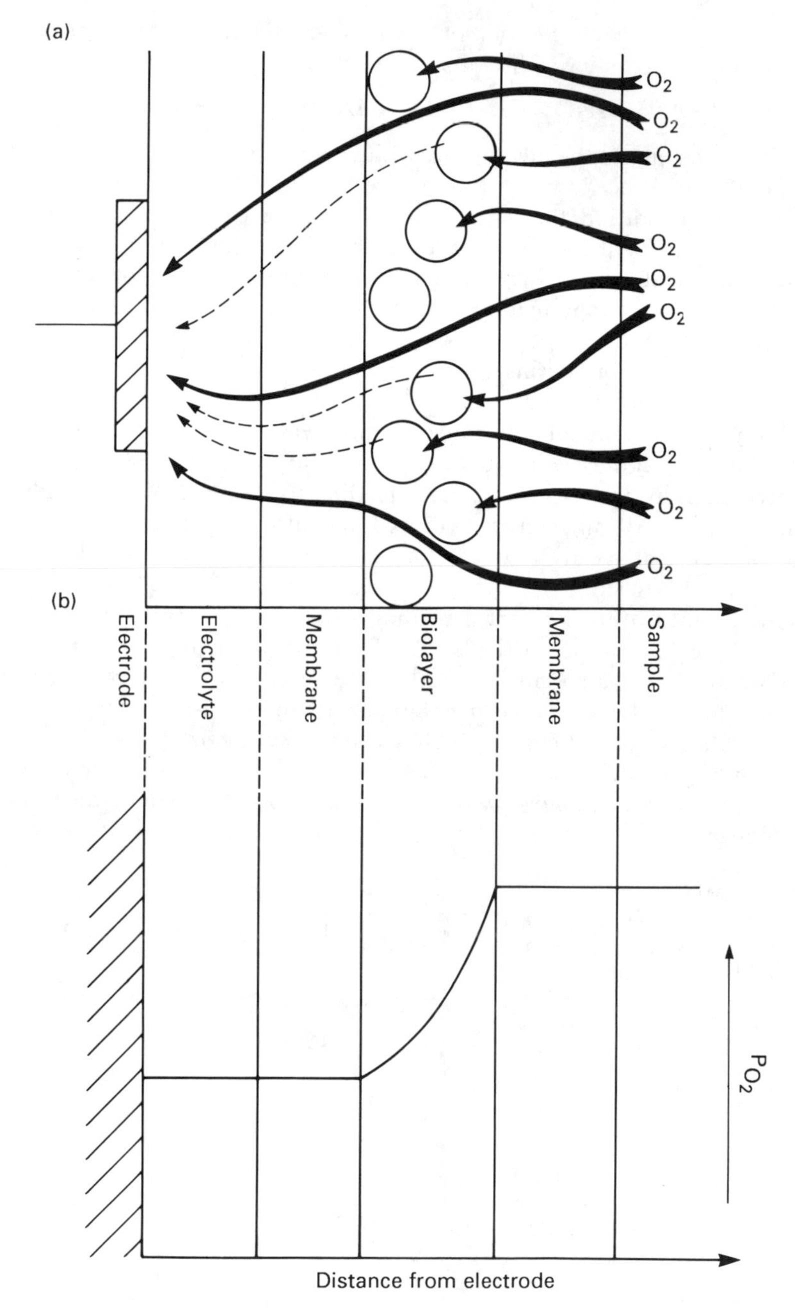

Fig. 7.3 (a) Schematic representation of the action of an oxygen electrode microbial sensor. (b) Typical concentration profile through the layers of the microbial sensor.

Total Sugars

The vast majority of fermentation broths utilize sugars as their major carbon source. Optimization of their concentration is necessary in order to prevent any inhibition in cell growth. A sensor for total sugar estimation can be constructed by employing *Brevibacterium lactofermentum* in the same configuration as already described, in conjunction with an oxygen electrode. This sensor shows similar responses for glucose, sucrose and fructose (relative sensitivity 1.0 : 0.92 : 0.80, respectively) but a sensor more specific to glucose can be realized through the use of *Pseudomonas fluorescens* where only slight responses are noted for sugars other than glucose.

Vitamin B_1

In the successful microbial biosensor, the organism is 'tuned' to the target analyte. Screening of readily available microorganisms revealed that baker's yeast was an excellent choice for the assay of thiamine (vitamin B_1) (Mattiasson *et al.*, 1982). The basis of this assay is the fact that a dried yeast preparation will grow in a glucose medium in the absence of thiamine, but that growth is considerably enhanced (and thus oxygen uptake increased) in the presence of thiamine. The microorganisms were immobilized on a nylon net, consolidated by the formation of insoluble calcium alginate, and the respiratory activity of the microorganism monitored in a standard glucose culture medium before and after exposure to thiamine. The procedure requires that the membrane be used for 'one-off' measurements only, since subsequent exposure to thiamine results in decreasing changes in O_2 uptake. This multistep single use method is not attractive for large numbers of samples, so that although the detection limit of 10^{-8} g cm^{-3} is well within the requirements of most applications its use would be limited to small numbers of even single samples—perhaps the system is more usefully employed as a glucose sensor, since it showed a linear response to glucose below 15 mM!

As may be seen from this example, the usefulness of any particular successful research system is strongly influenced by the *ease of use and potential real applications*.

Escherichia coli 215 also requires vitamin B_{12} for growth. Karube *et al.* (1987a) have prepared a microbial sensor with the *E. coli* 215 immobilized on a porous acetylcellulose membrane. The thiamine response of this electrode was enhanced by the presence of glucose and limited for concentrations of glucose > 1%. Glucose (1%) was therefore introduced into the assay solution, and vitamin B_{12} could be estimated in the range 5×10^{-9}–25×10^{-9} g cm^{-3}. The system could be used for multiple assays over a 25-day period with storage at −25°C between use. The response decreased by about 8% in this period.

Oxides of Nitrogen

An increasing need for environmental monitoring is found in the control of pollution. Although nitrous oxide, N_2O, is the only naturally produced major oxide of nitrogen, nitric oxide (NO) and nitrogen dioxide (NO_2) are produced in waste gases. NO, for example, is emitted by vehicle exhausts and oxidized to NO_2 it produces a toxic gas.

In what may perhaps be one of the most economic sources for a biorecognition system for a biosensor, unidentified nitrifying bacteria were obtained from active sludges in waste-water treatment plants (Okada *et al.*, 1983). The required organisms could be isolated from a culture rich in nitrite without contamination by other microorganisms, since in general nitrite is toxic to other microorganisms.

The bacteria were immobilized in a porous acetylcellulose membrane sandwiched between two gas-permeable Teflon membranes in front of an oxygen electrode. The assay of NO_2 is based on the formation of nitrite ion in aqueous media:

$$2NO_2 \rightleftharpoons N_2O_4$$
$$N_2O_4 + H_2O \rightleftharpoons HNO_2 + HNO_3$$
$$2NO_2^- + O_2 \xrightarrow{\text{nitrite-oxidizing bacteria}} 2NO_3^-$$

so that NO_2 concentration could be related to oxygen concentration, as measured by the oxygen electrode, so long as sufficient bacteria were immobilized in the 'biolayer'. *Conditions of pH and temperature must be optimized for maximum sensitivity of the sensor.* As might be expected the sensitivity of this system is related to the permeability of the NO_2 through the gas-permeable membrane to the 'biolayer' (Fig. 7.4), and optimization of the temperature and pH for maximum activity of the microorganism. The sensor showed increasing activity up to 30°C and then a decrease with further increases in temperature, due to the bacteria. This was also the case at pH values below pH 8.5, so that the internal buffer solution was maintained at high pH. However, permeability of the NO_2 through the gas-permeable membrane retaining the 'biolayer' reached a maximum at *low* pH, so that these two conflicting requirements were achieved by using an outer buffer solution (Fig. 7.4) held at pH 2, to inhibit the formation of the nitrite ion.

Ammonia

In conjunction with a second microorganism, the system for NO_2 may be extended to assay ammonia (Hikuma *et al.*, 1980):

$$2NH_3 + 3O_2 \xrightarrow{\textit{Nitrosomonas europaea}} 2NO_2^- + 2H_2O + 2H^+$$
$$2NO_2^- + O_2 \xrightarrow{\textit{Nitrobacter sp.}} 2NO_3$$

Comparison of microbial biosensors employing cultured *Nitrosomonas europaea* and the nitrifying bacteria obtained from activated sludge showed that the latter were more stable and could be used over longer periods. However, in extended use, contamination by other microorganisms reduced selectivity and introduced a response for glucose. This could be overcome by the use of the antibiotic chloramphenicol, which showed selective inhibition of the non-nitrifying bacteria.

This sensor which has its roots in waste waters, also finds application in the determination of ammonia in waste waters. Minimal recalibration was necessary in a two-week trial and the results compared favourably with conventional assay methods. Similar correlation of assay methods was found for the determination of NH_3 in human urine with the microbial sensor, showing none of the inherent interference effects by metal ions and volatile amines.

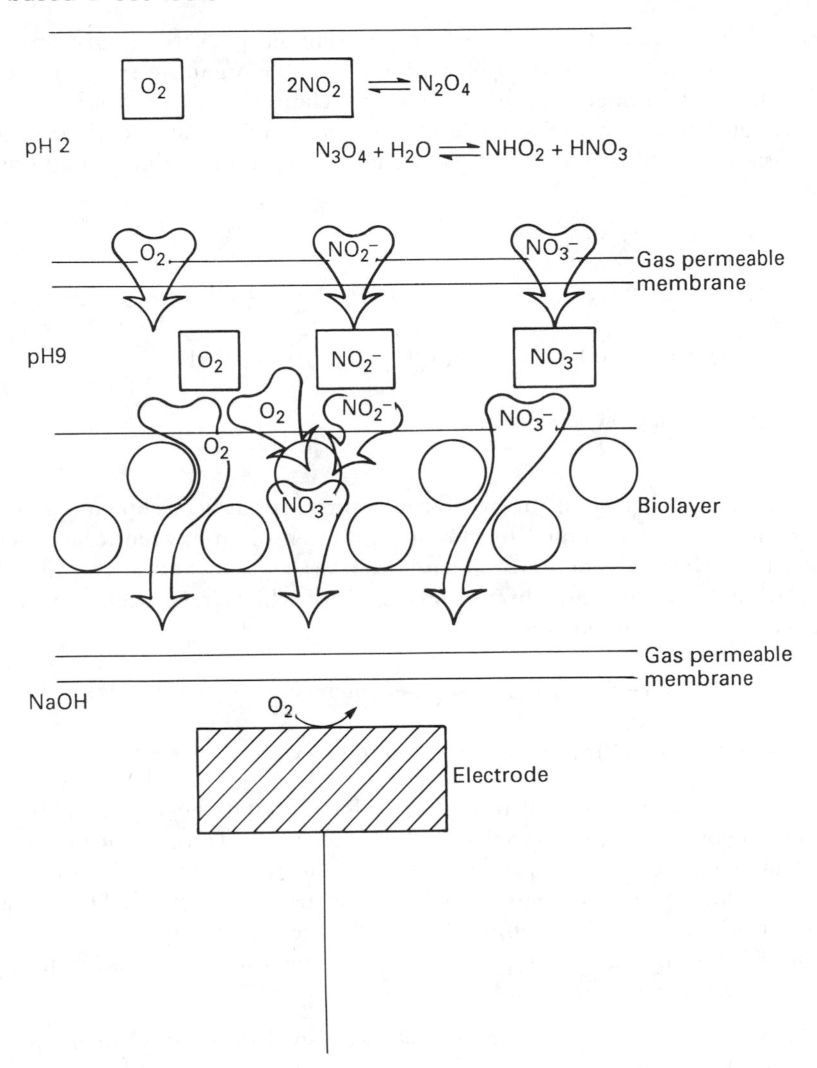

Fig. 7.4 Schematic representation of an optimized nitrite microbial-sensor based on an oxygen electrode.

It is apparent from this device that careful culturing of a specific microorganism for a biosensor does not always give as robust a sensor as the 'impure' mixture isolated from 'natural sources'.

Methane

Also produced by the treatment of waste is the fuel gas, methane. This is an attractive energy source and its monitoring allows the control of waste treatment to give maximum production. Its rapid determination is also relevant to such applications as coal mining, where escapes from gas pockets can give explosive mixtures with air. (While the first report of a whole organism sensor is cited as 1975, should the miners' canary perhaps not claim this attribution?)

Methane assay is a particularly good example of the application of whole organisms, since although the route for the oxidation of methane is an obvious one,

$$CH_4 \xrightarrow[\text{NADH}+H^+ \to NAD^+]{O_2 \to H_2O} CH_3OH \xrightarrow{\to PQQH_2} HCHO \xrightarrow{\text{'2H'}} HCOOH \xrightarrow{NAD^+ \to NADH + H^+} CO_2$$

methane mono-oxygenase

the enzymic mechanism is subject to controversy. It has been shown that growth on methane is accompanied by the incorporation of an oxygen atom, yielding methanol. Methane mono-oxygenase from *Methylococcus capsulatus* has been isolated and purified into three proteins, all of which have been shown to be necessary for methane oxidation:

$$CH_4 + O_2 \xrightarrow{*} CH_3OH; \quad \text{Protein } A_{(red)} \to \text{Protein A}; \quad \text{Protein C} \to \text{Protein } C_{(red)}; \quad NADH + H^+ \to NAD^+$$

The role of Protein B is not clear although it is essential to the overall pathway (its possible involvement is marked * above). Development of an enzyme biosensor for methane would therefore require isolation, purification and immobilization of all three of these proteins, together with a means by which NADH could be continuously generated or supplied in stoichiometric amounts.

On the other hand, three types of organism grow with the C_1 source methane. These methanotrophs are:

(1) Methane-oxidizing bacteria (obligate)—will grow only on methane or methanol, e.g. *Methylococcus* sp., *Methylosinus* sp., *Methylomonas* sp. and *Methylocystis* sp.
(2) Methane-oxidizing bacteria (facultative)—can also grow on organic compounds such as glucose, e.g. *Methylobacterium organophilium*.
(3) Methane-oxidizing yeasts.

Clearly the first group are most selective for methane and therefore more suitable for exploitation in a sensor. Karube *et al.* (1982b) have reported a methane sensor based on immobilized bacterium, *Methylomonas flagellata*. In order to eliminate fluctuations in oxygen concentration, the measurement is made with respect to background oxygen level, measured at a second oxygen electrode. The response time was <60 s and the minimum detection limit 5 μmol (linear range 0–6.6 mmol).

Carbon Dioxide
pCO_2 is commonly determined potentiometrically with the Severinghaus electrode (see Chapter 3). However Suzuki *et al.* (1987) have employed a CO_2-utilizing autotropic bacteria S-17, thought to be the genus *Pseudomonas*, in a CO_2 bacterial sensor linked to an oxygen assay. The bacteria cells were retained on a cellulose nitrate membrane in front of an oxygen electrode. The response was linear to 8 mM carbonate which at the operating pH of 5.5 corresponded to 5 mM CO_2 (200 mg l^{-1}), i.e. the saturation concentration for CO_2 in the buffer solution. It was estimated that the sensor could detect a minimum change of 5 mg l^{-1} CO_2, and was insensitive to the presence of organic acids, other than acetic acid, which may have offered an alternative carbon source. The sensor showed some decay in output with time, which levelled off during the first week of operation, to give a device stable for at least 3 weeks.

Mutagens
A rather different application of microbial sensors has been devised for detecting various types of chemical mutagens. *Salmonella typhimurium* TA100 is a histidine-requiring bacterial mutant which can be reverted back to histidine-free wild type with chemical mutagens. In histidine-free medium, therefore, growth (or oxygen uptake) is related to the rate of mutagenesis.

Assembly of a screening test for mutagens requires the same components as any other microbial sensor. In this example (Karube *et al.*, 1982a) a membrane filter retaining the *S. typhimurium* mutant was attached to the surface of an oxygen electrode. The electrode was equilibrated in histidine-free phosphate/glucose buffer and then incubated in the mutagen sample. Maximum sensitivity was achieved after a 10 h incubation period, and measurement taken at this time showed an increase in electrode response for increasing mutagen concentration.

Although such sensors may not replace long-term carcinogenicity tests with animals, they do provide a fast preliminary test with a high degree of prediction. This particular device has been tested for 2-(2-furyl)-3-(5-nitro-2-furyl)acrylamide, *N*-methyl-*N*′-nitro-*N*-nitrosoguanidine, nitrofurazone, methylmethane sulphonate and ethylmethyl sulphonate. Other organisms have been used for other mutagens. Obviously, broad-band mutagenic screening would employ a multiple array of organisms.

These examples demonstrate the application of a microbial sensor to the detection of a genetic mutation which gives a nutrient-induced secondary activity in a microorganism.

BOD
One of the most general uses of the oxygen electrode in conjunction with a microorganism is that of biological oxygen demand (BOD). BOD is one of the most widely used assessments of organic pollution, and by conventional means a five-day incubation period is required.

Two organisms have been reported by Karube *et al.* (1977b) for employment in microbial electrode BOD sensors at 25–30°C: *Clostridium butyricum* and *Trichosporon cutaneum*. The microorganisms were mounted in front of an oxygen electrode in the standard configuration and flushed with oxygen-saturated buffer solution until a steady-state electrode current response was reached and a sample could be injected. The new current output reflected the BOD of the sample solution. A linear relationship was found for the difference between these two current values and the conventional BOD assay. The latter organism has been employed in a continuous monitoring system for BOD, commercialized in Japan.

However, the temperature of waste water is often higher than the optimum temperatures for these microorganisms. Karube *et al.* (1987b) have developed a BOD sensor utilizing a thermophilic bacterium isolated from hot springs, and which retained high-temperature stability, even after prolonged treatment above 60°C.

One of the major advantages of microbial biosensors is that the biorecognition system is maintained in its natural environment, i.e. within the cell—*thus stabilizing its activity*. This condition can also be achieved through the use of tissue slices from plant and animal cells.

TISSUE-BASED SENSORS

These tissue sensors are more recent additions to the family than the microbial sensors, but nevertheless in many cases they have proved to be particularly useful. Bovine liver tissue, for example, contains a high concentration of the enzyme required for the reaction,

$$2H_2O_2 \rightleftharpoons O_2 + 2H_2O$$

so that, unlike previous microbial examples (where analyte presence has been indicated by a decrease in the partial pressure of oxygen), this assay correlates increase in electrode current due to oxygen with the peroxide concentration (Mascini *et al.*, 1982). The use of this differentiated tissue reveals a most satisfactory selectivity; even in the presence of high concentrations of sugars, alcohols and acids specificity for hydrogen peroxide is demonstrated.

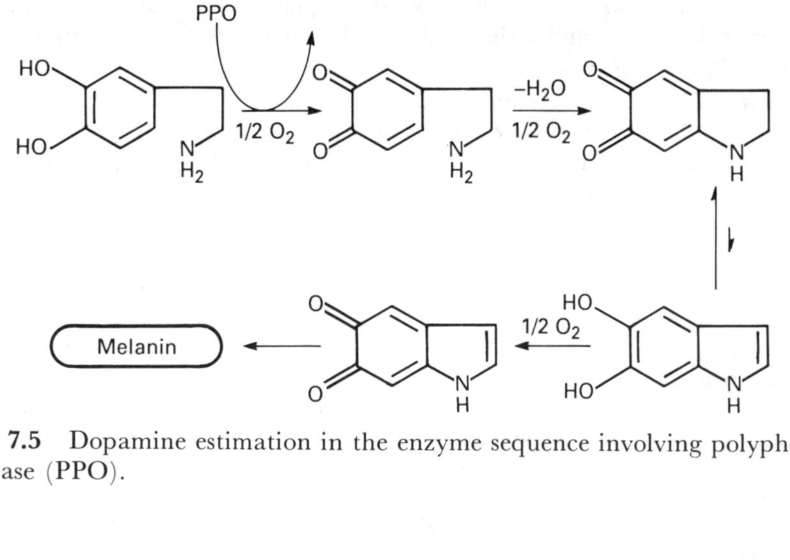

Fig. 7.5 Dopamine estimation in the enzyme sequence involving polyphenol oxidase (PPO).

Table 7.2 Potentiometric cell-based sensors. Sensing element in all cases is NH_3 except where indicated

Analyte	*Biolayer*	*Analyte*	*Biolayer*
Asparate	*Bacterium cadaveris*	Nitrate	*Azotobacter minelandii*
L-arginine	*Streptococcus daecium*	NAD^+	*Escherichia coli* + NADase
	Streptococcus lactis	Glutamate	*Escherichia coli*
	Streptococcus faecium		Yellow squash
Arginine	Bovine liver	Pyruvate*	Corn Kern
Glutamine	Porcine kidney cells		*Streptococcus faecium*
	Sarcina flava	Antibodies**	*Citrobacter freundii*
Adenosine	Mouse small-intestinal mucosal cells	Histidine	*Pseudomonas* sp.
		Tyrosine	*Aeromonas phenologenes*
Guanine	Rabbit liver		
Urea	*Proteus mirabilis*		
	Jack-bean meal		

*Sensing element = CO_2. **Sensing element = H^+.

The number of types of differentiated tissue may not be quite as numerous as the number of microorganisms, but nevertheless the sources and their applications appear limited only by the imagination: banana pulp immobilized on an oxygen electrode provides a source of polyphenol oxidase (PPO) for the catalytic oxidation of dopamine (Fig. 7.5), allowing dopamine to be estimated via oxygen consumption (Sidwell and Rechnitz, 1985). Similarly, tyrosinase contained in sugar-beet tissue can be employed to monitor tyrosine (Schubert *et al.*, 1983), but specificity by this system was not as good, with responses shown to phenols and aromatic amines. Obviously further tissue screening might reveal a more specific source.

Potentiometric Sensors

AMMONIA-LINKED SENSORS

Arginine

The overwhelming majority of cell-based biosensors are linked with potentiometric electrodes. Unlike the collaboration with the oxygen electrode, these systems do not monitor general respiratory activity but can be more specific to a particular enzyme reaction within the cell.

The first potentiometric microbial biosensor (Table 7.2) was reported by Rechnitz *et al.* (1977) for L-arginine. Using a bacterium *Streptococcus faecium* coupled to an ammonia gas-sensing electrode, arginine could be assayed via the metabolism of arginine by the microorganism in the sequence:

$$\text{L-arginine} + H_2O \xrightarrow{\textit{arginine deiminase}} \text{citrulline} + NH_3$$

$$\text{citrulline} + H_3PO_4 \xrightarrow{\textit{ornithine transcarbamylase}} \text{ornithine} + \text{carbamoylphosphate}$$

$$\text{carbamoylphosphate} + \text{ADP} \xrightarrow{\textit{carbamate kinase}} \text{carbamic acid} + \text{ATP}$$

$$\text{carbamic acid} \rightarrow CO_2 + NH_3$$

A linear response (i.e. 40–45 mV/decade) was obtained for arginine in the range 10^{-4}–6.5×10^{-3} M. This organism is specific for arginine, not producing significant amounts of ammonia in the presence of glutamine or asparagine, and it was shown that a 20 day-old electrode that had begun to lose its activity could be regenerated by placing it in growth medium for 48 h. This suggested that the bacteria were still viable and indicated that, if this regeneration process was generally applicable to cell-based sensors, these devices could offer certain advantages in terms of sensor lifetime. However, *without careful control of the storage conditions to discourage the growth of contaminants, the specificity is lost* (Corcoran and Kobos, 1983).

The first report of a tissue based biosensor is also claimed for arginine by Rechnitz (1978), who co-immobilized urease with tissue slices of bovine liver,

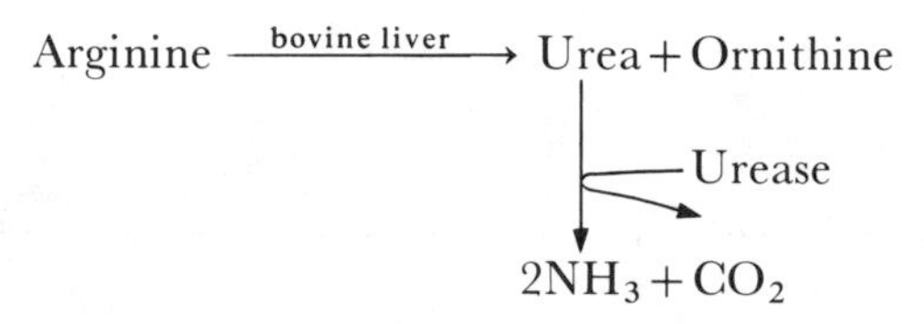

and tissue slices from other sources have been 'tuned' to asparagine, glutamine, etc.

Glutamine

Porcine kidney cortex tissue containing the enzyme glutaminase, immobilized on an ammonia electrode renders the electrode specific for glutamine. This sensor appears to be highly selective, showing no response for urea or other L-amino acids and even though a high concentration of D-amino acid oxidase is present in the tissue slice, no response was found for D-amino acids. This has been correlated with an absence of the flavin adenine dinucleotide (FAD) required by this enzyme (Guilbault and Hrabankova, 1971), and *both sensitivity and selectivity can be altered by additives in the buffer solution.*

This sensor has been compared with a microbial sensor based on the microorganism *Sarcina flava*. The response properties are of similar order, the tissue-based system offering some advantages over the microorganism-based sensor, due to the risk of contamination of cell line in the latter biosensor. Indeed, most of these ammonia-linked biosensors have both microbial and tissue-based counterparts, and in common with all these cellular systems some absolute selectivity must be traded against their extended lifetime compared with the isolated enzymes.

Adenosine

The enzyme adenosine deaminase is frequently isolated from intestinal mucosa suggesting that immobilization of mouse small-intestinal mucosal cells against an ammonia electrode would provide a sensor capable of measuring adenosine. However, as with many tissue preparations, this source contains enzymes directed at many different metabolic pathways, so that the electrode also exhibits a response to certain adenosine-containing nucleotides. *Elimination of this interference has been achieved with the use of inhibitors and blocking agents* (Arnold and Rechnitz, 1981).

Mucosal cells were immobilized on the surface on an ammonia gas-sensing electrode via glutaraldehyde–BSA crosslinking. The resulting electrode showed a comparable response for adenosine, AMP, ADP and ATP (54.5, 50.8, 50.2 and 46.8 mV/decade, respectively) with similar linear ranges. It would obviously be desirable to selectively suppress reactivity with the adenosine nucleotides. The optimum pH for reaction with adenosine (pH 9) is higher than for the nucleotides, so that the relative responses can be optimized by efficient buffering at high pH.

Total elimination of interfering activities requires identification of the metabolic pathways involved. Deamination of the adenosine nucleotides can occur via two possible routes:

(i) *Direct deamination of the nucleotide*

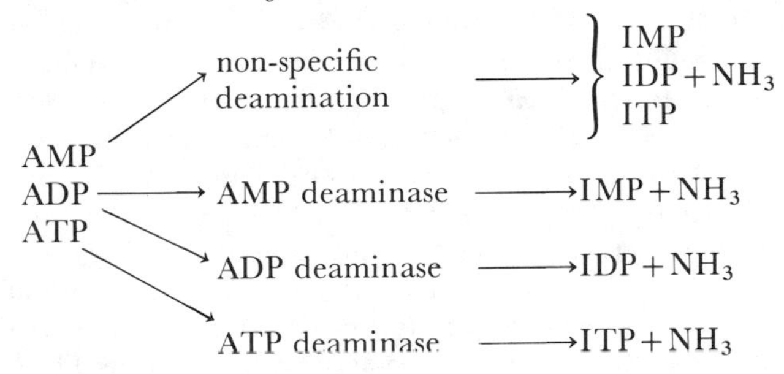

(ii) *Cleavage of adenosine followed by deamination of adenosine*

$$\left.\begin{array}{l}\text{AMP}\\ \text{ADP}\\ \text{ATP}\end{array}\right\} \xrightarrow{\textit{alkaline phosphatase}} \text{adenosine} + \text{phosphate}$$

$$\text{adenosine} \xrightarrow[H_3O]{\textit{adenosine deaminase}} \text{inosine} + NH_3$$

Phosphate has been shown to inhibit both alkaline phosphatase and AMP deaminase, whereas potassium ion is necessary for AMP deaminase activity. The presence of phosphate ions were shown to inhibit nucleotide interference, although potassium showed no effect. These findings were consistent with the second route, a finding that is substantiated by the high AMP deaminase activity commonly associated with muscle tissue and the maximum alkaline phosphatase activity in intestinal mucosal cells.

Although this tissue-based sensor does not show inherent selectivity for adenosine, an understanding of the interfering metabolic pathways can allow the introduction of suitable inhibitors to manipulate the specificity.

Guanine

A guanine biosensor employing rabbit liver tissue demonstrates the importance of optimization of *both operating and storage* conditions. The microbial sensor for NO_2

described earlier in this chapter has shown how each buffer layer must be pH optimized for maximum biocatalytic activity, gas diffusion, solution, etc. For the guanine biosensor, maximum ammonia production—i.e. maximum biocatalytic activity (pH 9.5) is sacrificed in preference for the stability of the biocatalyst and extended electrode lifetime that is achieved at pH 8, where only 50% enzyme activity is recorded.

Nitrate

Because of its importance many different methods have been devised to measure nitrate. The use of the dual enzyme pathway

$$NO_3^- + NADH \xrightarrow{\text{nitrate reductase}} NO_2^- + NAD^+ + H_2O$$
$$NO_2^- + 3NADH \xrightarrow{\text{nitrite reductase}} NH_3 + 3NAD^+ + 2H_2O$$

relates its concentration to ammonia and suggests a potential link with an ammonia electrode. Screening of microorganisms for appropriate enzyme activity has revealed a strain of *Acetobacter vinelandi* as suitable. However the electrode showed a super-Nernstian response (85 mV/decade) and a very narrow linear concentration range (7×10^{-5}–30×10^{-5} M). This behaviour was characteristic of fluctuations in enzyme activity caused by pH changes in the 'biolayer'. Although this would be conceivable for such an ammonia producing reaction, it was found that deviation in pH of the buffer solution (pH 7.4) was minimal, and the actual cause of the fluctuation lay in the supply of nutrients and the activation of metabolic pathways other than those directed at nitrite and nitrate. The carbon source was supplied by a 1% sucrose solution, but the only nitrogen source available was the varying ammonia production from the sample nitrate. Identification of this variable allowed its correction with an ammonia assimilation inhibitor such as isonicotinic acid.

In order to obtain a linear concentration response, therefore, the substrates and products of the targeted enzyme sequence must be directed uniquely towards that sequence and inhibited from participation in other metabolic or synthetic processes that might alter the output signal.

NAD

Screening of microorganisms or tissue for selective activity does not always reveal a satisfactory cell system for the desired analyte. Estimation of the coenzyme NAD^+, for example, could be linked to an ammonia electrode via the enzyme sequences shown in Fig. 7.6 (Kobos *et al.*, 1979). However although the bacterium *Escherichia coli* is a rich source of nicotinamide deamidase and the fungus *Neurospora crassa* contains NADase, neither organism alone can catalyse the complete deamination of NAD^+. Further screening might reveal a suitable single cell-based source of the two enzymes but an alternative solution is found in the development of a hybrid NH_3 electrode of immobilized *Escherichia coli* together with NADase isolated from *Neurospora crassa*. This combination does not however include all the advantages of extended lifetime associated with whole cells, since although the immobilized bacteria are stable for more than a week, the enzyme activity is lost continuously over this period, causing the limit of detection to deteriorate.

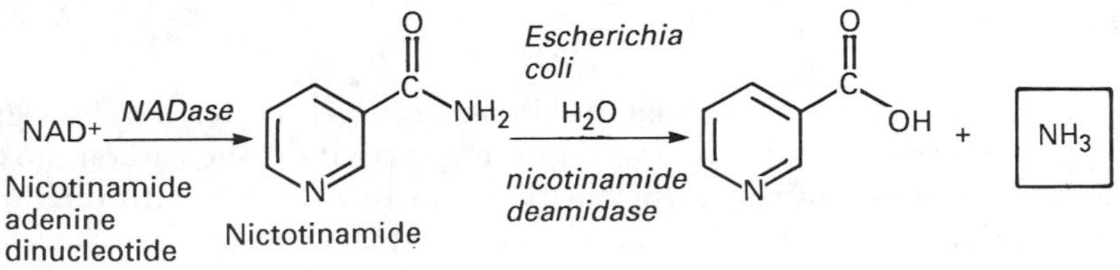

NAD+ —NADase→ Nictotinamide —Escherichia coli, H_2O, nicotinamide deamidase→ nicotinic acid + NH_3

Nicotinamide adenine dinucleotide

Fig. 7.6 Estimation of NAD^+. *E. coli* only provides one of the enzymes required in the sequence.

HYDROGEN-SULPHIDE LINKED SENSORS

Cysteine

Cysteine is metabolized by the bacterium *Proteus morganii* according to the reaction:

$$\text{cysteine} \xrightarrow{\textit{cysteine desulphydrase}} \text{pyruvate} + NH_3 + H_2S$$

so that the construction of a biosensor based on ammonia is indicated. Jensen and Rechnitz (1978) however have devised a bacterial membrane electrode coupled to a hydrogen sulphide-sensing electrode. A response of 25 mV/decade was recorded for the 5×10^{-5}–9×10^{-4} M range of cysteine, but applications of the device are limited by the efficiency of the base-sensing device, which is unable to distinguish between carbon dioxide and hydrogen sulphide. Background levels of carbon dioxide had to be removed from the sample solution in order to perform the assay.

Sensor efficiency can thus frequently be limited by both the base transducing device and the biorecognition component.

CARBON DIOXIDE-LINKED SENSORS

Glutamate and Pyruvate

Glutamate and pyruvate are two of the analytes that can be monitored via carbon dioxide emission, caused by the decarboxylase:

$$\text{glutamate} \xrightarrow{\textit{glutamate decarboxylase}} \text{4-aminobutyrate} + CO_2$$
$$\text{pyruvate} + H_2O \xrightarrow{\textit{pyruvate decarboxylase}} \text{acetaldehyde} + CO_2$$

The source of glutamate decarboxylase was the first plant material reported in a tissue biosensor. A glutamate sensor with excellent selectivity has been constructed with a thin slice of yellow squash tissue immobilized by BSA–glutaraldehyde crosslinking on the surface of a CO_2 electrode (Kuriyama and Rechnitz, 1981).

A source of pyruvate decarboxylase has been identified in corn kernels for the corresponding pyruvate biosensor (Kuriyama *et al.*, 1983) and in comparison with a similar electrode using the isolated enzyme, the tissue electrode displayed longer response times but reliable operation over a one-week period, in contrast with one day for the enzyme electrode. Indeed for the latter system, the response decreased from about 35 mV/decade to 12 mV/decade over three days, while the tissue electrode remained essentially unchanged.

Antibiotics

We have already discussed the relationship between O_2 uptake and biological activity. An alternative indication of respiratory activity is the carbon dioxide production. Inhibition of respiratory activity can be achieved with the use of selective antibiotics and this has been mentioned as a means of controlling contamination in microorganism-based sensors. This phenomenon can also be employed, however, as a means of assaying the antibiotics themselves.

Tetracycline hydrochloride (Simpson and Kobos, 1982) and the aminoglycosides gentamicin, streptomycin and neomycin (Simpson and Kobos, 1983) have been estimated via their inhibition of carbon dioxide production in *Escherichia coli*. The inhibition is irreversible and the method does not appear to be sufficiently promising for the general assay of pharmaceutical and plasma preparations to be exploited, in the development of 'single use' biosensor devices for antibiotic assay.

An antibiotic biosensor assay has been devised for cephalosporin around a pH electrode. Cephalosporinase isolated from *Citrobacter freundii* catalyses the ring-opening reaction of cephalosporin with production of H^+ (Matsumoto *et al.*, 1979):

R'CONH, S, N, O, CH_2R, COOH —*Cephalosporinase*, H_2O→ R'CONH, S, N, O, O^-, CH_2R, COOH

Whole cells of *Citrobacter freundii* were immobilized in a collagen membrane in front of a pH electrode, and in test in a fermentation broth of *Cephalosporium acremonium*, the microbial sensor showed a relative error of 8% compared with HPLC. Since this assay does not rely on an irreversible inhibition, it may be used continuously over several days.

Cell-based FETs

Chapter 4 described how FET devices can be applied to sensor applications. In particular H_2- and NH_3-sensing CHEMFETs would be suitable for use in combination with a 'biolayer' in a way similar to that already described. However one of the major problems with FET devices is encapsulation and their use in aqueous solutions. Although examples exist where FET sensors are used to monitor biological processes, the FET is usually placed behind a gas-permeable membrane, remotely from the sample solution. For example, hydrogen is produced by many microorganisms, particularly under anaerobic conditions. *Enterobacteriaceae* responsible for many urinary-tract infections produces hydrogen as the end product of acid fermentation. The susceptibility of such bacteria to treatment by antibiotics could be tested via the inhibition of hydrogen production. Analysis of the 'head space' from a sealed culture, by injection of a 2 ml sample into a PdMOSFET cell produced a quick estimate of antibiotic susceptibility (Hörnsten *et al.*, 1985).

Mediated Cell-based Sensors

ENERGY TRANSDUCTION

So far the whole cell sensors that have been described have functioned in an indirect mode, that is to say, in combination with an amperometric or potentiometric electrode probe, which measures a secondary analyte. An alternative approach would be to transduce the specific cellular reaction directly. In order to achieve this measurement, it is necessary to 'tap into' the energy transfer involved in the biological process in such a way that it can be transduced as an electrical signal. Biological energy-transducing systems can operate in two different energized states, and coupling of energy between one system and the next apparently originates from a common intermediate point which occurs when cycling between these different energized states, i.e. when cycling between two alternative redox potentials (E^{o} and E^{o*}) overlaps with cycling between two pK_a values (pK_a and pK_a^*) or two phosphate potentials (PTP and PTP*). This is particularly well demonstrated by the photosynthetic process. Light energy, acting on water, is transduced to redox energy at the chlorophyll reaction centres (Fig. 7.7). Subsequently in photosynthesis this is converted to acid–base energy and then a 'phosphate bond energy storc', which can be used to increase the previous redox energy level.

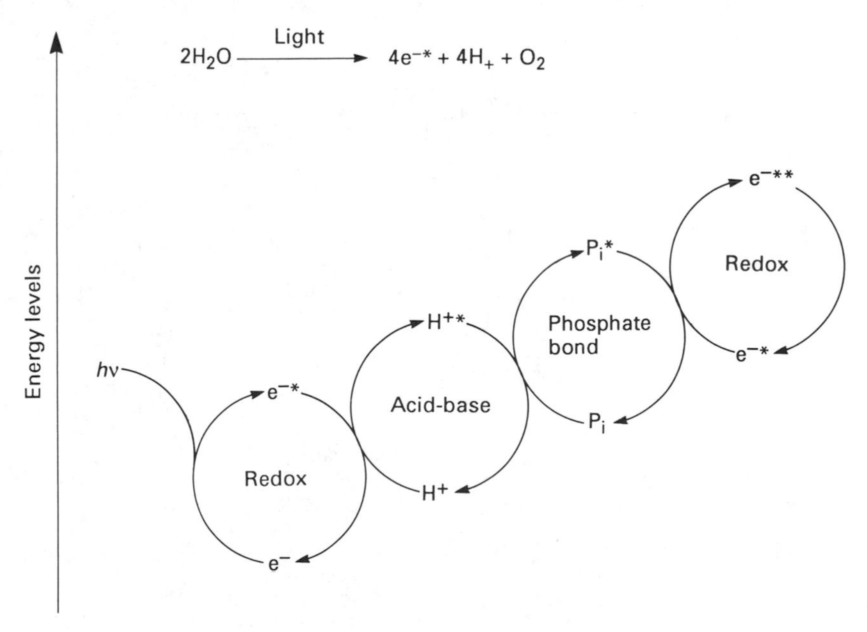

Fig. 7.7 Energy transduction system.

ARTIFICIAL MEDIATORS

Artificial entrance into these electron-transport chains is usually achieved at the redox cycle via *redox mediators*. The use of redox mediators in biosensors has become particularly important in amperometric biosensors, as will be discussed in Chapter 8, but electron transduction from a whole cell presents rather different problems from an isolated enzyme, since transport is restricted by the cell walls and the membrane. However, much of thc effort that has been devoted to achieve this has come from the development of biofuel cells (Fig. 7.8).

The efficiency of electron transfer using intermediates can be high and since the redox potential seen by the electrode is that of the last component in the chain, this usually reduces the potential difference between the two half-cells. The first direct biofuel cells are attributed to Cohen (1931) using bioelectrochemical cells in series, but the first reports of mediated electron transfer to whole organisms came half a century later (Bennetto *et al.*, 1980). These fuel cells have revealed information concerning the efficiency of different redox mediators. A considerable variation in the efficiency of reduction of similar mediators by the same organism was noted. Roller *et al.* (1983) reported a biofuel cell based on *Escherichia coli* for the utilization of lactate wastes. This involved thionine-mediated electron transfer at the cathode. However subsequent work (Roller *et al.*, 1984) showed that methylene blue was a better mediator for *E. coli B/r*, whereas outstanding mediator efficiency with *Proteus vulgaris* was demonstrated by thionine.

The performance of *E. coli* cells growing on glucose with ferric chelate reagents as mediators (Tanaka *et al.*, 1983a) correlated mediator efficiency with the reaction kinetics at both electrode and organism. The studies attributed fast electron transfer at the electrode with mediator efficiency. Thionine, for example

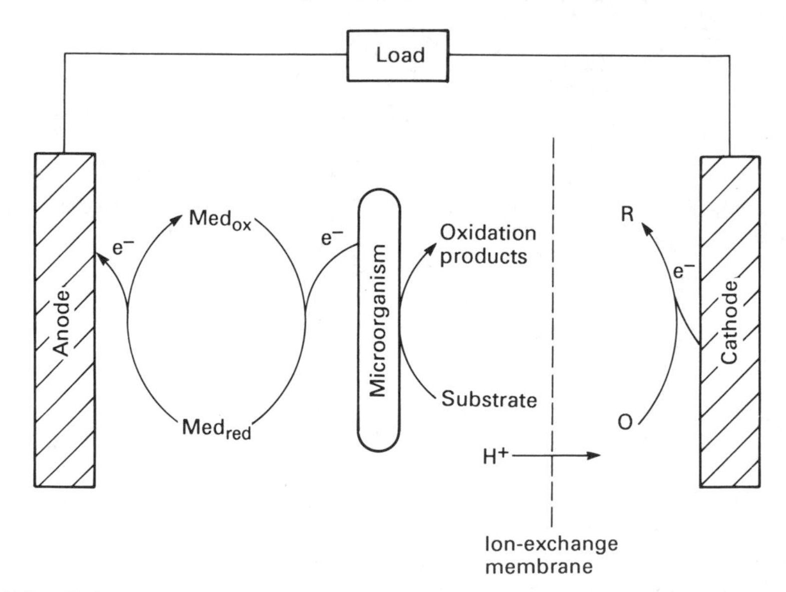

Fig. 7.8 Schematic representation of a bio-fuel cell.

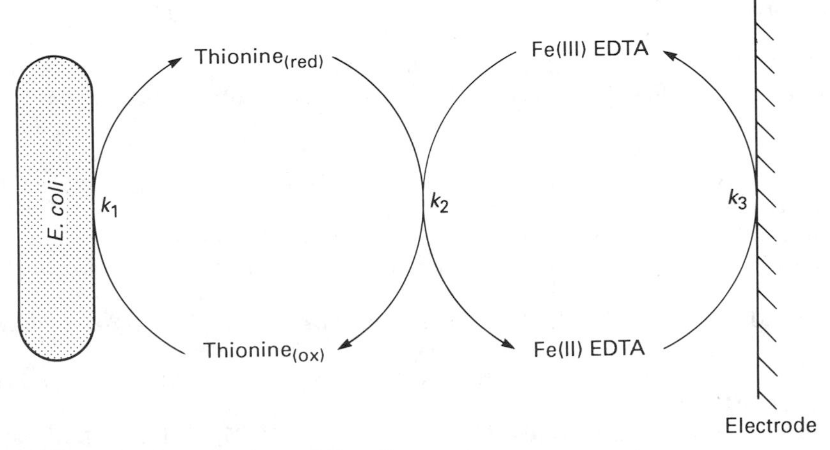

Fig. 7.9 Mixed mediation of *E. coli*.

showed a rate constant for reaction with *E. coli* which was 100 times faster than the ferric chelates. With the latter systems, reduction rates were slower than with thionine, although coulombic yields were comparable. The results suggested that a mixed mediator is required (Fig. 7.9); for example, combining thionine for interaction with the organism, and EDTA for electrode interaction might be most successful! Indeed, this mixed mediator has been proved to be more efficient than the individual components (Tanaka *et al.*, 1983b).

Turner *et al.* (1986) have reported a 'biocheck' analyser using ferri- or ferrocyanide in combination with benzoquinone, to mediate the cell reaction with the electrode and thus give a sample organism count, with a detection limit of 10^6 organisms/ml.

The question as to mediator design for a particular organism is a complicated one. It is not simply a matter of redox potential but lipophilicity and surface interactions play an important role in penetration of the cell membrane. Retention of the mediator at the sensor surface is also critical, and in recent years many immobilization procedures tailored to redox-mediator systems have been investigated. One of the potential problems with immobilization of the mediator is that it is no longer 'free' to shuttle electrons between the electrode and the immobilized biolayer. Arrangements have therefore been envisaged to overcome this. For example, mediators anchored by a 12-carbon chain to a polymer support would theoretically be able to transfer electrons across a 10 nm gap by rotation on their anchorage. Alternatively, direct immobilizing association of the mediator with the organism itself would give a molecular conductor to channel electrons between organism and electrode. Although this might involve the introduction of 'foreign bodies' into the cell (which may reduce the lifetime of the organism), some such cellular modifications have been achieved. If this can be successfully accomplished then, together with electron-hopping mediator complexes, tuned for redox and surface interaction properties, directly transducing cellular molecular conductors could be feasible.

References

Arnold, M.A. and Rechnitz, G.A. (1981). *Anal. Chem.* **53**, p. 515.

Bennetto, H.P., Stirling, J.L., Tanaka, K. and Vega, C.A. (1980). *Soc. Gen. Microbiol.* **8**, p. 37.

Cohen, B. (1931). *J. Bacteriol.* **21**, p. 18.

Corcoran, C.A. and Kobos, R.K. (1983). *Anal. Lett.* **16**(B16), p. 1291.

Divies, C. (1975). *Ann. Microbiol.* **126A**, p. 175.

Guilbault, G.G. and Hrabankova, E. (1971). *Anal. Chim. Acta* **56**, p. 285.

Hikuma, M., Kubo, T., Yasuda, T., Karube and Suzuki, S. (1979a). *Anal. Chim. Acta* **109**, pp. 33–38.

Hikuma, M., Kubo, T., Yasuda, T., Karube, I. and Suzuki, S. (1979b). *Biotechnol. Bioeng.* **21**, p. 1845.

Hikuma, M., Kubo, T., Yasuda, T., Karube, I. and Suzuki, S. (1980). *Anal. Chem.* **52**, p. 1020–1024.

Hörnsten, G., Elwing, H., Kihlström, E. and Lundström, I. (1985). *J. Antimicrob. Chemother.* **15**, 695–700.

Jensen, M.A. and Rechnitz, G.A. (1978). *Anal. Chem.* **46**, p. 246.

Karube, I., Matsunaga, T., Mitsuda, S. and Suzuki, S. (1977a). *Biotechnol. and Bioeng.* **XIX**, pp. 1535–1547.

Karube, I., Mitsuda, S., Matsunaga, T. and Suzuki, S. (1977b). *J. Fermentation Technol.* **55**, pp. 243–248.

Karube, I., Suzuki, S., Okada, T. and Hikuma, M. (1980). *Biochimie* **62**, p. 567.

Karube, I., Nakahara, T., Matsunaga, T. and Suzuki, S. (1982a). *Anal. Chem.* **54**, p. 1725.

Karube, I., Okada, T. and Suzuki, S. (1982b). *Anal. Chim. Acta* **135**, p. 61.

Karube, I., Wang, Y., Tamiya, E. and Kawarai, M. (1987a). *Anal. Chim. Acta* **199**, pp. 93–97.

Karube, I., Yokoyama, K. and Tamiya, E. (1987b). *Appl. Env. Microbiol.*, p. 179.

Kobos, R.K., Rice, D.J. and Flournoy, D.S. (1979). *Anal. Chem.* **51**(8), p. 1122.

Kuriyama, S., Arnold, M.A. and Rechnitz, G.A. (1983). *J. Membr. Sci.* **12**, pp. 269–278.

Kuriyama, S. and Rechnitz, G.A. (1981). *Anal. Chim. Acta* **131**, pp. 91–96.

Mascini, M., Jannelle, M. and Palleschi, G. (1982). *Anal. Chim. Acta* **138**, p. 65.

Matsumoto, K., Seijo, H., Waranabe, T., Karube, I. and Suzuki, S. (1979). *Anal. Chim. Acta* **105**, pp. 426–432.

Mattiasson, B. Larsson, P.-O., Lindahl, L. and Sahlin, P. (1982). *Enzyme Microb. Technol.* **4**, p. 153.

Okada, T., Karube, I. and Suzuki, S. (1983). *Biotechnol. and Bioeng.*, **XXV**, p. 1641.

Rechnitz, G.A. (1978). *Chem. Eng. News*, **56**, p. 16.

Rechnitz, G.A., Kobos, R.K., Riechel, S.J. and Gebauer, C.R. (1977). *Anal. Chim. Acta* **94**, pp. 357–365.

Roller, S.D., Bennetto, H.P., Delaney, G.M., Mason, J.R., Stirling, J.L., Thurston, C.F. and White, Jr., D.R. (1983). *World Biotec Report 1983*, p. 655. London, Online Publications.

Roller, S.D., Bennetto, H.P., Delaney, G.M., Mason, J.R., Stirling, J.L. and Thurston, C.F. (1984). *J. Chem. Tech. Biotechnol.* **34B**, p. 3.

Schubert, F., Wollenberger, U. and Scheller, F. (1983). *Biotech. Lett.* **5**, pp. 239–242.

Sidwell, J.S. and Rechnitz, G.A. (1985). *Biotech. Lett.* **7**, pp. 419–422,

Simpson, D.L. and Kobos, R.K. (1982). *Anal. Lett.* **15**, p. 1345.

Simpson, D.L. and Kobos, R.K. (1983). *Anal. Chem.* **55**, pp. 1974–1977.

Suzuki, M. and Tamiya, E. and Karube, I. (1987). *Anal. Chim. Acta* **199**, pp. 85–91.

Tanaka, K., Vega, C.A. and Tamamushi, R. (1983a). *Bioelectrochem. Bioenergetics* **11**, pp. 135–143.

Tanaka, K. Vega, C.A. and Tamamushi, R. (1983b). *Bioelectrochem. Bioenergetics* **11**, pp. 289–297.

Turner, A.P.F., Cardosi, M.F., Ramsay, G., Schneider, B.H. and Swain, A. (1986). *Biotechnology in the Food Industry*, p. 97. Pinner, Online Publications.

Chapter 8

Amperometric Biosensors

The 'Enzyme Electrode'

The major disadvantage with the use of whole-cell sensors is that of specificity. Very often, the cell employed can utilize analyte-related, or even totally unrelated substrates to give a similar signal. Methods for eliminating this interference have already been discussed in the previous chapter, but in many instances the enzymes can be isolated and used outside their natural environment, so that an alternative approach is to employ immobilized enzymes in conjunction with an electrochemical assay.

In considering the treatment of an amperometric enzyme electrode it is necessary to identify the controlling reaction step (Mell and Maloy, 1975). The device may be limited by the enzyme kinetics, or if this is sufficiently fast, it may be under diffusion control.

The rate of an enzyme-catalysed reaction is given by (see Chapter 2):

$$\frac{-\mathrm{d}[\mathrm{S}]}{\mathrm{d}t}=\frac{k_2[E_\mathrm{o}][\mathrm{S}]}{K_\mathrm{M}+[\mathrm{S}]}$$

The steady-state current for an amperometric electrode under diffusion control is given by:

$$i_\mathrm{d}=\frac{nFAD[\mathrm{S}]}{d}$$

where d is the diffusion layer thickness and D is the diffusion coefficient in that layer. Assuming a boundary condition that half the enzyme generated product diffuses away into solution and not to the electrode, then for an enzyme electrode under kinetic control, the current will be:

$$i_k = \frac{nFAdk_2[E_o][S]}{2(K_M + [S])}$$

which gives a maximum current response i_{max} (when $[S] \gg K_M$):

$$i_{max} = \frac{nFAdk_2[E_o]}{2}$$

This will of course also be the maximum expected current, even for an electrode under diffusion control, so that when $K_M > [S]$, i_d and i_k can be written as:

$$i_d = \frac{2i_{max}D[S]}{k_2[E_o]d^2}$$

$$i_k = \frac{i_{max}[S]}{K_M}$$

i_d therefore is dependent on both d and D and decreases with d, whereas i_k is independent of D and increases with d.

Chapter 9 describes the treatment of enzyme kinetics with respect to diffusion through the enzyme layer and identifies a dimensionless parameter V, the diffusion modulus,

$$V = \frac{k_2[E_o]d^2}{DK_M}$$

Inspection of this parameter essentially compares rate of enzyme reaction with diffusion through the layer. $V < 1$ if $[E_o] \leqslant K_M$ or $d^2 < D$ and under these conditions enzyme kinetics predominate.

It can be seen from the evaluation of i_k and i_d, that deviations from enzyme kinetic control will occur as d_2 compares with D, and at high substrate concentrations. If $V > 1$, the electrode is under diffusion control. Under these conditions the linear response of the electrode may extend to $[S] \gg K_M$, since the current is no longer controlled by the enzyme catalysis.

Empirically, for enzymes with a low activity, the electrode is likely to be under enzyme control, irrespective of substrate concentration (Varfolomeev and Berezin, 1978). Similarly, for enzymes with a high activity, the electrodes will be under diffusion control, even at concentrations comparable with K_M. At intermediate activities there will be mixed control, depending on conditions.

Membrane-limiting Diffusion

The diffusional barrier introduced in a membrane-covered electrode has been discussed in Chapter 5. The effect of membrane thickness on response time (Tse and Gough, 1987) and the depletion of the sample of the electroactive species was considered there.

For an enzyme-coupled system, the current may be limited by three main processes:

(i) diffusion of substrate from bulk solution

$$i = nFAV_{max}d[S]$$

where $[E_o]$ is high and $K_M \gg [S]$

(ii) diffusion of substrate within the enzyme layer

$$i = \frac{nFAD_s[S]}{d}$$

where $K_M \gg [S]$ and D_s diffusion coefficient in the enzyme layer

(iii) enzyme reaction

$$i = \frac{nFAV_{max}d[S]}{2K_M}$$

Deviations from ideal behaviour in assay measurements can often be rationalized by consideration of these limiting cases. Film permeability and thickness are parameters that may be adjusted to 'tune' the electrode response and alter the linear concentration range. The apparent K_M for an enzyme is dependent on the immobilization environment, for example, and often differs from the solution species.

OXYGEN-LINKED ASSAY

Continuing the analogy with cell-based sensors, systems can also be devised for enzymes linked to the amperometric oxygen electrode (see Chapter 5). One of the first analytical devices to adopt the name '*enzyme electrode*' (Updike and Hicks, 1967) related changes in the partial pressure of oxygen caused by the following enzyme-catalysed reaction:

$$\text{glucose} + O_2 \xrightarrow{\textit{glucose oxidase}} \text{gluconic acid} + H_2O_2$$

to the substrate concentration.

Glucose oxidase was immobilized in a layer of acrylamide gel on a gas-permeable membrane over an oxygen electrode (Fig. 8.1). In the presence of glucose, the diffusion of oxygen through the gas-permeable membrane to the electrode was reduced. For concentrations of glucose below the apparent K_M for the immobilized glucose oxidase and in an excess of oxygen, there is a linear relationship between decrease in the electrode current due to oxygen and glucose concentration.

Similarly, oxalic acid whose determination in urine is clinically important in the diagnosis of various forms of hyperoxaluria, is specific to two enzymes which have been isolated in the pure form: oxalate decarboxylase and oxalate oxidase.

$$(COOH)_2 \xrightarrow{\textit{oxalate decarboxylase}} CO_2 + HCOOH$$
$$(COOH)_2 + O_2 \xrightarrow{\textit{oxalate oxidase}} 2CO_2 + H_2O_2$$

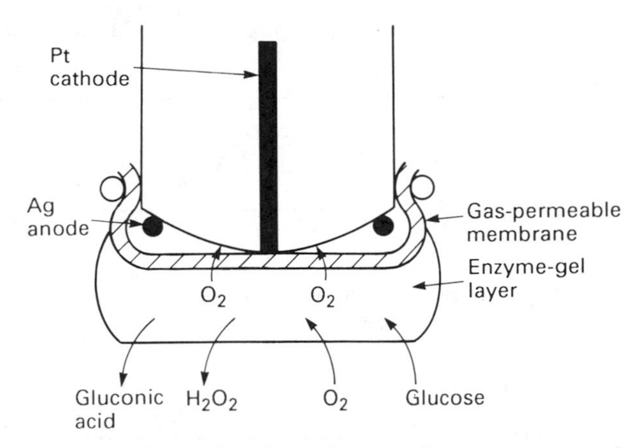

Fig. 8.1 The enzyme electrode, devised around an amperometric oxygen electrode.

The latter enzyme-catalysed reaction has been followed with an oxygen electrode, behind a membrane containing oxalate oxidase (Nabi Rahni *et al.*, 1986). The enzyme was immobilized on a pig intestine membrane by BSA–glutaraldehyde crosslinking, and the signal was optimized in a succinate buffer solution, by the introduction of EDTA and 8-hydroxyquinoline in sufficient concentration to bind free metal ions (e.g. Ca^{2+}), and thus ensure a maximum of free oxalate. High concentrations of these metal complexing agents cause inhibition of the enzyme so that *the use of such 'signal enhancers' must be optimized to prevent 'signal inhibition'*.

In a test carried out on urine samples, comparison of the measurements obtained by this technique with a spectrophotometric method gave a correlation coefficient of 0.95, but in contrast to the spectrophotometric method, the immobilized enzyme electrode assay was a one-step procedure, stable for two months and >200 assays.

Other substrates can also be linked to oxidase enzymes and assayed by a similar enzyme electrode. The relatively recent recognition that sulphiting agents, used as 'freshness' preservatives on raw fruit and vegetables and in the processing of other foods and beverages, are not as 'desirable' as previously considered, has demanded some interest in the development of analytical methods capable of determining sulphite at low levels.

Enzyme-linked reaction of sulphite can be described by:

$$\text{sulphite} + O_2 \xrightarrow{\textit{sulphite oxidase}} \text{sulphate} + H_2O_2$$

Sulphite oxidase was immobilized in a gel matrix on a dialysis membrane at the surface of an oxygen electrode (Smith, 1987). Sample sulphite was estimated via the oxygen uptake. The lower limit of detection for this system was 10 ppm in food samples, but with amplification techniques, as described later in this chapter, a lower limit of detection could probably be anticipated.

Table 8.1 Examples of some enzyme-linked assays involving oxygen and reported in the literature

Analyte	*Enzyme*	*Typical response time (min)*	*Stability (days)*
Glucose	Glucose oxidase	2	>30
Cholesterol	Cholesterol oxidase	3	7
Monoamines	Monoamine oxidase	4	14
Oxalate	Oxalate oxidase	4	60
Sucrose	Invertase	6	>14
Hydrogen peroxide	Catalase	2	30
Uric acid	Uricase	2	>14

MODES OF OPERATION

In principle, this enzyme electrode arrangement could be employed for any of the enzyme-catalysed reactions involving oxygen (e.g. Table 8.1), but in many cases alternative methods are available, so that the final choice of sensor configuration will depend on manufacturing, application and many factors other than the basic analytical principle. Considering, for example, the glucose oxidase catalysed oxidation of its substrate glucose (Fig. 8.2). The reaction causes various secondary changes in parameters which could be monitored. For example, electron transfer from glucose to the enzyme results in the following:

(1) Conversion of glucose to gluconic acid—i.e. a local pH change which could be monitored at a pH electrode.
(2) Reduction of oxygen to hydrogen peroxide—i.e. decrease in partial pressure of O_2 and increase in concentration of H_2O_2. Hydrogen peroxide can be measured colorimetrically by Trinder's method (1969), or electrochemically at a platinum anode polarized at $+0.7$ V versus SCE.

It follows therefore that an enzyme electrode may be constructed with the same components as the oxygen-linked enzyme electrode, except that H_2O_2 is estimated at a Pt electrode polarized at $+0.7$ V versus SCE (the oxidation potential for H_2O_2), instead of -0.65 V versus SCE (the reduction wave for oxygen).

Oxygen is the principal electron acceptor for a large number of oxidases apart from glucose oxidase. Table 8.2 shows some of the target analytes that could be assayed since their respective oxidases can be re-oxidized by molecular oxygen, producing peroxide.

MEMBRANE-TRAPPED ENZYME ELECTRODE

One of the earlier devices which measured peroxide (Clark, 1973) used L-amino oxidase in a cellophane sandwich over a platinum anode. The electrode was

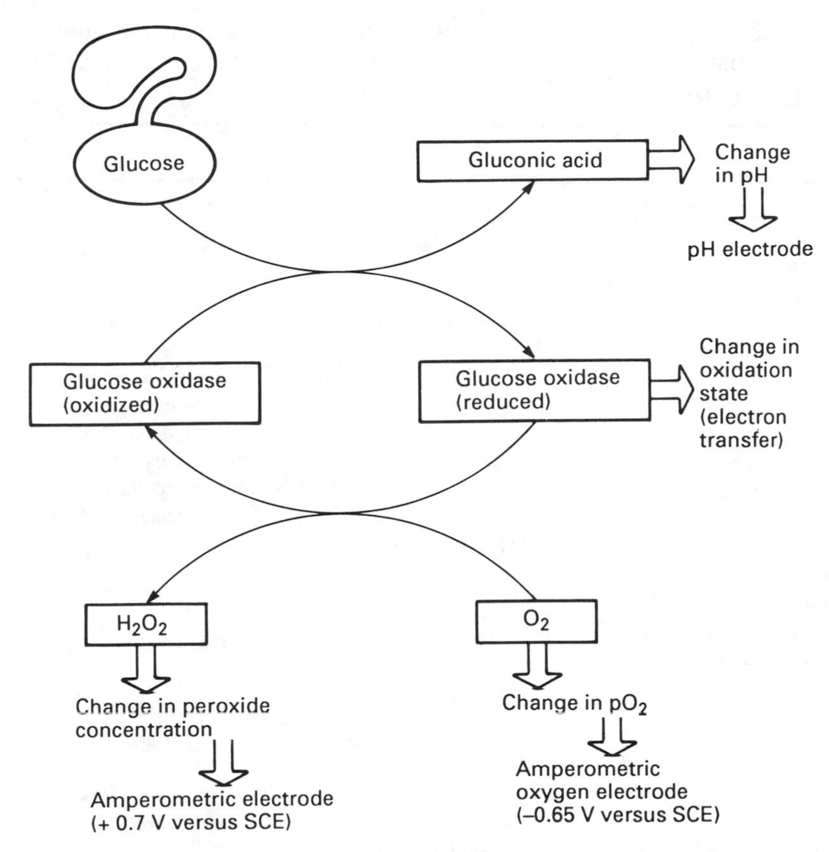

Fig. 8.2 Glucose oxidase catalysed oxidation of glucose: identification of measurement parameters for glucose assay.

specific for leucine, phenylalanine, methionine, tryptophan and nor-leucine, giving no response for alanine or glycine (or other unrelated compounds such as sucrose and glucose). As would be expected, the response time was dependent on the membrane thickness. A robust electrode made with 0.8 mm dialysis tubing had a response time of 8 min, while 0.1 mm cuprophane reduced the response time to 3 min. Placing the enzyme in a celite paste directly on the electrode reduced the response time still further to 30 s, but produced a more fragile enzyme electrode unsuitable for development as a biosensor device.

Glucose Assay

While applicable to many oxido-reductase enzyme systems, much of the work in this area has been devoted to the measurement of glucose. The reasons for this emphasis are multiple. Not only does it provide a low-cost research model, but

Table 8.2 A polarographic enzyme electrode for the measurement of oxidase substrates: examples of oxygen oxido-reductases which yield hydrogen peroxide

Analyte	*Enzyme*	*Source*
Glycollate L-Lactate D-Lactate (+)-Mandalate	Glycollate oxidase	Spinach, rat liver
L-Lactate	Lactate oxidase	*Mycobacterium smegmatis*
D-Glucose	Glucose oxidase	*Aspergillus niger*, *Penicillium amagasakienses*, honey (bee), *Penicillium notatum*
2-Dioxy-D-glucose 6-Dioxy-6-fluoro-D-glucose 6-Methyl-D-glucose	Glucose oxidase	
D-Glucose D-Galactose D-Mannose	Hexose oxidase	
L-Gulono-λ-lactone L-Galactonolactone D-Manonolactone D-Altronolactone	L-Gulonolactone oxidase	Rat liver
D-Galactose Stachyose Lactose	Galactose oxidase	*Dactylium dendroides*, *Polyporus circinatus*
L-2-Hydroxy acid	L-2-Hydroxy acid oxidase	Hog renal cortex
Formaldehyde Acetaldehyde	Aldehyde oxidase	Rabbit liver, pig liver
Purine Hypoxanthine Benzaldehyde Xanthine	Xanthine oxidase	Bovine milk, porcine liver
Pyruvate	Pyruvate oxidase	
Oxalate	Oxalate oxidase	
L-4,5-Dihydroorotate NAD	Dihydroorotate dehydrogenase	*Zymobacterium orioticum*
D-Aspartate D-Glutamate	D-Aspartate oxidase	Rabbit kidney
L-Methionine L-Phenylalanine 2-Hydroxy acids L-Lactate	L-Amino acid oxidase	Diamond rattlesnake, cotton mouth moccasin, rat kidney
D-Alanine D-Valine D-Proline	D-Amino acid oxidase	Hog kidney

Table 8.2 *(cont.)*

Analyte	*Enzyme*	*Source*
Monoamine Benzylamine Octylamine	Monoamine oxidase	Beef plasma, placenta
Pyridoxamine phosphate	Pyridoxamine phosphate oxidase	Rabbit liver
Diamines Spermidine Tyramine	Diamine oxidase	Bovine plasma, pea seedings, porcine plasma
Sarcosine	Sarcosine oxidase	*Macaca mulatta*, rat liver mitochondria
N-Methyl-L-amino acids	*N*-Methyl-L- amino acid oxidase	
Spermine Spermidine	Spermine oxidase	*Neisseria perflava*, *Serratia marcescens*
Nitroethane, aliphatic nitro urate	Nitroethane oxidase Urate oxidase	Hog liver, ox kidney
Sulphite	Sulphite oxidase	Beef liver
Ethanol and methanol	Alcohol oxidase	Basidiomycetes
D-Glucose	Carbohydrate oxidase	Basidiomycetes
D-Glucopyranose D-Xylopyranosc 1-Sorbose D-gluconolactone		*Polyporus obtusus*
NADH	NADH oxidase	Beef heart mitochondria

glucose is one of the most common substrates for the fermentation industry and as such a potential market for glucose monitoring is revealed. Perhaps more importantly, however, is the search for better methods to control the insulin-dependent diabetics. Diabetes mellitus is a world-wide health problem, involving an insulin deficiency which results in an inability to control blood glucose levels. About 2% of the population (higher for the over 40s) suffer from insulin deficiency. In the more chronic cases this can be treated by insulin therapy, usually involving a discontinuous administration of insulin a number of times a day by injection. More desirable would be an implantable glucose sensor/insulin delivery device, but whatever the technique of insulin therapy, measurement of the blood glucose levels are essential (see Chapter 1).

OXYGEN LIMITATION

Many of thc carly rcports wcre directed towards investigating glucose levels in the brain and tissue. Clark and Clark (1973) studied variations in glucose levels on the surface of the brain, following intravenous injection of insulin, and Silver (1976)

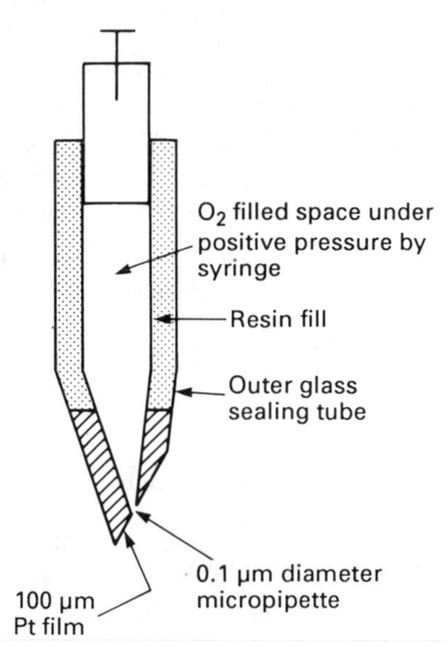

Fig. 8.3 Needle enzyme electrode with oxygen reservoir.

developed glass needle Pt-electrodes (Fig. 8.3) to which enzyme was immobilized by dipping in a 1% solution of glucose oxidase and drying. The electrode was polarized at +0.6 V versus Ag/AgCl to monitor the peroxide, and a constant oxygen supply was maintained from the reservoir in the centre of the needle. This latter refinement ensured that the enzyme electrode could be deployed in hypoxic tissue without falsely low readings. Indeed in a study of healing in connective tissue, an earlier conclusion, that the glucose concentration decreased with blood flow was reviewed. *It was apparent that fluctuations in the oxygen tension could lead to a situation where hydrogen peroxide production did not reflect the glucose concentration, but was limited by the oxygen supply.*

The problem of oxygen independence and interference of the signal by other electroactive species, notably ascorbate, has probably demanded most attention in the development of a stable glucose sensor.

Once a parameter has been identified whose measurement can be related to the concentration of target analyte, its dependence on the sampling environment must be fully tested. Clearly, both of the techniques already discussed for oxidase-linked electrodes (O_2 consumption or H_2O_2 production) are dependent on a constant excess oxygen supply. The following sections will discuss possible solutions to this supply.

'OXYGEN-STABILIZED' GLUCOSE ANALYSIS

Some fermentation broths contain oxygen at concentrations equivalent to air saturation. Many, however, are at much lower concentrations.

The enzyme catalase causes the reaction

$$H_2O_2 \xrightarrow{catalase} H_2O + \tfrac{1}{2}O_2$$

and in an attempt to 'stabilize' the oxygen supply of the enzyme electrode, catalase (together with glucose oxidase) was introduced into the enzyme layer of an oxygen electrode linked device, thus replacing up to 50% of the oxygen consumed by the enzyme reaction with glucose (Fig. 8.4a). Nevertheless, in an aerated culture of *Candida utilis*, the linear range was $<0.3\,\mathrm{g\,l^{-1}}$, with a saturation limit of $0.75\,\mathrm{g\,l^{-1}}$—clearly far short of the 2–3 $\mathrm{g\,l^{-1}}$ requirement (Enfors, 1981).

In an imaginative attempt to solve this problem Enfors (1981) proposed a cell based on a principle of regeneration of oxygen by feedback control of the O_2 measuring electrode (Fig. 8.4b).

The enzymes were immobilized on a Pt gauze, in front of an oxygen electrode. (A galvanic oxygen electrode was in fact used here, but the technique would be generally applicable.) The gauze formed part of an electrolysis cell (Fig. 8.4c) and the system was based on the principle that oxygen consumption measured at the oxygen electrode should provide feedback control to the electrolysis circuit, where an equivalent amount of oxygen was produced by the electrolysis of water from the catalase reaction.

$$H_2O_2 \xrightarrow{\text{catalase}} H_2O + \tfrac{1}{2}O_2$$
$$H_2O \xrightarrow{\text{anode}} \tfrac{1}{2}O_2 + 2H^+ + 2e^-$$
$$2H^+ \xrightarrow{+2e^-} H_2$$

The extent to which such a regeneration procedure can be employed depends largely on the avoidance of unwanted electrolysis reactions (such as destruction of the enzyme!), but under ideal conditions the net transfer of two electrons in the electrolysis circuit is equivalent to 1 mol of glucose oxidized by the enzyme.

Apart from side reactions, 'ideal conditions' here must include the condition where there is no exchange of oxygen between the electrode and the sample. At the potential required for O_2 production,

$$2H_2O \rightarrow O_2 + 4H^+ + 4e^- \; (E^\circ + 1.23\,\mathrm{V})$$

there are in fact likely to be many other interfering reactions, and even if these can be avoided, the production of H_2 or O_2 gas is likely to cause some considerable interference if bubbles form on the electrode. Obviously, while the authors demonstrated that the technique extended the measurement range of the sensor to $2\,\mathrm{g\,l^{-1}}$ glucose in the fermentation broth, the device is perhaps undesirably complex and electrolysis of water could be rather severe a requirement for many applications.

However, further manipulations can be imagined which might give greater improvements in oxygen independence for the oxygen-stabilized system:

(1) Oxygen is also regenerated during the electrochemical oxidation of hydrogen peroxide—i.e. during one of the possible assay procedures. Optimization of the collection efficiency for the Pt anode detector so that the electrochemically generated oxygen flux were equivalent to the total peroxide flux would achieve more efficient recycling of the O_2. With the electrode geometry discussed so far (see Chapter 8) at least 50% of the H_2O_2 is lost to the sample and is thus not available for recycling.

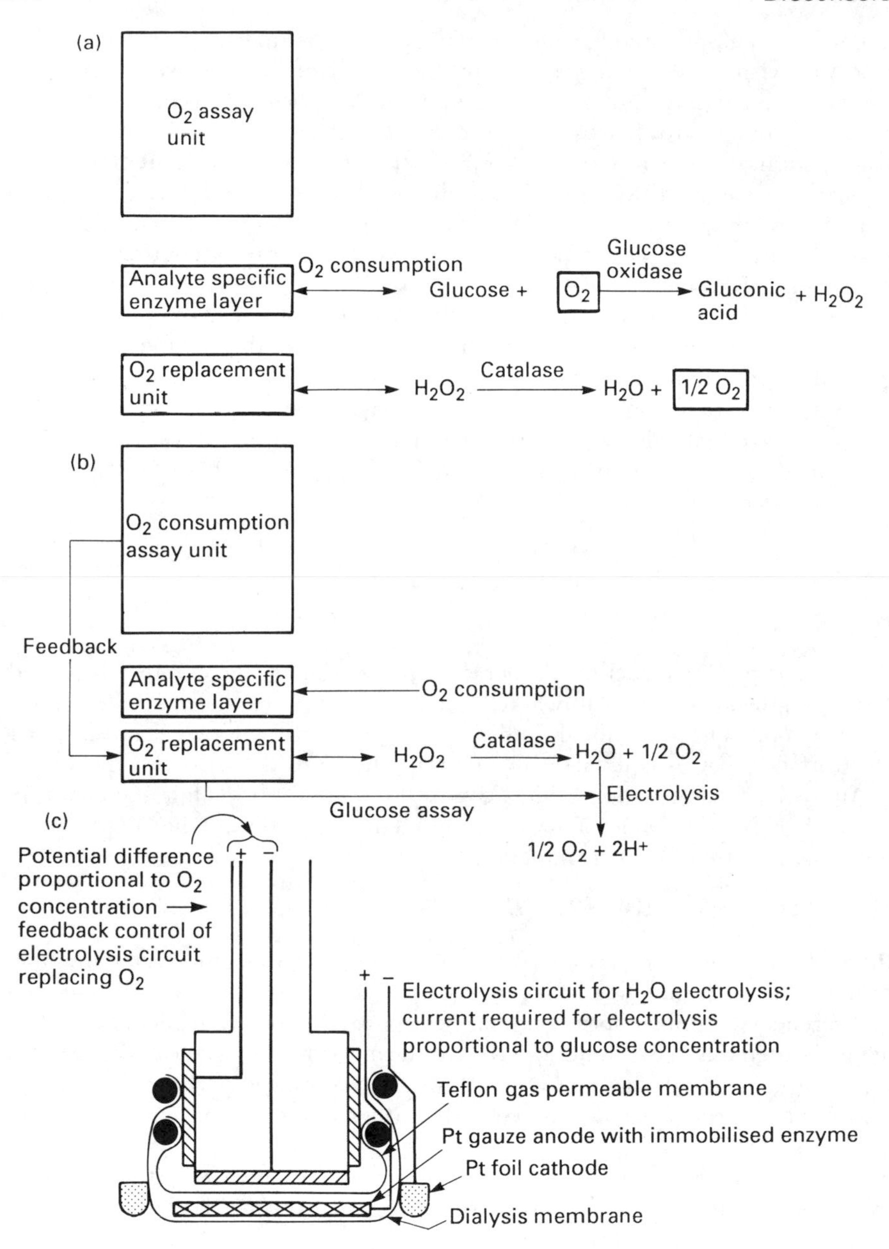

Fig. 8.4 (a) O_2-stabilized electrode unit construction involving catalase; (b) O_2-stabilized electrode unit construction involving electrolysis of H_2O; (c) schematic representation of O_2-stabilized enzyme electrode.

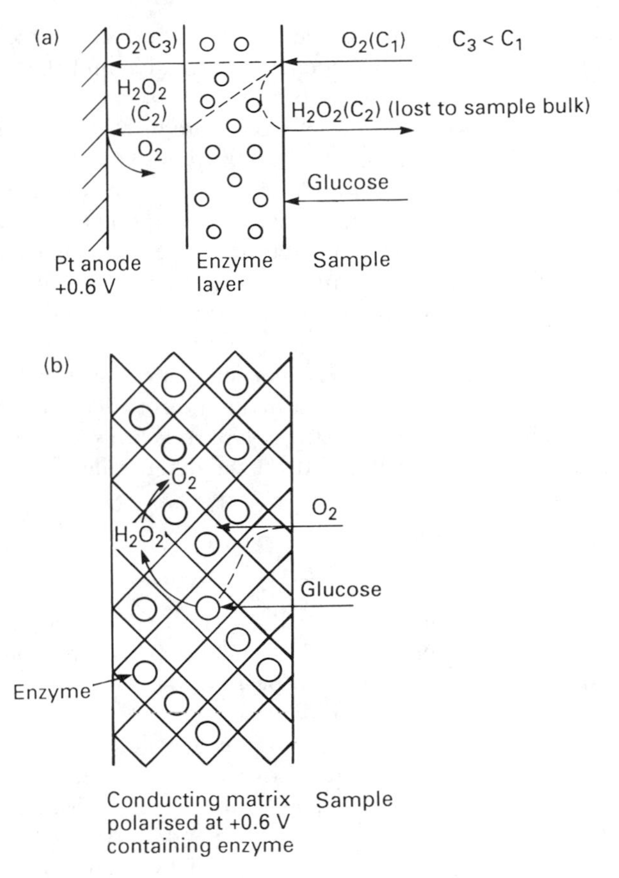

Fig. 8.5 Comparison of substrate and product exchange in (a) the enzyme membrane-covered electrode and (b) the 'ideal' conducting enzyme matrix.

(2) A conducting matrix, polarized at $+0.6$ V and containing the enzyme immobilized within the matrix (similar to Fig. 8.5), is a model where the diffusion path for the enzymatically produced peroxide would then be infinitely small, and it would be recycled essentially instantaneously at the conducting matrix electrode, without substantial loss of sample.

Electrochemically Deposited Immobilization Matrix

The use of electrochemically deposited conducting polymers has in fact been investigated in a rather different approach which assessed polypyrrole and polyaniline as a membrane material for entrapment of enzyme. These polymer films are particularly interesting, since they exist in both insulating and conducting forms, depending on oxidation state. The films are grown by *electro-*

oxidation of monomer (or a derivative) and the structure and redox properties of the polymer film can be altered by choice of derivative and the electrolyte and solvent of the electropolymerization solution.

Polypyrrole films (Diaz *et al.*, 1981) are known to be permeable and appear to offer an ideal matrix for enzyme immobilization. Electropolymerization of pyrrole from an aqueous solution containing glucose oxidase produced a polypyrrole film containing enzyme. The electro-oxidation is believed to proceed via a reactive π-radical cation, which reacts with neighbouring pyrrole species to produce a chain that is predominantly α,α'-coupled (Fig. 8.6). The resulting polymer incorporates anions from the supporting electrolyte and has a net positive charge.

The electrode responded to glucose concentration measured via peroxide oxidation at 0.8 V. However, at this potential in the presence of peroxide, Umana and Waller (1986) have reported that the polypyrrole films are degraded, so that an indirect strategy was adopted based on the determination of H_2O_2 via I_2 reduction:

$$H_2O_2 + 2H^+2I^- \xrightarrow{\textit{molybdate}} I_2 + 2H_2O$$
$$I_2 + 2e^- \xrightarrow{0.2\ V} 2I^-$$

At the potential required to measure I_2 (< 0.2 V versus SCE), the polypyrrole does not degrade, but introduction of this additional step requires that reagents be added to the sample solution in order to perform the peroxide estimation, so it then fails to achieve the reagentless biosensor.

Possibly a better solution would be to use a pyrrole derivative that did not degrade at the potential of peroxide oxidation. In fact in the similar device reported by Foulds and Lowe (1986) on a Pt–ink electrode base, no degradation is reported at the assay potential of 0.7 V (versus Ag/AgCl). It is well established that slight modifications in the polymerization conditions can alter the characteristics of the resultant polymer, so that these apparently contradictory findings may well be the result of different polymerization recipes.

Bartlett and Whitaker (1987a, b) have prepared a similar enzyme electrode, starting with the monomer *N*-methylpyrrole. At a rotating electrode polarized at 0.95 V versus SCE, the response to glucose was linear to 22 mmol l^{-1} and no film degradation due to operation at this potential was reported.

Polyaniline has also been successfully employed as an immobilization matrix for glucose. Shinohara *et al.* (1988) prepared a non-conducting polyaniline film containing the enzyme, by polymerization at pH 7. The resulting matrix allowed permeation by oxygen, so that glucose was estimated in the 10^{-4}–5×10^{-3} M range via the decrease in oxygen tension monitored at -0.5 V versus Ag/AgCl. The technique was also tested for Pt fibre microelectrodes (50 μm diameter) and found satisfactory.

In more acidic conditions conducting polyaniline films can be polymerized, but very low pH is detrimental to the enzyme, so that pH conditions must be compromised. J.C. Cooper and E.A.H. Hall (forthcoming) have produced a more conducting polyaniline entrapment matrix and monitored glucose concentration via peroxide assay.

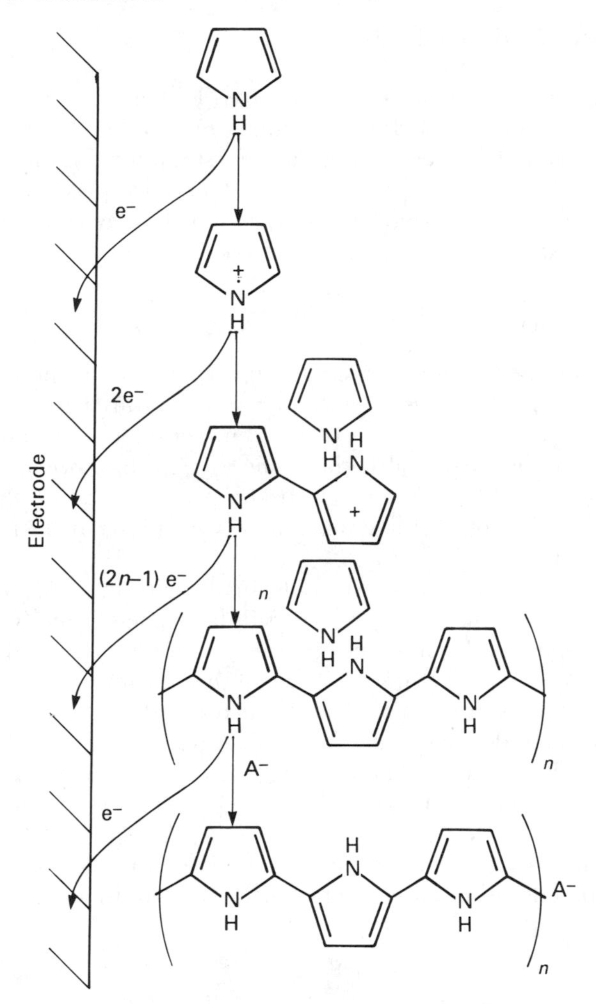

Fig. 8.6 Mechanism of polypyrrole film generation by electrolytic oxidation of pyrrole.

In view of the previous discussions on diffusion barrier membranes (see Chapter 5), it is also interesting to note that the polypyrrole device reported by Foulds and Lowe had a K_M for glucose/glucose oxidase of 31 mmol dm^{-3}, which is comparable with the solution enzyme. It would appear therefore that this film must be sufficiently porous not to act as a diffusional barrier. It is also possible that manipulation of the electropolymerization would give a film which extended the linear range for glucose and reduced the oxygen dependence.

In situ Glucose Monitoring

Implantable glucose sensors have been reported for consideration in closed-loop insulin therapy (Bessman *et al.*, 1981), but in animal studies they failed to achieve complete control, partly due to inaccurate estimation of the glucose levels (readings of the order of 50% were noted), a familiar observation for measurements made where the physiological concentration of oxygen is lower than the K_M values of the enzyme.

NEEDLE ELECTRODES

In developing a sensor suitable for *in vitro* and *in vivo* use, the possibility of local reaction to the sensor itself must be anticipated. Miniature needle-shape designs, as opposed to flat architectures have been considered as most ideal for implantation. In fact essentially the same design has been developed independently by two groups. Needle electrodes prepared from a Pt-wire anode, mounted in the middle of a cylindrical cathode were dip-coated with an 'enzyme sandwich' (Fig. 8.7).

Churchouse *et al.* (1986) mounted a 50 μm Pt wire in a 0.25 mm diameter stainless-steel tube reference. A polyethersulphone membrane was dip-coated from a dimethylsulphoxide solution to give a membrane 0.5–6 μm thick, covering the electrode cell. Enzyme was immobilized on this membrane by glutaraldehyde crosslinking and the resulting enzyme electrode responded to glucose, via peroxide assay at the Pt electrode polarized at +0.6 V. A response time of as little as 0.4 s could be achieved, depending on the membrane and enzyme thickness, but with an ambient pO_2 of 100 mmHg, an oxygen dependence was noted at >2 mM glucose.

However, mass-transfer effects meant that the introduction of a diffusion barrier, by dip coating an outer membrane of polyurethane from tetrahydro-

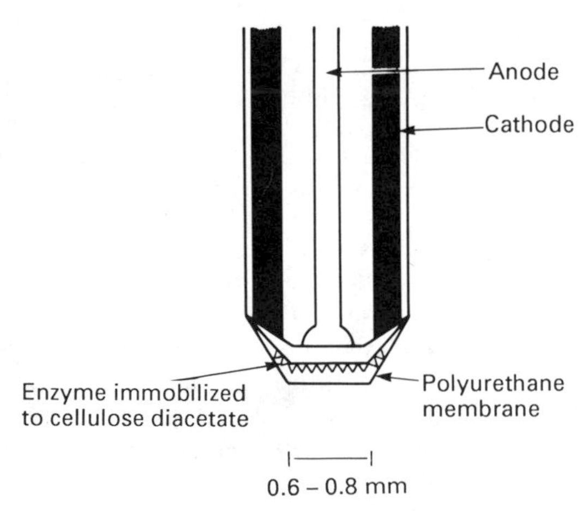

Fig. 8.7 Needle-type sensor for *in vitro* monitoring.

Table 8.3 Comparison of *in vivo* needle glucose electrodes

Parameter	*Scichiri* et al. (1984)	*Churchouse* et al. (1986)
Size		
diameter	0.4–0.8 mm	0.25 mm
length	20 mm	
Outer semi-permeable membrane	Polyurethane	Polyurethane
Enzyme immobilization	Glutaraldehyde crosslinking	Glutaraldehyde crosslinking
Inner gas-permeable membrane (on which enzyme is immobilized)	Cellulose diacetate	Polyethersulphone (0.5–6 μm)
Anode	Pt (0.2 mm)	Pt (50 μm)
Cathode	Ag	Stainless steel
Polarizing voltage	+0.6 V	+0.65 V versus Ag/AgCl
Electrolyte	Body fluid	Body fluid/phosphate buffer
Electrode body	Glass	Teflon
Response time	15–25 s	10–60 s
Concentration range (mmol/l)	0–27.5	15–70
Drift	0.8 ± 1.3%/24 h	20%/6 weeks
O_2 dependence	0.1%/mmHg pO_2	

furan, could extend the linear range of the sensor 10-fold. Although this was achieved at the expense of response time, the resulting response time of 10–60 s was adequate for use in whole blood. It was considered, however, that although suitable for use in venous blood, the oxygen independence had not been extended sufficiently for application in oxygen deficient tissue.

Oxygen dependence can thus be reduced by the introduction of a diffusion barrier. The thicker the membrane the better the oxygen independence—but the slower the response time.

A needle electrode with a comparable performance has also been investigated by Shichiri *et al.* (1982, 1984) (Table 8.3). Except that the inner gas-permeable membrane was dip-coated from cellulose acetate in acetone–ethanol, the construction details are essentially the same as the Churchouse system.

The use of a diffusion-limiting membrane with a greater permeability coefficient for oxygen than glucose can reduce the ambient pO_2 *dependence of the glucose measurement, and thus extend the linear range of the sensor. Applications involving implantable sensors require biocompatibility of this membrane, if the sensor is to be used successfully over long periods.*

On tests in subcutaneous tissue in diabetic and normal dogs and humans, and in the jugular vein of dogs, the device showed oxygen independence, but a decrease in electrode output was noted of the order of 25% in 3 days. This was accompanied by an increase in response time, and scanning electron micrographs suggested a build up of protein on the sensor surface. *Protein deposition on sensors will inhibit long-term use of implantable devices.* This deposition problem must also be solved before implantable devices can realistically be employed, since their longevity is a prerequisite of their application.

However, converted to a telemetry glucose monitoring system (Shichiri *et al.*, 1984), short-term exploratory experiments were carried out by insertion of a glucose sensor attached to a VHF transmitter in the subcutaneous tissue of the forearm of a diabetic patient. Considerable variation in the glycaemia of diabetics was demonstrated on a daily basis by this continuous monitoring, a situation which is obviously incompatible with the normal regimented insulin therapy. Indeed in conjunction with an artificial pancreas, the feedback control gained from the sensor to infuse insulin and glucose as required achieved better management in diabetic patients than manual discrete prescribed applications.

However, even under closed-loop control, the infusion of insulin provokes changes in the concentration of intermediate metabolites such as lactate, pyruvate, alanine, etc. and ideally signals other than just blood glucose are required to normalize the response. Mascini (1988) has reported, for example, that for a diabetic patient infusion of insulin results in a rapid return of the glucose concentration to normal levels, but an increase in the L-lactate level. Full insulin management is proposed by the development of algorithms based on multiple analyte signals.

Signal Enhancement

Apart from the assay error associated with glucose measurements made in tissues with a low oxygen tension, actual depletion of oxygen in the tissue by sensor operation can create an oxygen gradient in that tissue. It is therefore beneficial to make the sensor as small as possible, thus minimizing the O_2 consumption. However, since the current density (I) from an amperometric electrode is given by

$$I = i/A = nFkC$$

where A is electrode area, i is current, n is the number of electrons transferred, k is the rate constant and C is the concentration of electroactive species, it can be seen that any reduction in the size of the electrode will have a direct effect on the electrode current, thus decreasing the signal. Electrochemical techniques resulting in signal enhancement are particularly important, therefore, if miniature devices are to be employed. Chapter 5 describes the manipulation of control and measurement regimes for performing non-steady-state measurements for amperometric oxygen electrodes. This mode of operation results in a considerable increase in the current signal at the beginning of the applied potential pulse (see Fig. 5.12). The enzyme electrode presents some modification to the oxygen model, since in this case the electrochemically active substance that is to be monitored (H_2O_2 for example) is being produced by the enzyme reaction, so that in a static situation (i.e. when the electrode is out of circuit) where no removal of this electroactive compound is occurring, its concentration will increase with time. From $t = 0$, when substrate/analyte is introduced to the enzyme layer, there will be an increasing concentration of the 'product' in the enzyme layer until saturation is reached. Assuming therefore that the total reaction is dependent only on the concentration of analyte, then the size of the initial electrode signal will depend on the time after $t=0$ that the electrode was switched 'on'. If the *electrode* is under diffusion control (which it should be!) then a steady-state current will be reached, at time t_s after switching on, where the 'reservoir' of electroactive compound has been used up. Switching off the electrode will then allow the 'reservoir' to be refilled and the process can begin again.

In order to assess this treatment it is necessary to expand the classical treatment of enzyme kinetics to take account of the diffusional characteristics of the membrane in which the enzyme is immobilized. Two extreme cases can exist for both the diffusion of substrate/analyte from the sample solution into the membrane and for diffusion of the 'product' to the electrode surface.

In the former case, the dimensionless parameter V can be used to describe the relationship between the maximum rate for the enzyme reaction and diffusion through the membrane layer (Mell and Maloy, 1976)

$$V=\frac{(k_2[E_o]d^2)}{DK_M}$$

where k_2 is the rate constant for the enzyme-catalysed reaction, $[E_o]$ is the concentration of enzyme in the membrane layer, d is the membrane thickness and D is the diffusion coefficient of the substrate in the membrane layer. For large values of V, the rate of diffusion is slow compared with the enzyme reaction, while for small values of V, the enzyme kinetics become the rate-limiting process. Similar considerations are necessary for examination of product penetration through the membrane.

In addition to diffusion-limiting effects associated with immobilization, the non-steady-state mode proposed here must take account of accumulation of product with time during the electrode 'inactive' period and the interaction between electrode kinetics and enzyme/membrane kinetics during the 'active' period. Obviously the most straightforward electrochemical model is furnished by the case where diffusion within the membrane is slow compared with the enzyme reaction, and diffusion to the electrode is slow compared with electron transfer so that the 'active' electrode period is entirely under diffusion control. Models could however be considered for all extreme and intermediate cases, and concentration range responses could be 'tuned' by choice of membrane characteristics and measurement regime.

Mell and Maloy (1976) investigated this non-steady-state technique for an amperometric glucose electrode, assaying peroxide via the titration:

$$H_2O_2+2H^++2I \xrightarrow{\textit{molybdate}} I_2+2H_2O$$

This indirect assay introduces an additional step, but the application of a non-steady-state control mode, described by this group, is nevertheless relevant to the amperometric enzyme electrode techniques as a whole. They showed that increases in the ratio of 'inactive time' (τ) to 'active time' (t) led to successive enhancements of the current measurement until a point was reached when the 'inactive time' was sufficiently long (in their case >900 s) for saturation of the membrane by product to occur (corresponding to $\tau/t=100$). In non-steady-state operation, this enzyme electrode model is complicated. So long as t is sufficiently long for the electrode reaction to be diffusion controlled, and τ is sufficiently short for the rate of removal of product into bulk solution to be less than the rate of production, then it is the τ/t ratio that is critical. The τ and t values are associated with the concentration gradients emanating from the enzyme and the electrode respectively. They thus influence not only the current (present) measurement but also the environment for the next measurement. Figure 8.8 shows the product concentration profiles which develop for extreme values of t and τ when t is sufficiently long for the electrode reaction to be diffusion controlled. The actual concentration 'seen' by the electrode, and thus the size of the current measurement is a function of these parameters, in particular the τ/t ratio. However, with successive increases in t the system approaches the 'normal' steady-state case (see Fig. 8.8) where there are none of the advantages of increased current. For a given τ/t ratio therefore a maximum in the current enhancement will be found for small absolute values of t. Optimization of t is associated with diffusion through the membrane layer, and will thus be related to the diffusion coefficient D and the membrane thickness d.

Such constructional manipulations in electrode geometry and membrane characteristics that have been described are important in the optimization of any amperometric system. Similarly, mathematical manipulations of the control and processing procedures can significantly enhance the signal and increase the range and sensitivity of the sensor. Indeed we have discussed how the oxygen dependence of the signal due to glucose can be considerably varied using these techniques in a sensor employing glucose oxidase. Even so, improvements often fall short of ideal and in order to achieve any further advances it is necessary to return to Figure 8.2 and consider the enzyme reaction and the role of oxygen.

Oxygen By-pass Approach to Oxidase Sensors

Glucose oxidase is a redox enzyme whose function it is to accept electrons from the oxidation of glucose. The enzyme acts only as a temporary store for these electrons and normally the active oxidized form of the enzyme is regenerated and the electron-transport chain is continued through the action of the natural mediator oxygen (see Fig. 8.2). In order to completely eliminate oxygen dependence from this process, it would be necessary to intercept the chain before reaction with oxygen, and thus perform the electron transfer from the reduced enzyme by some artificial means (Fig. 8.9).

An obvious solution to this requirement would be to attach the enzyme to an electrode, and to perform direct electron transfer between electrode and enzyme. This approach is particularly attractive since the actual re-oxidation of the enzyme would also produce a current signal that was proportional to the target analyte (i.e. the enzyme's substrate), so that no additional secondary reagent would be required to produce the signal in order to perform the assay.

Electron transfer to oxido-reductase enzymes is discussed in Chapter 5. The majority of oxidases contain a flavin cofactor (Fig. 8.10) plus an apoenzyme

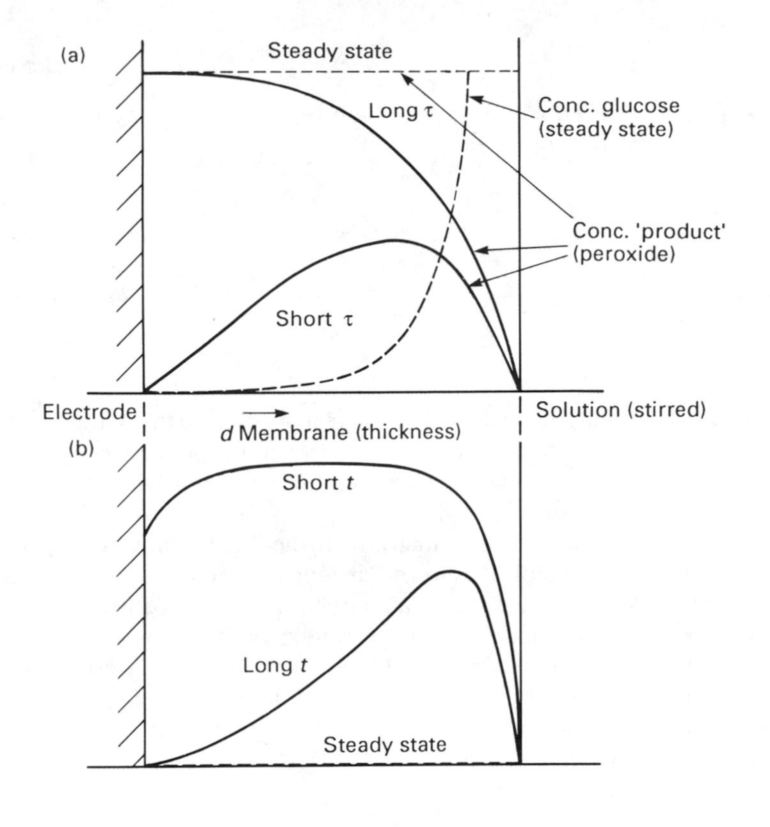

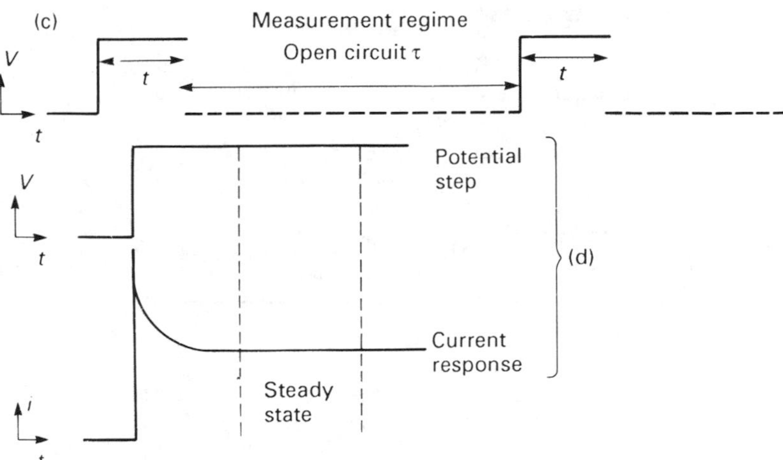

Fig. 8.8 Idealized concentration profiles, considering one phase only, emanating from a zero point at the electrode surface and imposing the boundary condition of membrane–solution interface concentration being zero. (a) Concentration profiles for given t and variable τ; (b) concentration profiles for given τ and variable t; (c) identification of the measurement regime; (d) relationship between applied potential and current response.

protein. The overall reduction of the flavoproteins involves the binding of two hydrogen atoms at the 'redox target' area:

$$+ H^+ + e^- \rightleftharpoons \qquad + H^+ + e^- \rightleftharpoons$$

Redox target

However, as was discussed in Chapter 5, electron transfer to the isolated prosthetic groups presents a very different problem to electron transfer between enzyme and electrode, and a limiting current is rarely obtained even when the enzyme is covalently attached to the electrode.

A possible solution would be to first attach the flavin redox coenzyme to the electrode and then reassemble the enzyme around the cofactor. The efficient redox behaviour of FMN at an electrode has already been established (see Fig. 5.18) and FAD adsorbed on graphite electrodes, by immersion in flavin solutions, showed similar behaviour (Fig. 8.11). In fact, in buffer solution these electrodes could be cycled repeatedly without any loss of material, but it was not possible to reconstruct the flavoenzyme from this base (Gorton and Johansson, 1980; Narasimhan and Wingard, 1985).

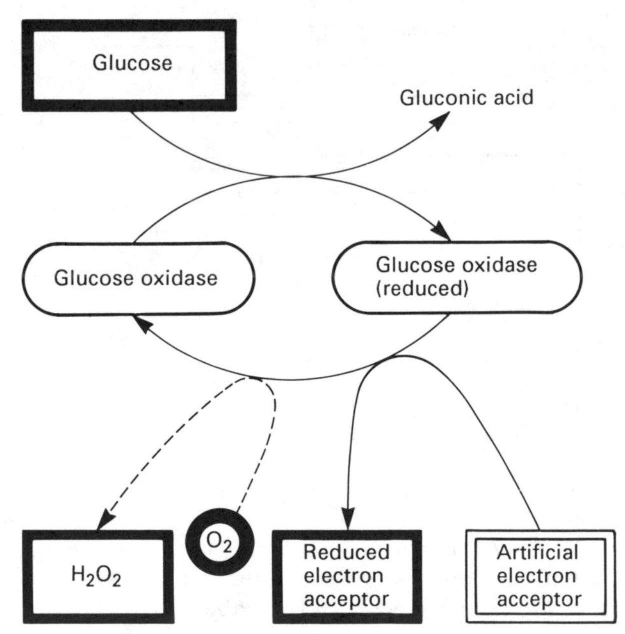

Fig. 8.9 Glucose oxidase catalysed oxidation of glucose: the introduction of artificial electron acceptors to act as mediators.

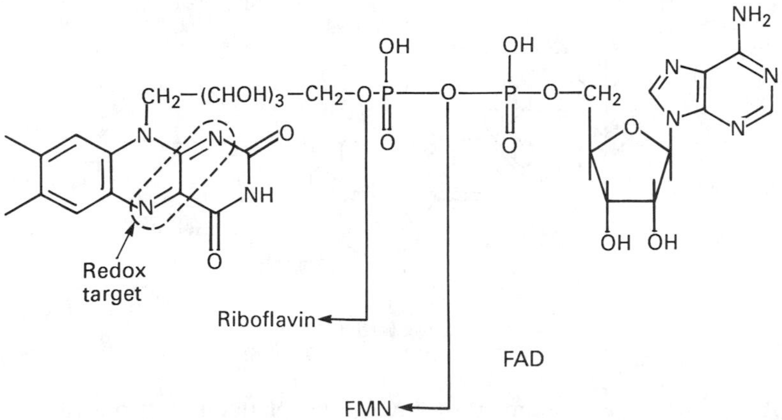

Fig. 8.10 Flavin cofactor for oxidases.

If covalent bonding of the cofactor to the electrode is considered, there are several positions on the isoalloxazine ring system where covalent coupling could occur. Since electron transport within and between molecules is facilitated by overlapping of π-orbitals, it is attractive to design an immobilization position where the π-system of the 'redox target' nitrogens (N-1 and N-5) of the flavin are in the same plane as the electrode ligand (Fig. 8.12).

Wingard (1982) proposed a Wittig reaction between aldehyde derivatized glassy carbon and 8-methyl-modified riboflavin, triphenylphosphonium riboflavin, to bring about the desired coupling. The redox-active riboflavin was then enzymically converted to FMN and then FAD, the ultimate electrode bound FAD-containing glucose oxidase finally being reported (Wingard, 1983). This technique has still however to make a great impact on biosensor technology,

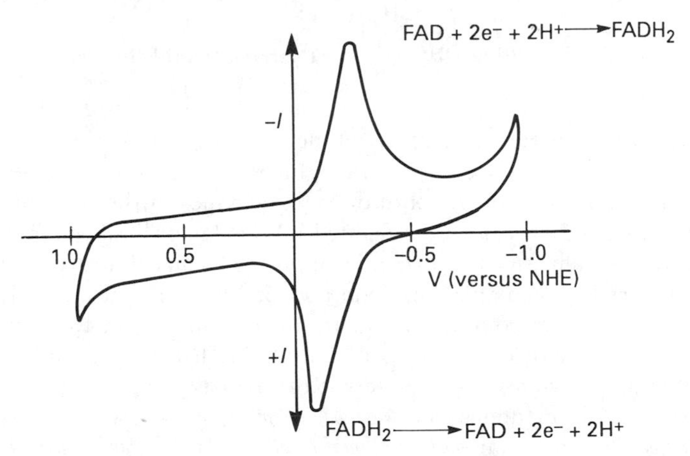

Fig. 8.11 Cyclic voltammogram for FAD adsorbed on a graphite electrode.

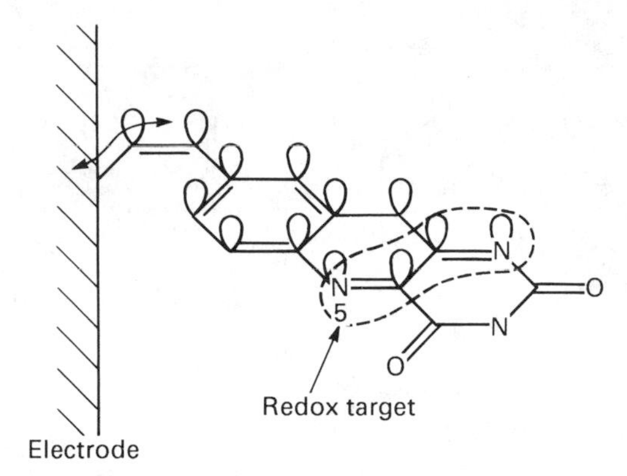

Fig. 8.12 Design for attachment of flavin cofactor to an electrode. The 'route' to the redox target (p–π orbitals shown above the plane only).

perhaps because it relies entirely on the ability to reconstruct the desired enzyme from cofactor and apoenzyme.

Another possibility would be to modify the electrode surface in such a way as to facilitate electron transfer between enzyme and electrode. Direct electron transfer has been claimed for glucose oxidase attached to graphite via a cyanuric chloride bridge (Ianniello *et al.*, 1982) or to glassy carbon modified with (aminophenyl)boronic acid (Narasimhan and Wingard, 1986):

-CONH- B—OH OH
Boronic acid bridge

—O— N N N Cl Cl
Cyanuric chloride bridge

However, in both examples the activity of the resulting immobilized enzyme was uncertain, and in the presence of substrate glucose, no current increase could be unambiguously assigned to re-oxidation of enzyme. Although this finding questions whether the enzyme or dissociated FAD cofactor has been immobilized at the electrode, evidence for the former is supported by an exhaustive comparison of the differences between this electrode and one prepared from FAD. It was postulated therefore that complexation with the enzyme occurs through the diol groups of the carbohydrate portion, and that this leads to a conformational change in the apoenzyme which renders it catalytically less active.

It is clear from this evidence that *the mode of immobilization of the enzyme must not occlude the active site or cause any conformational change that will alter its catalytic activity.*

These problems present certain limitations to attempting electron transfer between electrode and enzyme by this approach. Other solutions have also been sought to by-pass or solve the paradox, while enabling the efficient regeneration of oxidized enzyme, assay of the enzyme's substrate and independence from oxygen.

The central problem is finding routes for electron transfer and in principle there are two such pathways: the direct route already discussed and the indirect route, using low-molecular-weight *redox mediators* to shuttle electrons between electrode and enzyme. As may be seen from Fig. 5.17 (see Chapter 5) the redox potential for the $FMN/FMNH_2$ or $FAD/FADH_2$ couple is -0.22 (versus NHE). However, as already established, any attempt to perform direct electron transfer from the electrode to the flavoenzymes requires substantial overpotentials. The use of a 'messenger' to transport the electrons circumvents this failure and since a mediator with a redox potential close to the flavin couple is used, the electrode may be operated at considerably lower potentials than required in the assay via hydrogen peroxide, thus obviating possible interference of the signal by other electroactive species.

This is analogous to the redox catalysis described in Chapter 5, except that here the model must be extended to involve the reaction between enzyme and substrate:

$$\mathrm{Enz_{ox}} + \mathrm{S} \rightleftharpoons \mathrm{P} + \mathrm{Enz_{red}}$$

$$\mathrm{R} \rightleftharpoons \mathrm{O} + ne^- \qquad \text{(mediator)}$$

$$\mathrm{O} + \mathrm{Enz_{red}} \rightarrow \mathrm{R} + \mathrm{Enz_{ox}} \qquad \text{(mediated electron transfer to enzyme)}$$

For the case where the enzyme is saturated with substrate (i.e. $[S] \gg K_M$), all the enzyme is essentially involved in complex formation with substrate and the model can be approximated to that already described, thus allowing a pseudo-first-order rate constant to be obtained from a plot of i_k/i_d versus $(k/a)^{1/2}$.

Many such electron acceptors for glucose oxidase have been proposed, but without doubt the most successful class of mediators to date remain those based on the ferrocene (η^5-bis-cyclopentadienyl-iron) molecule, a transition metal π-arene complex consisting of an iron sandwich of cyclopentadienyl (Cp) rings:

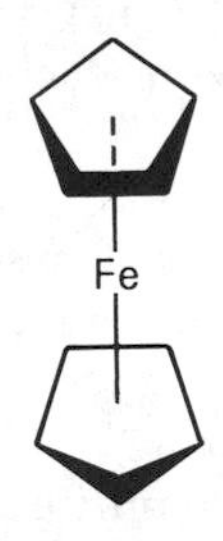

The well-behaved redox electrochemistry of ferrocene has traditionally made it an electrochemical standard. A redox potential of 165 mV versus SCE can be varied by derivatization to give both more anodic and more cathodic redox species, with a wide range of solubilities. This means that the choice of mediator employed becomes dependent on:

(i) ease of immobilization; stability of species in use
(ii) kinetics of reaction with enzyme
(iii) redox potential (low to reduce interference)

As may be seen from Fig. 8.13, the trends in redox potential do not necessarily reflect the kinetics of the reaction with enzyme, so that the final choice of mediator is a compromise between these competing factors. However, whichever derivative is chosen, the general scheme for the system remains the same:

$$\text{glucose} + \text{GOD}_{(ox)} \rightarrow \text{gluconolactone} + \text{GOD}_{(red)}$$
$$\text{GOD}_{(red)} + 2\text{Fecp}_2\text{R}^+ \rightarrow \text{GOD}_{(ox)} + 2\text{Fecp}_2\text{R} + 2\text{H}^+$$
$$2\text{Fecp}_2\text{R} \rightarrow 2\text{Fecp}_2\text{R}^+ + 2e^- \rightarrow \text{electrode current}$$

It should be noted from this scheme that the whole assay is a cyclic reversible process, with *no* net depletion of the electroactive species. It therefore offers significant advantages over the irreversible electrochemistry of oxygen. However, while solving the problem of oxygen dependence by the use of an artificial mediator, the construction of a biosensor requires the immobilization of an additional component—i.e. the attachment of the mediator at the electrode surface, together with the enzyme.

Ferrocene-mediated Electrodes

If immobilization of the ferrocene is to be achieved by adsorption onto an electrode surface, then the solubility of not only the ferrocene derivative but also its *ferricinium ion* is critical in determining the lifetime of the electrode. The rate constant (k) for the reaction of glucose oxidase with its natural mediator oxygen is quoted as $1.5 \times 10^6\,\text{l}\,\text{mol}^{-1}\,\text{s}^{-1}$ (25°C, pH 7) (Cass *et al.*, 1984), which is greater than most of the observed rates with the ferricinium ion derivatives. In solution however, a rate constant of $k = 2.01 \times 10^5\,\text{l}\,\text{mol}^{-1}\,\text{s}^{-1}$ can be obtained for ferrocene monocarboxylic acid, giving a reaction competitive with the oxygen mediated kinetics. Adoption of this derivative in a biosensor device is also favoured by its low redox potential (275 mV versus SCE), but is inhibited by the solubility of the ferricinium ion, allowing leaching of the mediator into the sample solution, away from the electrode.

1,1′-Dimethyl ferrocene provides a more suitable compromise to these competing factors. Different types of such electrode systems have been proposed. Based on a graphite electrode, for example, are the following:

(*i*) *Carbon Paste-mediated Electrode*
Carbon paste electrodes normally comprise a paste of graphite powder in liquid paraffin. Inclusion of the mediator in this paste gives a mediating electrode, upon which the enzyme is retained with a suitable membrane.

(*ii*) *Graphite Foil-mediated Electrode with Covalently Immobilized Enzyme* (Cass *et al.*, 1984)
Graphite foil sealed with the edge plane exposed is doped with 1,1′-dimethylferrocene, by evaporation from a solution of the ferrocene compound

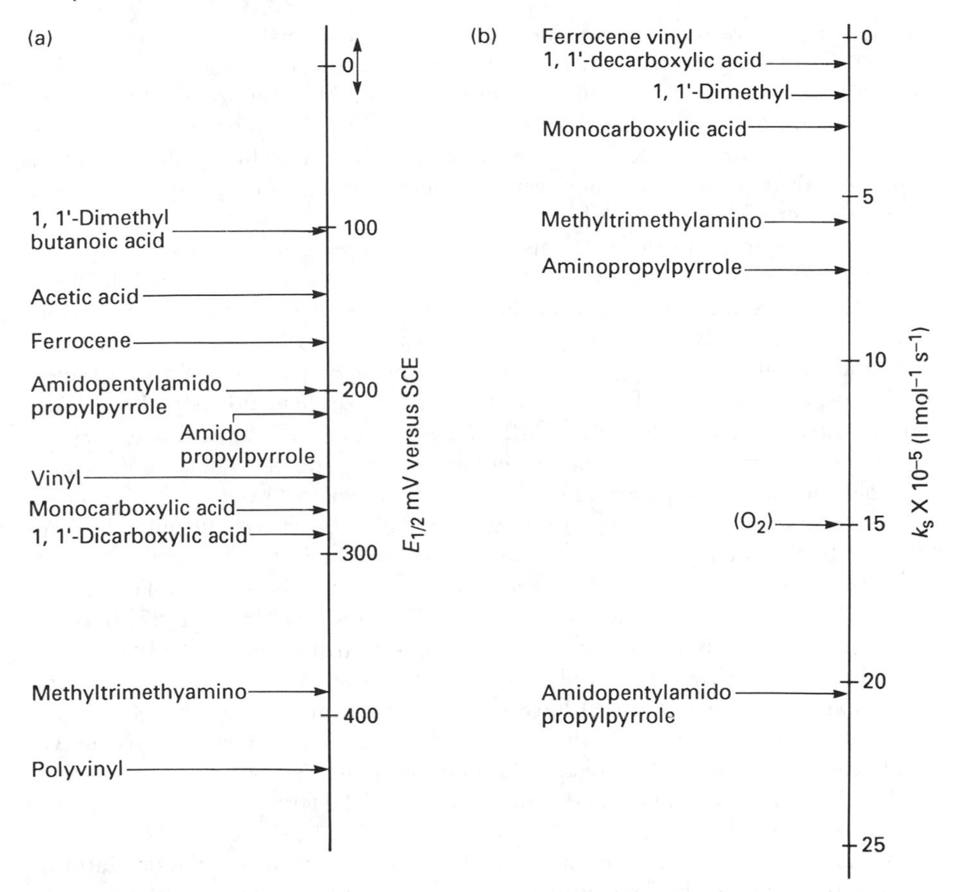

Fig. 8.13 (a) Redox potential of ferrocene derivatives; (b) kinetics of glucose oxidase oxidation by ferricinium ion derivatives.

> on the surface. Advantages over the previous system are obtained for this construction by the covalent immobilization of the enzyme, for example, via a carbodiimide-type reaction with surface carboxyl groups (see Fig. 1.12f). The enzyme reaction is reported to occur predominantly at the edge plane, so that it is important that this should remain exposed in the electrode mounting.

These electrodes were operated at 160 mV (versus SCE) (E° 1,1′-dimethylferrocene 100 mV versus SCE) and gave a linear response to glucose over the range 1–30 mM. In plasma samples from diabetic patients the limits for hyper- and hypoglycaemia usually lie within this range, thus demonstrating the relevance of the device in clinical analysis. Placing a membrane over the electrode introduces a diffusional barrier to the system, so that the linear range can be extended to 80–100 mM.

One of the pertinent questions, however, in the adoption of mediators that substitute for oxygen is whether oxygen dependence has been eliminated. This is particularly relevant where the kinetics of the artificial mediator are slower than oxygen. The system described here showed a decrease in current of the order of 5% between anaerobic and air-saturated conditions. It is however proposed that since venous blood and plasma have lower concentrations of oxygen, the likely error would be considerably less and would compare favourably with the errors obtained from ascorbate interference when making the measurement via hydrogen peroxide at +0.7 V!

In endeavouring to develop this electrode for *in vitro* use in fermentation monitoring, it was found that even using the carbodiimide immobilization procedure, a significant amount of the enzyme was immobilized by adsorption and thdrefore easily lost during use (Brooks *et al.*, 1987/88). An improved immobilization method was developed where the enzyme, modified with sodium periodate to produce surface carbonyl groups, was covalently attached to an amine-activated electrode via a Schiff's base; then additional modified enzyme was bound using the bifunctional reagent adipic dihydrazide. The authors report a 6-fold increase in response and a lifetime extended to 14 days.

Improvement of the competition with oxygen by more efficient electron transfer between enzyme and mediator might be achieved if the mediator were integral with the enzyme, i.e. attached covalently to the surface. Ferrocene-modified glucose oxidase has been reported (Delgani and Heller, 1987; Bartlett and Whitaker, 1987b) to show direct electro-oxidation at an unmodified electrode, sometimes with an improved catalytic current over the solution species mediation.

Lenhard and Murray (1978) have reported that immobilization of ferrocene derivatives to a modified platinum electrode (Fig. 8.14) gives a redox active electrode in acetonitrile showing *very slow* decay kinetics, second order with respect to the ferricinium ion. This infers some mobility of the ferricinium species which might, in principle, reduce the lifetime of the device.

With the mediator–enzyme conjugate described above, where the mediator is fixed to the enzyme, the ferricinium ion would presumably no longer be mobile, and the decay characteristics could be reduced. Although enzyme modification is required to produce these conjugates, the actual production of the biosensor derived from these modified enzymes would only require immobilization of this single species rather than a multistage immobilization of enzyme and separate mediator.

If manufacturing considerations are taken into account then multistep chemical modifications and immobilizations are unfavourable. It has already been noted, however, that considerable improvements can be made over the adsorption model in the sensor operation. Ultimately, it is the relevance of a particular refinement to an assay environment which is critical. 'One off' disposable discrete assays are not concerned with long-term operational stability nor leaching of the mediator due to the solubility of its oxidized form. They do however require good storage stability characteristics, as indeed do all biosensors. Long-term stability during use is more relevant to on-line or continuous monitoring and the economics concerned with the development of sensors for the different applications are considerably different.

In addition to immobilization of the biorecognition molecule, localization of

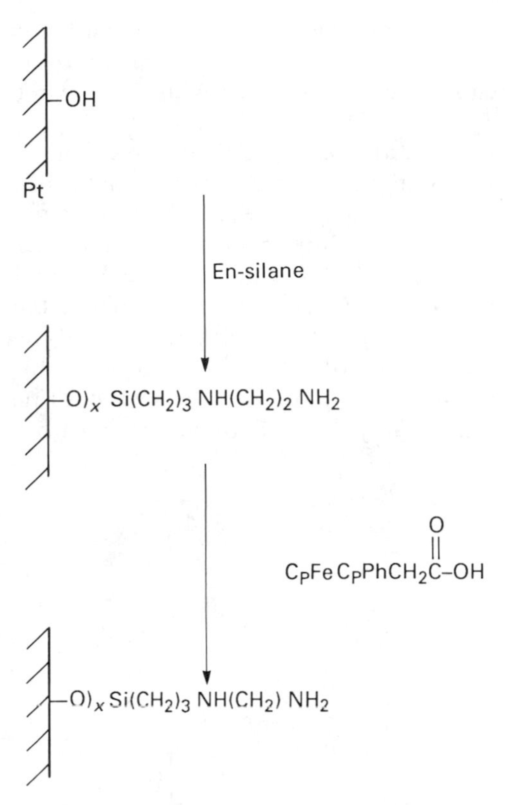

Fig. 8.14 Covalent attachment of ferrocene (CpFeCp) derivatives to platinum electrodes.

the redox couple in the vicinity of the electrode surface, while maintaining its reversible electron transfer activity causes some problems. Several lines of attack have been developed in addition to those already discussed, which produce polymeric coatings at the electrode surface and contain the redox centre as part of the polymer or retained within the polymer by electrostatic binding. Charge can be transported through the polymer layer by interaction between redox centres, diffusion of counterions or diffusion of the electroactive species (see Chapter 5).

Redox polymers have also been considered, but since the heterogeneous charge-transfer rate constant is typically 10^3 smaller than the monomeric solution species, they do not appear to offer an ideal phase for mediation between enzyme and electrode.

If the electroactivity of the redox centres can be increased by altering the conductivity of the retaining matrix, then a case presents itself for the adoption of modified conducting polymers of polyacetylene, polypyrrole, polyaniline, polythiopene, etc. The use of a polypyrrole or polyaniline matrix has already been discussed as an immobilization medium for the enzyme, glucose oxidase. These polymers are particularly suitable since it may be formed by electro-oxidation from *aqueous* medium, thus enabling biochemicals to be dissolved in the polymerization solution and become incorporated in the polymer matrix during polymer growth.

The possible film degradation highlighted by Umana and Waller (1986), caused by polarization of the polypyrrole electrode in the presence of peroxide at +0.8 V, would be circumvented by the use of a mediated system with a lower redox potential. However, an additional requirement of this mediated electrode would be the co-immobilization of the mediator (Foulds and Lowe, 1988). Redox mediator derivatives of polypyrrole were prepared from ferrocene–pyrrole conjugates, to give ferrocene amidopropylpyrrole (FAPP) and ferrocene amidopentylamidopropylpyrrole (FAPAPP) (Figs 8.13 and 8.15). In solution these mediator conjugates showed second-order homogeneous rate constant estimates which compared favourably with ferrocene monocarboxylic acid. Indeed

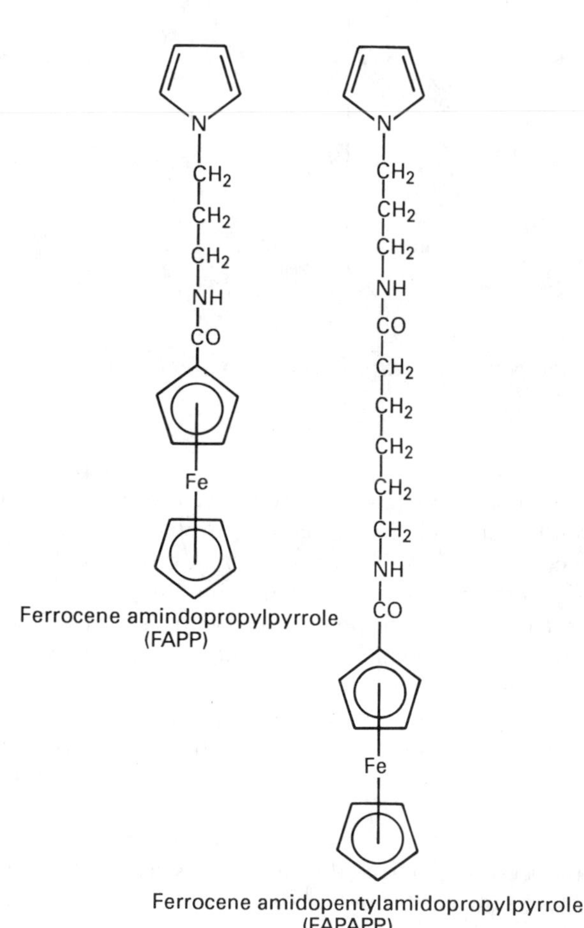

Fig. 8.15 Ferrocene-modified pyrrole derivatives for enzyme mediation.

FAPAPP gave a value of $2.1 \times 10^6\,l\,mol^{-1}\,s^{-1}$, which is faster than oxygen-mediated kinetics. However, by analogy with the polyvinylferrocene polymer described, the heterogeneous charge-transfer rate constants for the polymeric films were somewhat reduced. In fact, electropolymerization of these derivatives was self-limiting, probably due to the low conductivity of the *N*-substituted pyrroles and a better film was formed as a 1 : 1 copolymer with pyrrole. This copolymer could be formed in aqueous solution in the presence of enzyme and mediator, but the catalytic efficiency of the pyrrole–ferrocene mediator was reduced compared with the solution species. Clearly, although this system realizes all the components of a reagentless biosensor, the requisition of a heterogeneous charge transfer in these films for the enzyme–electrode mediation is not altogether straightforward. Indeed, a copolymer of FAPP–pyrrole containing glucose oxidase showed a 38% increase in current response between N_2-saturated and air-saturated solutions for 30 mM glucose, while in solution FAPP is competitive with the oxygen mediation.

Glucose Dehydrogenase-linked Glucose Assay

In the quest to eliminate oxygen dependence from the enzyme-linked glucose assay, modification of most of the constructional and measurement parameters has been considered. However, throughout these modifications the enzyme glucose oxidase has been retained, although perhaps the ideal enzyme for an amperometric glucose sensor would not be oxygen-dependent. The NAD^+-independent enzyme glucose dehydrogenase contains the prosthetic group pyrroloquinoline quinone (PQQ) and is thus included in the class of enzymes known as *quinoproteins*. Glucose dehydrogenase is reported to show an enzyme-substituted reaction mechanism (D'Costa *et al.*, 1984), and is capable of catalysing the *glucose-dependent* reduction of various redox systems but excluding flavins (nor does it reduce O_2). Indeed, in solution the second-order rate constant for the ferrocene monocarboxylic acid-mediated glucose dehydrogenase reaction was an order of magnitude greater than that for glucose oxidase (D'Costa *et al.*, 1986). Biosensors were prepared with this enzyme using 1,1′-dimethylferrocene adsorbed on porous graphite foil electrodes and covered with an enzyme layer, covalently bound via a carbodiimide reaction (Fig. 8.16a). This configuration was comparable with that already described by Cass *et al.* (1984) for glucose oxidase, but higher enzyme loadings were obtained with an additional layer, formed by glutaraldehyde crosslinking with enzyme (Fig. 8.16b). These electrodes showed a 95% response time within 25 s, which is significantly faster than the corresponding glucose oxidase electrode, and a K_M of 1.5 mM for the 'low-loaded' electrode and K_M of 8 mM for the high loading.

Duine and Frank (1981) have reported the possibility of reconstructing the active enzyme of a quinoprotein by combining the prosthetic group with the apo-enzyme. Earlier in this chapter, attempts to reconstruct enzymes with the prosthetic group attached to the electrode were discussed. It would appear therefore that this quinoprotein, glucose dehydrogenase, may be a suitable

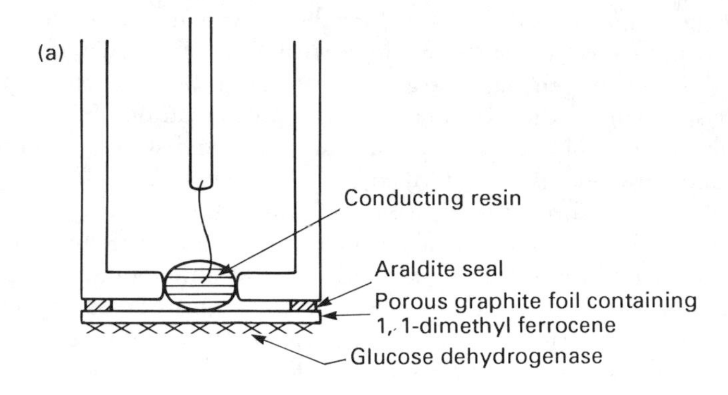

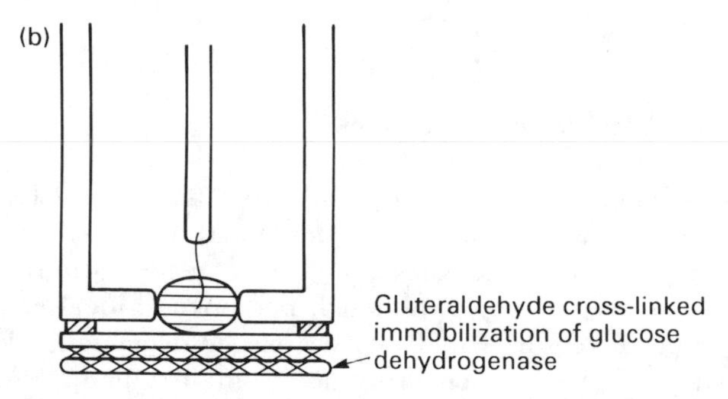

Fig. 8.16 Glucose dehydrogenase-linked glucose assay (D'Costa *et al.*, 1986). (a) Basic electrode with adsorbed enzyme; (b) electrode with additional enzyme layer, immobilized by glutaraldehyde crosslinking.

candidate for this approach, particularly in view of the fast electron transfer kinetics that have been reported.

In general, reports on the use of this enzyme indicate its suitability for adoption in enzyme-linked glucose assay. Commercial devices however must consider more than research results in their design. Availability and cost of individual components are important factors and, where glucose oxidase can *adequately* be employed, it is unlikely that it will be substituted by a considerably more expensive enzyme, glucose dehydrogenase.

Enzyme-label Linked Assays (Glucose)

The solution to the enzyme biosensor allows the estimation of not only the enzyme's substrate, but also opens up a whole family of enzyme-linked assays. The glucose oxidase–ferrocene system, for example, has been applied to an immunoassay which was tested for the antigen lidocaine (DiGleria *et al.*, 1986). As discussed earlier, not only ferrocene but many of its derivatives act as electron acceptors for glucose oxidase. For this application, the hybrid lidocaine ((α-diethylamino)-2,6-dimethylacetanilide) derivative of 1′,3-dimethylferrocene carboxylic acid was employed:

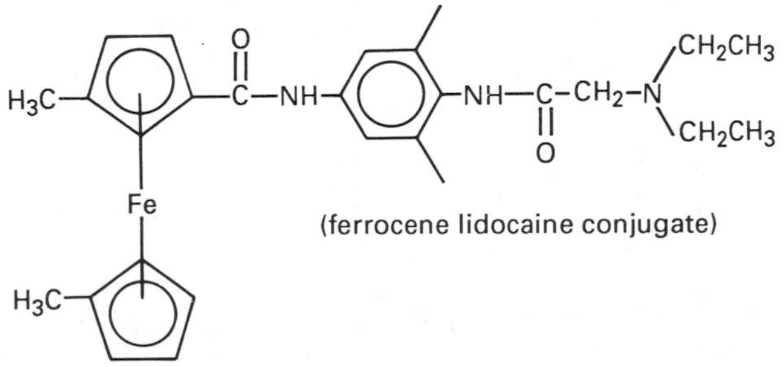

(ferrocene lidocaine conjugate)

The principle of the assay is summarized in Fig. 8.17. In the presence of anti serum, anti-lidocaine, a complex is formed (Fc: Ag–Ab) with the ferrocene–lidocaine conjugate (Fc: Ag). Unlike the free conjugate, this complex does not act as a mediator for the glucose oxidase reaction. However, on addition of a sample containing lidocaine, a competition is established between the ferrocene–lidocaine (Fc: Ag) and the sample lidocaine (Ag) for the antiserum, thus releasing a proportion of the Fc: Ag from the Fc: Ag–Ab complex, and allowing it to mediate the enzyme reaction. This example has been reported for a solution assay (Fig. 8.17), where glucose oxidase, glucose, ferrocene–lidocaine mediator and anti-lidocaine must all be added to the assay solution in known concentrations and the current increase due to the catalytic activity of the 'released' ferrocene–lidocaine mediator measured at an unmodified electrode. Development of these assay methods in biosensor devices should however ultimately be feasible.

Another 'competitive' type application for the glucose oxidase–ferrocene electrode is in conjunction with other enzyme systems that also use glucose as a substrate—thus establishing a competition for the available glucose. For example ATP can be detected by employing the pathways shown in Fig. 8.18(a). In the presence of the enzyme hexokinase and ATP, glucose is diverted from the glucose oxidase pathway to glucose-6-phosphate. This competition can therefore be employed to assay ATP (Davies *et al.*, 1986), or be further extended to include creative kinase activity, as shown in Fig. 8.18(b).

While ferrocene and its derivatives appear to offer most of the requirements of a mediator, other redox couples have also been proposed and tested. Gyss and Bourdillon (1987) have investigated an enzyme label immunoassay model of anti-rabbit IgG—rabbit IgG—where the enzyme label, glucose oxidase, activity is measured by the catalytic current due to

$$\text{glucose} + \text{benzoquinone} \xrightarrow{\textit{glucose oxidase}} \text{gluconic acid} + \text{hydroquinone}$$
$$\text{hydroquinone} \rightarrow \text{benzoquinone} + 2H^+ + 2e^-$$

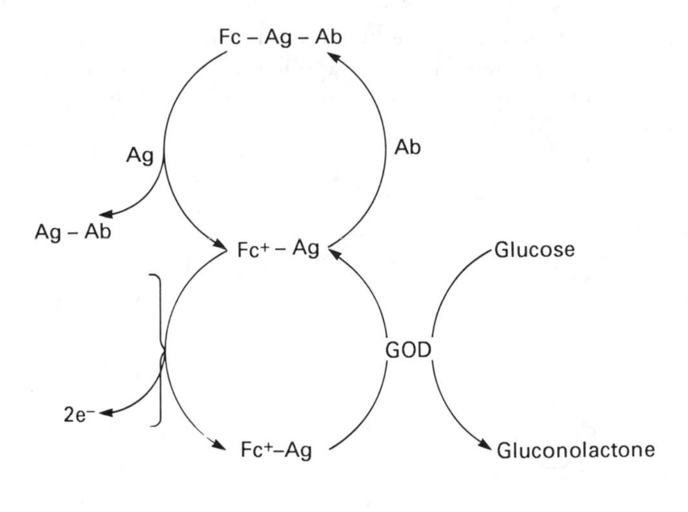

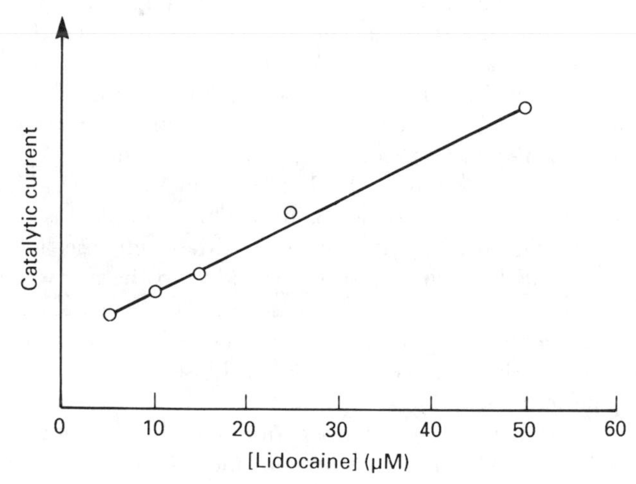

Fig. 8.17 Amperometric immunoassay of lidocaine (DiGleria *et al.*, 1986).

The antibody was adsorbed on an electrochemically pretreated carbon electrode to give the selective immunoreactive sensor surface. A sandwich-type assay was performed, so the resultant electrode was incubated first with sample antigen and then the glucose oxidase labelled antibody. The bound glucose oxidase could be estimated at 350 mV versus SCE in a solution containing benzoquinone and a saturating solution ($[S] \gg K_M$) of glucose. The estimated limit of detection for this model, *assuming minimal non-specific binding* was around 10^{-12} mol l^{-1}. As with all such immunoassay techniques, the efficient 'blocking' of the sensor surface, in order to prevent non-specific binding of one of the reagents involved, is paramount to the accuracy of the determination.

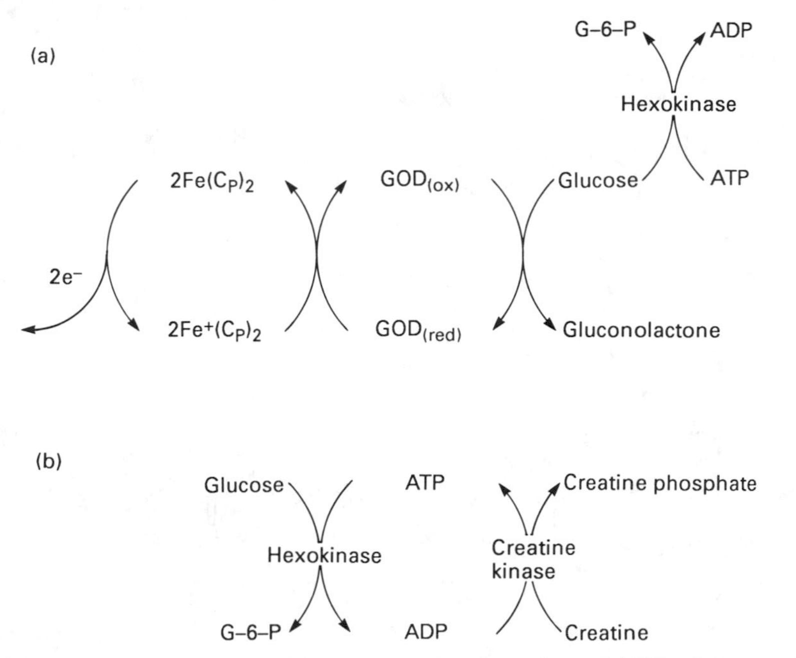

Fig. 8.18 (a) Competitive assay scheme for detection of ATP; (b) extension of (a) to give an assay of creatine kinase activity.

The spectrum of redox mediators for enzymes other than glucose oxidase is also wide and these will be discussed later in the chapter. However, even the glucose system has been explored for further viable alternatives to ferrocene.

Among other mediators, tetrathiafulvane (TTF) has been demonstrated to compare favourably with ferrocene (Turner *et al.*, 1987), but this compound is more usually associated with the conducting salts, based on stable charge-transfer (CT) complexes formed by the partial transfer of an electron from a donor to an acceptor. Typical donors and acceptors are shown in Fig. 8.19. The donor–acceptor complexes are metallic in nature at room temperature and Kulys *et al.* (1980) demonstrated that direct electron transfer from the flavin redox group of glucose oxidase to the CT complex could be achieved (Kulys *et al.*, 1980; Cenas and Kulys, 1980).

Provided that the donors and acceptors are packed into segregated stacks of alternating donors and acceptors, typical conductivities for these CT complexes at 300 K are $< 10^{-6}\,\Omega^{-1}\,cm^{-1}$. These electrodes are attractive not only because they achieve 'direct' electron transfer but also for their relative ease of preparation: finely divided charge-transfer complex is mixed with powdered enzyme and compressed into a disc electrode.

Electrodes based on glucose oxidase adsorbed on NMP^+TCNQ^- or NMA^+TCNQ^- are reported to retain their activity for > 100 days over a good linear range for glucose (Kulys *et al.*, 1980). Albery *et al.* (1985, 1987) found that

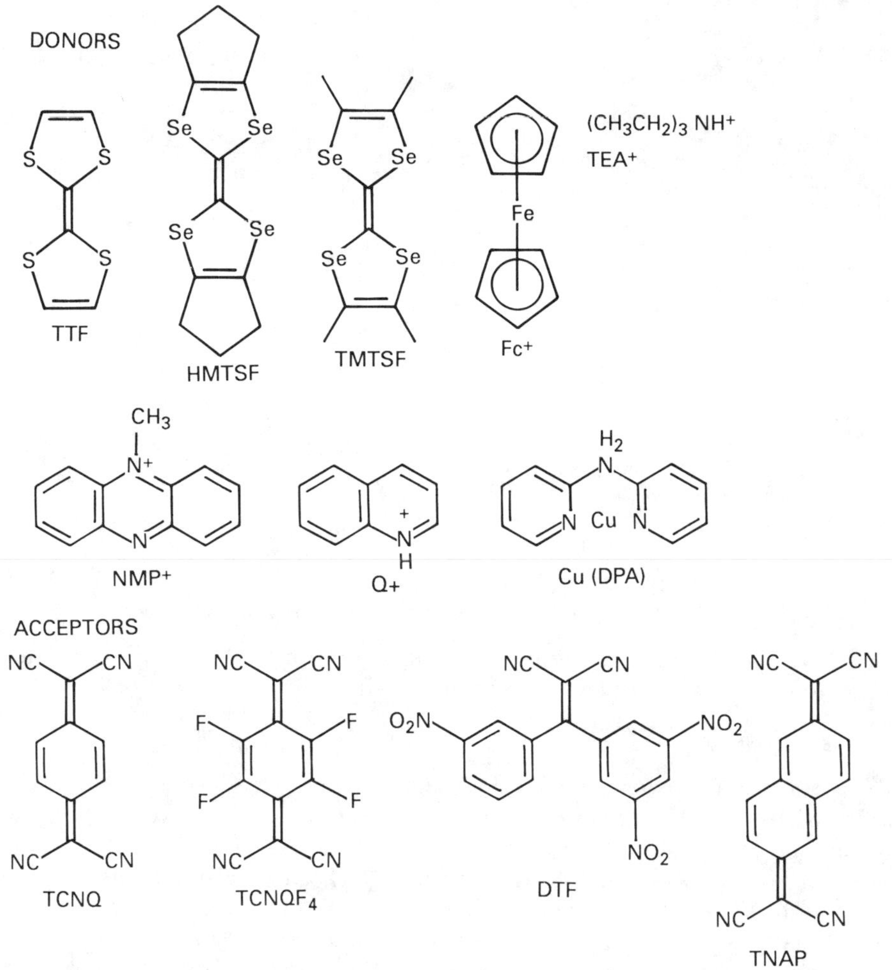

Fig. 8.19 Typical donors and acceptors for charge-transfer complexes.

nearly all the salts showed electrochemical activity with glucose oxidase, but confirmed that the $TCNQ^-$ salts of NMP^+, TTF^+ and Q^+ had the best electrochemical characteristics in terms of oxidation voltage and background current. They also concluded that the electrode kinetics were slower than the enzyme kinetics and established that in a membrane-covered electrode, the reaction was under diffusion control by the membrane and was independent of the enzyme or the electrode kinetics.

The mechanism involved in the action of these conducting salts has been hotly debated since one can imagine both *mediated* and '*direct*' electron transfer mechanisms. Kulys and Cenas (1983) have shown that TCNQ in solution is an efficient oxidant for glucose oxidase and TTF^+ has already been discussed as a mediator. Since the electron-transfer process with the flavin oxidases *only* occurs at the redox potential, Kulys and co-workers (Kulys and Razumas, 1983; Kulys and Samalius, 1983; Razumas *et al.*, 1984) interpret the reaction as mediated, due to slight dissolution of the organic metal at the electrode surface. Other groups and other enzyme systems have interpreted experimental evidence in terms of 'direct' electron exchange between redox centre and the electrode. In rotating-disc electrode experiments Albery *et al.* (1985) could find no evidence for loss of the conducting salt from the electrode. This negative evidence supported a more direct electron-transfer mechanism, rather than one where a soluble mediator exists for a finite lifetime.

NADH/NAD$^+$ Systems

In comparison with the flavo enzymes, the electrochemistry of nicotinamide cofactors is rather better behaved. Oxidation of NADH to NAD$^+$ can be carried out on a graphite electrode. Even so, oxidation occurs at potential substantially more positive than the standard electrode potential.

Blaedel and Jenkins (1976) developed an amperometric lactate electrode where lactate dehydrogenase (LDH) was co-immobilized with NAD$^+$ by glutaraldehyde crosslinking on a dialysis membrane (Fig. 8.20). The lactate-dependent reaction was

$$CH_3CHOHCOO^- + NAD^+ \xrightarrow{LDH} CH_3COCOO^- + NADH + H^+$$

and the NAD$^+$ was recycled by electrochemical oxidation at 0.75 V (versus Ag/AgCl) at glassy carbon, giving a current that was proportional to lactate concentration (0–5 mM). However, operation at such high overpotentials is undesirable and decreased the lifetime of the system. In fact, the electrode could be adequately operated at 0.45 V (versus Ag/AgCl) but the response was considerably reduced and did not compensate for the extension in operational lifetime.

The need to overcome these difficulties has led to the investigation of suitably modified surfaces. Like the other electrode systems already discussed, the

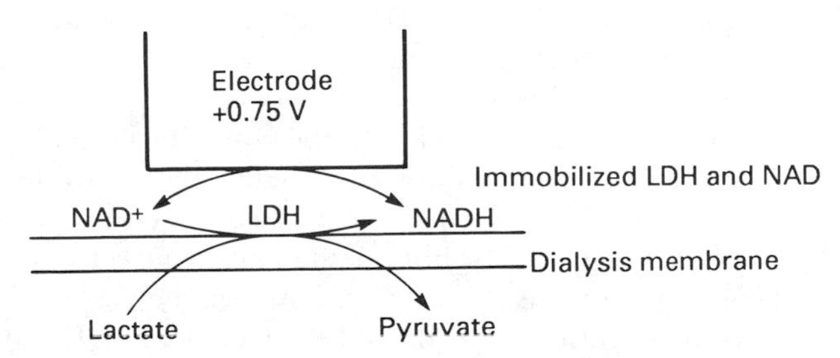

Fig. 8.20 Lactate electrode using the NAD-dependent lactate dehydrogenase (LDH) system.

Table 8.4 Enzyme electrodes using conducting salts

Salt	*Enzyme*	*Substrate*
TTF^+ $TCNQ^-$	Xanthine oxidase	Xanthine
	Glucose oxidase	Glucose
	D-Amino acid oxidase	D-Alanine
	L-Amino acid oxidase	Phenylalanine
	Choline oxidase	Choline betaine aldehyde
NMP^+ $TCMQ^-$	Xanthine peroxidase	H_2O_2
	Glucose oxidase	Glucose
	NAD^+ (>250 possible examples)	NAD^+-dependent enzymes and substrates

conducting salts are also suitable for many other redox enzyme-linked assays. Some of the possibilities that have been explored are shown in Table 8.4. A particularly large family is the NAD^+-dependent enzymes. NMP^+TCNQ^- is able to re-oxidize NADH to NAD^+ (Fig. 8.21), sothat it can potentially be applied in any of the 250 or more NAD^+/NADH-dependent dehydrogenases! Appled to the analysis of bile acids, and employing 3-*o*-hydroxysteroid dehydrogenase, the electrode demonstrated a considerable time improvement over conventional spectrophotometric assay (Albery *et al.*, 1987). Used inconjunction with alcohol dehydrogenase (Kulys an Razumasv, 1983) the system detected ethanol, and with yeast-ethanol dehydrogenase (Albery and Bartlett, 1984) a number of alcohols could be detected, including ethanol, butan-1-ol and propan-2-ol.

Often the preparation of modified surfaces has also been achieved very simply, by the passive adsorption of the modifier on the surface of the electrode. Much of the early work mentioned in Chapter 5 was based, for example, on organic species such as hydroquinones, catechols and redox dyes—Medola's Blue has proved particularly successful (Jaegfeldt *et al.*, 1981; Huck *et al.*, 1984; Gorton *et al.*, 1984).

Oxidation of the NADH at these modified electrodes is catalysed by the immobilized redox couple. If mediation is fast, then the conversion of NADH to NAD^+ occurs near the redox potential for the modifier, and at a considerably reduced overpotential for the NAD^+/NADH couple. An additional advantage of this indirect mechanism is that NADH can be oxidized with a greater efficiency, thus increasing the effective lifetime of the electrode.

The commonest problem encountered with these modified electrodes has been their poor stability, due to loss of mediator adsorbed on the electrode surface. For example hexacyanoferrate adsorbed on an electrode or in solution acts as a redox agent for NADH (Powell *et al.*, 1984), but does not give an electrode with a long-term stability. Inorganic polymer films may, however, be electro-deposited on nickel electrodes from a solution containing hexacyanoferrate(II) ions (see also Chapter 5). These films will oxidize NADH with a decrease in overpotential of the order of 150 mV compared with unmodified nickel (YonHin and Lowe, 1987).

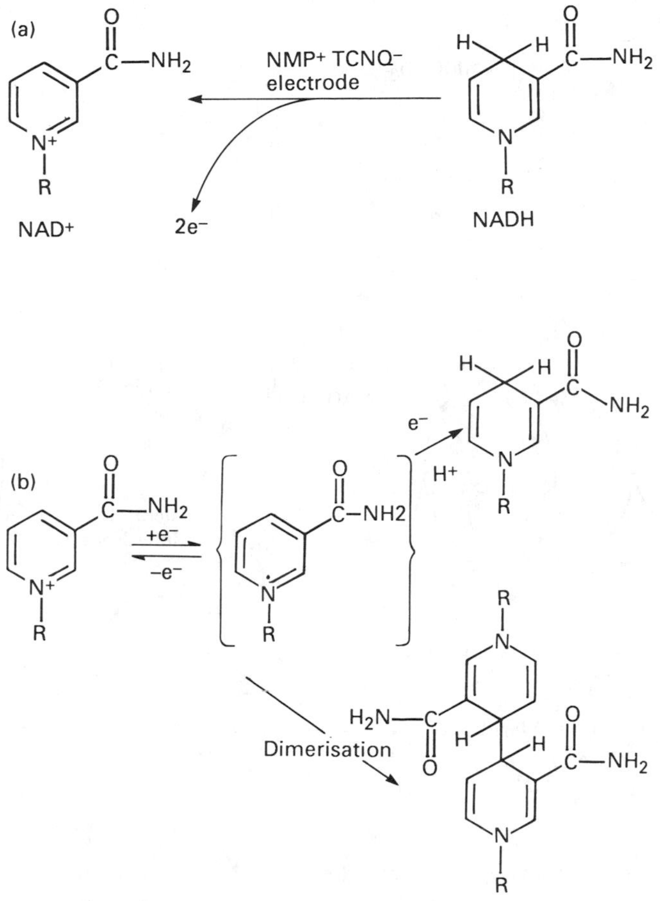

Fig. 8.21 (a) Oxidation of NADH via the conducting salt electrode NMP^+TCNQ^-; (b) electrochemical reduction of NAD^+ giving the dimer rather than NADH as the major product.

The redox reaction involves the migration of alkali-metal ions into and out of the film, since charge neutrality is maintained in the film by cationic doping (Fig. 8.22a). The films are, however, stable in storage and use, and their feasibility has been demonstrated (Fig. 8.22b) for the detection of ethanol using the NAD-dependent enzyme, alcohol dehydrogenase, immobilized on the nickel hexacyanoferrate by glutaraldehyde crosslinking. The electrode performed satisfactorily in both ethanol samples in buffer solution and ethanol-doped urine samples, showing no loss of catalytic activity over a period of several days. This system requires the addition of NAD^+ to the sample solution, so that further development of the device as a biosensor would clearly also involve the immobilization of the nicotinamide cofactor.

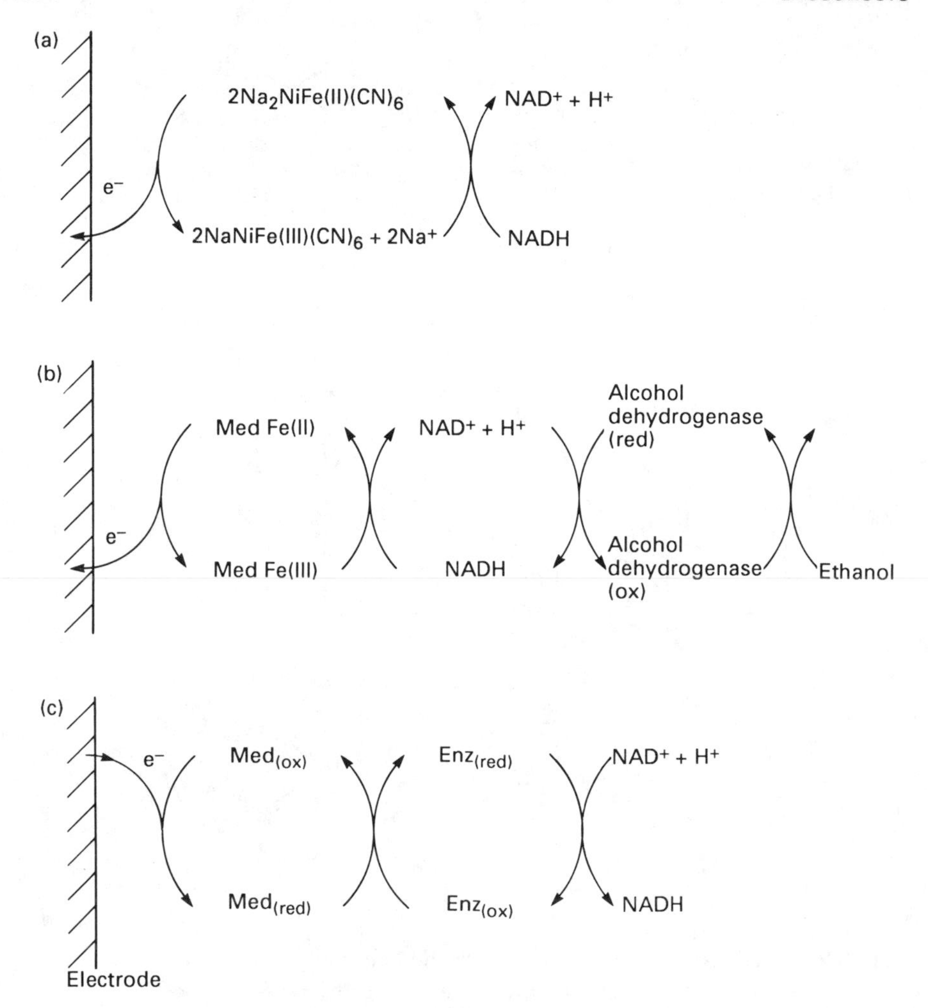

Fig. 8.22 (a) Hexacyanoferrate-mediated oxidation of NADH; (b) assay of ethanol via a hexacyanoferrate-mediated NAD-dependent alcohol dehydrogenase system; (c) mediated enzyme regeneration of NADH.

Comtat *et al.* (1988) have also proposed an assay for L-carnitine from the sequence

$$(CH_3)_3N^+CH_2CHOHCH_2COO^- + NAD^+ \rightleftharpoons (CH_3)_3N^+CH_2COO^- + NADH + H^+$$

$$NADH + 2[Fe(CN)_6]^{3-} \xrightarrow{\textit{diaphorase}} NAD^+ + 2[Fe(CN)_6]^{4-} + H^+$$

where the NADH is determined amperometrically via the oxidation current for hexacyanoferrate(II). L-Carnitine transports long-chain fatty acids through the mitochondrial membrane. Human carnitine deficiency leads to intracellular lipid accumulation and heart and muscle myopathies.

The reverse reaction, that is the reduction of NAD^+ to NADH would also be a useful transformation to be able to perform electrochemically. Direct electron transfer to NAD^+ occurs easily but reaction with proton is inhibited by such side reactions as free-radical dimerization (see Fig. 8.21) and adsorption onto the electrode surface. The dimerization process is partially deleterious since the reaction kinetics are so fast (the second-order rate constant has been estimated to be $>10^{-7} mol^{-1} s^{-1}$), and some considerable effort has been expended in trying to direct this process away from the dimerization to produce NADH. However, none of the electrode modifiers that successfully reduce the overpotential for the NADH oxidation process shows sufficient improvement for the reverse reaction.

Owing to these difficulties, attention has been directed at mediated enzyme regeneration (Fig. 8.22c). The enzyme that is employed must have a redox centre capable of reducing nicotinamide cofactor and the mediator must be able to transfer electrons between the electrode and the enzyme. One such mediator is methyl viologen,

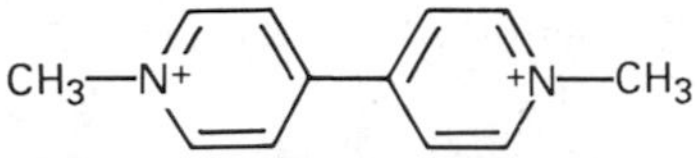

CH_3—N^+ ⟨ring⟩—⟨ring⟩ ^+N—CH_3

which has a redox potential of -0.45 V, and in the presence of diaphorase (lipoamide dehydrogenase) the electrochemical reduction of NAD^+ can be effected producing NADH. Similarly, ferredoxin NADP reductase can be employed with methyl viologen to regenerate NADPH (DiCosmo *et al.*, 1981).

Redox Proteins

CYTOCHROME C

In the same way that the nicotinamide redox couple acts as a cofactor for some redox enzymes, the redox protein, cytochrome C (see Fig. 2.10), is a protein mediator for a number of oxido-reductases. The Fe(II/III) centre of cytochrome C can be reduced/oxidized at unmodified electrodes, but the electrode kinetics are too slow to be included in a biosensor. One of the first modified electrodes, however, employing gold modified with a monolayer of 4,4′-bipyridyl (Eddowes and Hill, 1977) facilitated the electron transfer between electrode and cytochrome C.

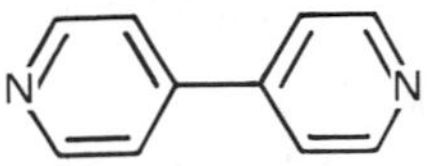

N⟨ring⟩—⟨ring⟩N (4, 4'-bipyridyl)

Unlike mediated electrode transfer, 4,4′-bipyridyl is not electrochemically active at the potential employed to achieve electron transfer to cytochrome C. The mechanism involved in the electron transfer at this modified electrode is not therefore altogether clear, but there is evidence that the protein becomes adsorbed and desorbed at the electrode surface (Albery *et al.*, 1981) and that the orientation of the bipyridyl molecules is important. It is suggested that they are orientated vertically with respect to the electrode and that excessive potentials cause re-orientation to a horizontal alignment destroying the catalytic activity of the modified electrode (Uosaki and Hill, 1981).

Azurins constitute another class of redox proteins with a copper redox centre and showing very slow electron transfer at gold and graphite electrodes. Carbon paste and 4,4′-bipyridyl ground together and compacted into an electrode, however, gives a modified electrode capable of achieving electron transfer to azurin close to the redox potential (+312 mV) (Dhesi *et al.*, 1983).

As with the cytochrome C redox protein, it is suggested that the adsorption–desorption process at the modified electrode involves hydrogen bonding between specific lysine residues on the surface of the protein and the nitrogen atoms of the promoter. Indeed modification of the surface side chains in cytochrome C inhibited the electrochemical charge transfer step at the modified electrode (Higgins and Hill, 1985).

The gold–bipyridyl cytochrome C system has been employed in enzyme-based sensors directed at various analytes. For example, incorporated with carbon monoxide oxido-reductase, reaction with carbon monoxide produces an electrode current (Turner *et al.*, 1984) (Fig. 8.23).

Many other electron-transfer *promoters* have also been examined and Allen *et al.* (1981) have concluded that a successful promoter requires:

- a surface active group with 'acceptor' properties, e.g. involving N, S or P
- weakly basic or anionic functional group suitably orientated

Electron transfer to redox proteins such as cytochrome C has also been facilitated with electrochemically *active* modifiers. An interesting approach to this solution is found in the benzadiamine based organics with $-Si(OMe)_3$ functionalities in the side chains, capable of exploitation in order to covalently anchor the mediator to the electrode by reaction with surface –OH groups. N,N,N′,N tetrakis-(trimethoxysilyl-3-propyl)-1,4-benzenediamine(**I**) and N,N-dimethyl-N′-ethyl-(trimethoxysilyl-4-butyl)-1,4-benzenediamine(**II**) have been shown to

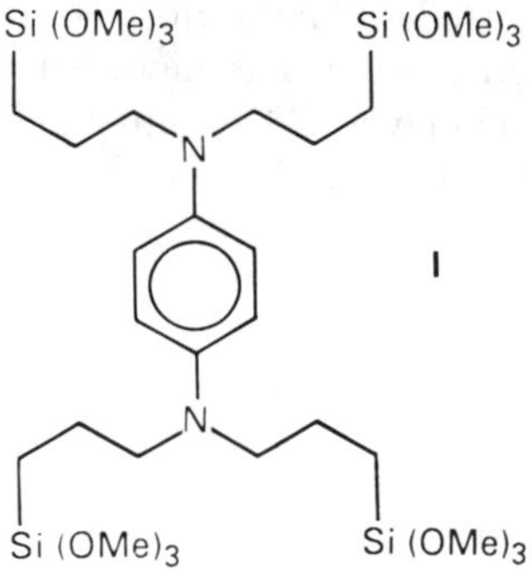

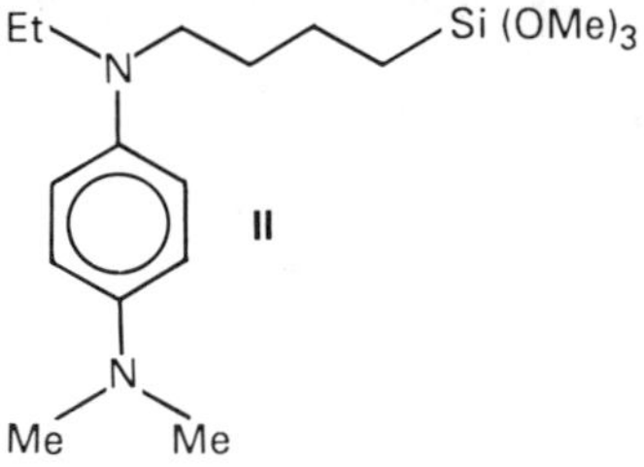

effect the interconversion of cytochrome C Fe(III/II) close to the standard redox potential (+0.04 V; E°+0.02 V versus SCE) with an initial mediation efficiency

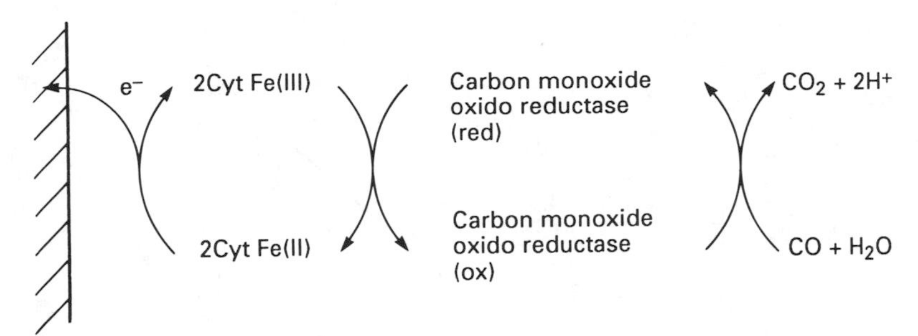

Fig. 8.23 The cytochrome C redox protein employed at a gold bipyridyl electrode together with the enzyme carbon monoxide oxido-reductase to monitor carbon monoxide.

of 100%. However rapid degradation of the electrocatalysis precludes application of these modifiers.

A redox mediator based on ferrocene with a silane anchoring point

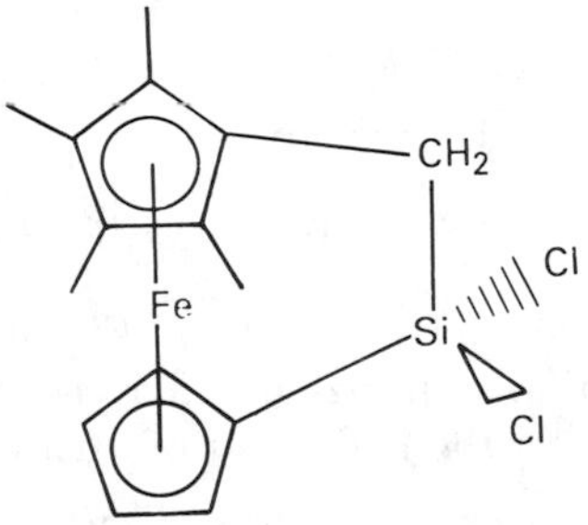

provides a considerably more durable system with the same redox potential. Either this monomer or its polymer, anchored to Pt (Chao *et al.*, 1983) are capable of providing better long-term improvement in the electrochemical response of cytochrome C.

Amperometric Determination of Enzyme-linked pH Changes

The use of pH-sensitive electrochemical redox processes was investigated by Kirstein *et al.* (1983), as a means of following hydrolase-catalysed reactions. The authors report hydrazine to be a particularly suitable pH indicator since $\log I$ is directly proportional to pH in the Tafel region of the electrochemical oxidation. This compares favourably with the activity of an ideal hydrogen ion-sensitive electrode showing a response of 59.14 mV/decade.

The mechanism of the electro-oxidation of hydrazine has been found to be both

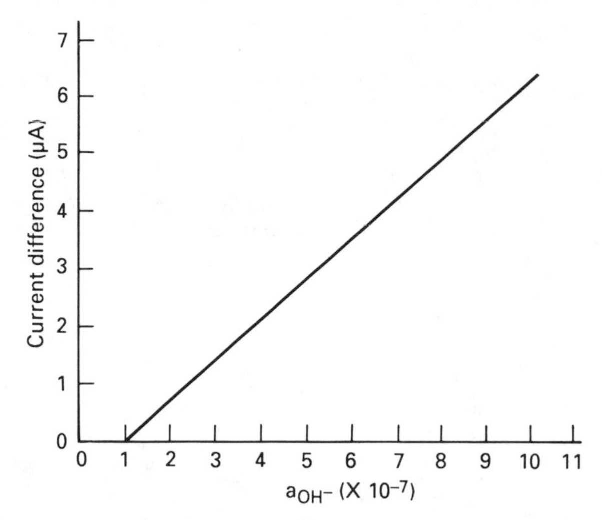

Fig. 8.24 Amperometric pH response at a hydrazine electrode (from data in Kirstein *et al.*, 1983).

pH dependent and electrode-material dependent and has been simply described by two reactions

$$N_2H_5^+ \rightarrow N_2 + 5H^+ + 4e^- \qquad (pH < 7)$$
$$N_2H_4 + 4OH^- \rightarrow N_2 + 4H_2O + 4e^- \qquad (pH > 7)$$

where the electroactive 'starting material' varies between $N_2H_5^+$ and N_2H_4 depending on pH.

The linear pH dependence of this system (Fig. 8.24) makes it particularly suitable for use in conjunction with enzyme reactions that involve a change in pH. For example urease catalyses the hydrolysis of urea giving an increase in pH:

$$3H_2O + (NH_2)_2CO \xrightarrow{urease} 2NH_4^+ + HCO_3^- + OH^-$$

Cast in polyvinylalcohol, a urease membrane may be formed, which is mounted over the end of a Pt electrode in a buffer solution containing hydrazine and EDTA as chelating agent to prevent inhibition of the urease by silver ions from the reference electrode (Kirstein *et al.*, 1985). The electrode showed a 20 s response time, with a linear response to urea between 0.8 and 35 $mmol\,l^{-1}$ at the pH optimum for the enzyme (pH 7.0–7.4). The authors report the successful operation of this sensor in urea monitoring of dialysis patients (Scheller *et al.*, 1987). Estimation of urea in dialysate and serum allows more effective adjustment of dialysis time and better patient handling.

Amperometric Determination of Bioaffinity Reactions

Bioaffinity interactions between binding proteins and their counterparts involve non-covalent forces and may be described by an equilibrium reaction with a characteristic association constant, indicating the specificity of the complexation.

For example biotin and dethiobiotin show similar affinity for avidin:

biotin + avidin $\rightleftharpoons$ biotin–avidin ($K_a = 10^{15}$ M)
dethiobiotin + avidin $\rightleftharpoons$ dethiobiotin–avidin ($K_a = 2 \times 10^{12}$ M)

However, the biotin analogues, lipoic acid and 2(4-hydroxyphenyl)azobenzoic acid (HABA), have a much reduced affinity for avidin:

lipoic acid + avidin $\rightleftharpoons$ lipoic acid–avidin ($K_a = 1.4 \times 10^6$ M)
HABA + avidin $\rightleftharpoons$ HABA–avidin ($K_a = 1.7 \times 10^5$ M)

Biotin or dethiobiotin would therefore displace HABA or lipoic acid from an avidin complex.

This type of competitive replacement can be employed to devise a bioaffinity sensor. For example, avidin immobilized at an electrode is complexed with an *enzyme-labelled* biotin analogue. The sample biotin will displace the labelled analogue in proportion to its concentration. The remaining complexed enzyme-labelled analogue can be assayed by addition of the enzyme's substrate. This is the principle that has been employed by Ikariyama *et al.* (1985) for a biotin sensor. A copolymer of 4-(aminomethyl)-1,8-octanediamine and glutaraldehyde was cast in a cellulose triacetate support to provide a reactive surface for the attachment of the protein ovalbumin (Fig. 8.25). HABA was immobilized via a carbodiimide reaction to the amino group of the protein, and the resulting membrane complexed with catalase-labelled avidin and mounted on a Clark-type oxygen electrode. The affinity electrode could then be incubated with the sample containing biotin, thus displacing some of the catalase-labelled avidin from the membrane. Subsequent removal of the electrode from the sample and estimation of the remaining complexed enzyme-labelled avidin could be achieved by addition of peroxide (see Fig. 8.25):

$$2H_2O_2 \xrightarrow{catalase} O_2 + 2H_2O$$

The rate of increase in oxygen tension is proportional to the concentration of the enzyme, giving a linear response in the range 10^{-9}–10^{-7} g mol^{-1} biotin. Incubation of the membrane complex with biotin was found in this system to proceed to completion within 10 min, and the reaction of the remaining enzyme-labelled avidin with the added peroxide, reached completion within 1 min. After each determination the bioaffinity sensor could be regenerated by incubation in an avidin–catalase conjugate solution, thus re-establishing the protein–biotin analogue complex.

This type of competition analysis requires that the analyte and the chosen analogue should show bioaffinity characteristics for the binding protein separated by several orders of magnitude; i.e. an affinity ratio for the two equilibria concerned probably $\geqslant 10^5$.

Amplification

Methods of electrode control and data capture have been discussed as a means of increasing the signal and prolonging the life of the sensor. Amplification may however be achieved by the analyte itself, it if causes access to a cycling process, so

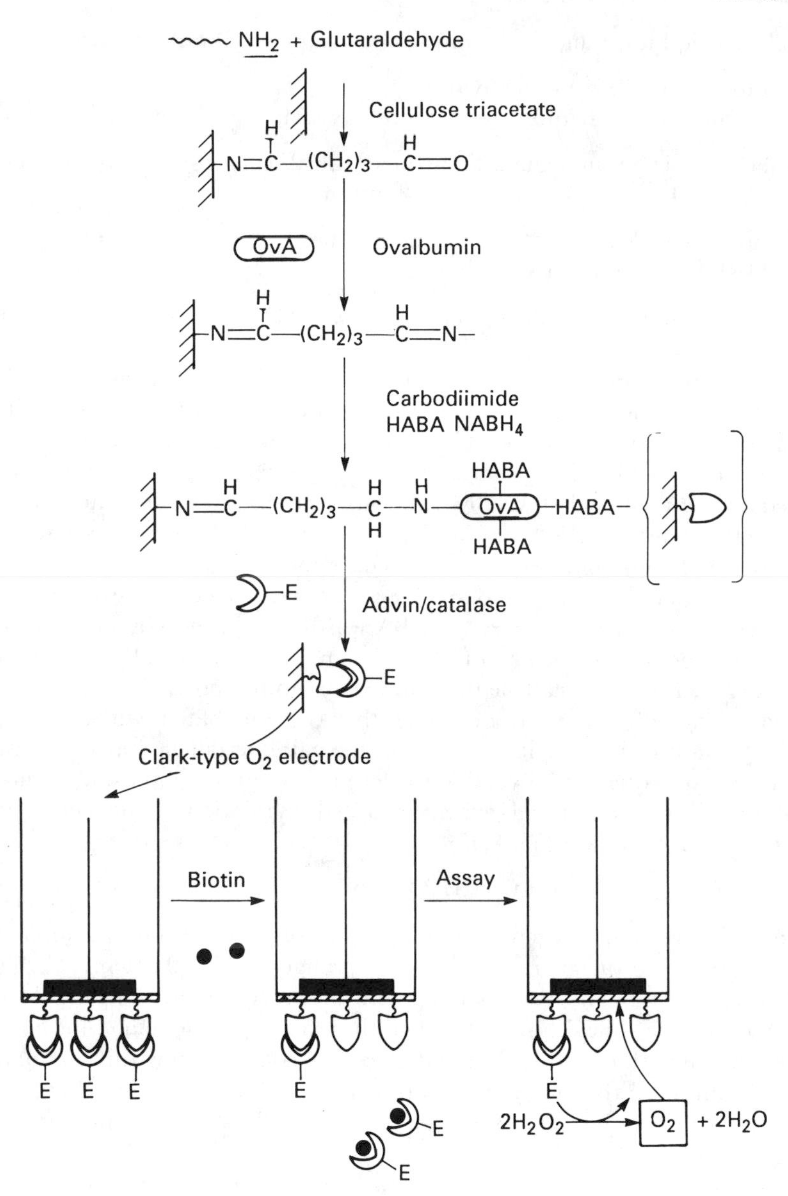

Fig. 8.25 Preparation of a bio-affinity electrode. The biotin analogue HABA is attached to the electrode and reacted with enzyme-labelled avidin. Sample biotin competes for the labelled avidin and the remaining labelled avidin, still complexed at the electrode, is assayed by addition of enzyme substrate.

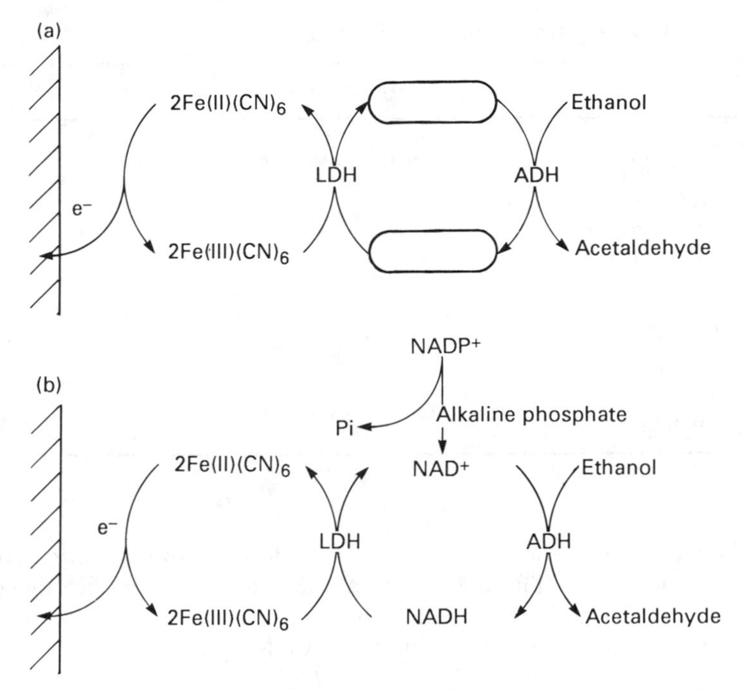

Fig. 8.26 (a) Amplification system for NAD^+; (b) linking $NADP^+$ assay to NAD^+ amplification system.

that a stoichiometric amount of product is formed and accumulated each time the analyte is turned over in the cycle. For example in Fig. 8.26(a), the system involves the two NAD-dependent enzymes, alcohol dehydrogenase (ADH) and lipoamide dehydrogenase (LDH). Entry of the missing NAD into the system will therefore drive the cycle, so that in the presence of excess ethanol, catalytic cycling of reduced and oxidized nicotinamide occurs, generating ferrocyanide which can be electrochemically oxidized at +0.45 V (versus SCE). The cycle will therefore be activated directly as an assay of NAD^+, or indirectly when NAD^+ is the product of another reaction. NAD^+, for example, is the product of the alkaline phosphatase-catalysed dephosphorylation of $NADP^+$ (Cardosi *et al.*, 1986) (Fig. 8.26b).

Similarly, the scheme proposed earlier for creatine kinase activity (see Fig. 8.18b) could be employed in an amplified ADP/ATP determination; Wollenberger *et al.* (1987) have also proposed a similar combination, replacing the creatine kinase with pyruvate kinase, and then following the pyruvate production with a third dual-enzyme system,

$$\text{pyruvate} + \text{NADH} \xrightarrow{\textit{lactate dehydrogenase}} \text{lactate} + \text{NAD}^+$$

$$O_2 \xrightarrow{\textit{lactate monooxygenase}} \boxed{\text{monitor } O_2 \text{ uptake}}$$

Table 8.5 Enzyme amplification systems

Enzymes	*Target analyte*	*Amplification*
Cytochrome B_2 Lactate dehydrogenase	Lactate/pyruvate	10
Glucose oxidase Glucose dehydrogenase	Glucose	10
Glutamate dehydrogenase Alanine aminotransferase	Glutamate	60
Pyruvate kinase Hexokinase	ADP/ATP	200
Lactate oxidase Lactate dehydrogenase	Lactate pyruvate	4000

where, in contact with the oxygen electrode, a decrease in O_2 tension could be related back to ADP/ATP concentration. Even greater amplification can be achieved by replacing the enzyme lactate monooxygenase with lactate oxidase and thus introducing a second amplification cycle:

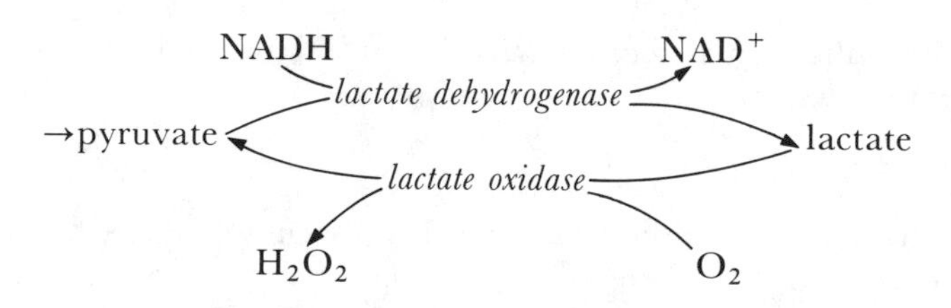

The technique has also been reported for many other multi-enzyme cyclic combinations, often achieving claims of up to a 4000-fold increase in sensitivity (Schubert *et al.*, 1985). The degree of signal enhancement varies from system to system, but the principle is always the same: *The specific substrate/cofactor for the first enzyme gives a product which is the substrate/cofactor for a second enzyme. Reaction with the second enzyme produces a product which once again is substrate for the first enzyme.* Some of the systems are highlighted in Table 8.5.

The use of the mediator in recycling schemes has also been proposed. An amperometric immunoassay for morphine has been developed by Weber and Purdy (1979). The assay was based on the competitive binding of ferrocene-labelled morphine and sample morphine or codeine with morphine antibody (Fig. 8.27). The remaining unbound morphine–ferrocene conjugate could be estimated by oxidation at a Pt electrode. However, as already discussed, ferrocene is a mediator for electron transfer between the electrode and the redox centre of various enzymes. It is apparent, therefore, that the signal could be enhanced by recycling of the ferrocene with glucose oxidase in the presence of an excess of glucose (Fig. 8.27b).

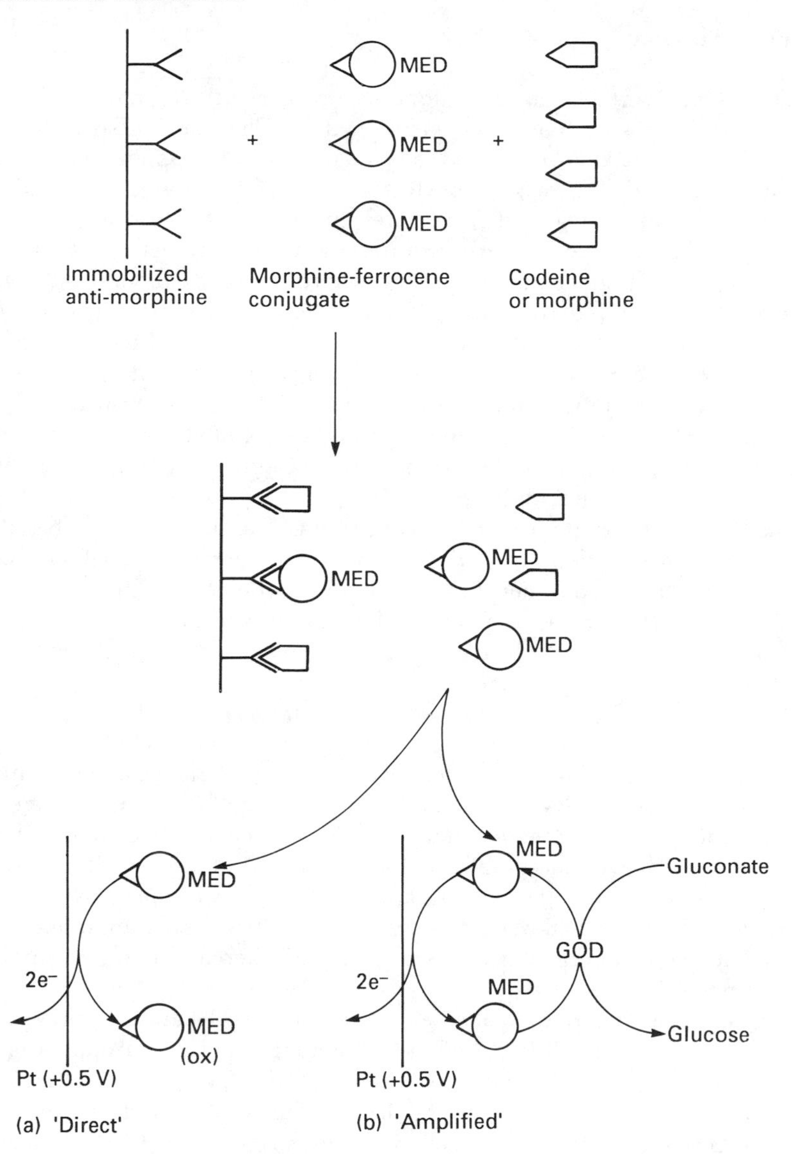

Fig. 8.27 Amperometric immunoassay for morphine using ferrocene-labelled morphine in a competition with sample morphine for antibody. (a) Direct measurement and (b) amplified measurement.

Miniaturization

Some of the advantages and techniques associated with the miniaturization of amperometric sensors have been mentioned in Chapter 5. Especially in the medical field, microfabrication is a prerequisite for implantable or *in vivo* biosensors. Fabrication of such devices is seemingly suitable for solution via currently available integrated circuit and thin film technology.

Often the assay environment requires multiple parameter determination, demanding closely spaced micro probes, where improved monitoring could be achieved by arranging the sensing surface close to the integrated electronic circuitry . . . ideally laying the sensor directly on the circuit. Modern technology allows precisely defined microstructures to be fabricated with enormous accuracy.

This is still very much an area of exploration, but is nevertheless making a significant impact on the fabrication and utilization of electrochemical sensors. Many *in vivo* enzyme systems, for example, can be anticipated extending the thin film electrodes described by Prohaska (Chapter 5).

Takatsu and Moriizumi (1987) have produced a glucose sensor based on a differential measurement of hydrogen peroxide, and produced entirely by photolithographic techniques. Enzyme was immobilized in a polyvinylalcohol (PVA) matrix which has been photosensitized with a stilbazolium group, so that on irradiation with UV light the enzyme becomes trapped in the cross-linked polymer.

The device was fabricated with two anodes and a common cathode, so that a differential measurement of hydrogen peroxide could be made between (i) the anode coated with PVA-containing active enzyme and (ii) the anode containing deactivated enzyme in PVA. The use of this 'accurate' blank rather than a bare electrode meant that elimination of reducing interferents such as ascorbate or uric acid could be achieved successfully, reducing their effect to $<3\%$. This then is an efficient solution to the 'ascorbate problem', so long as the total ascorbate + glucose signal does not then enter electrical saturation and require that the device be operated at a lower gain (and therefore lower sensitivity) in order to accommodate it.

Karube has also reported microelectrodes for H_2O_2 and O_2 produced by integrated circuit technology (Chapter 5). Vacuum deposition of a silane onto the electrode surface provides a reactive film for the covalent attachment of a biorecognition molecule. In view of the capability of this electrode to assay peroxide, this is suitable for the attachment of glucose oxidase (GOD) by a Schiff's base, for glucose determination. The GOD microelectrode is reported to show a linear response to glucose over the range 0.01–1 mg ml^{-1}, and a selectivity consistent with the enzyme selectivity (Karube, 1988).

In the same way that macro-glucose oxidase electrodes may be tuned to the potential required for peroxide or oxygen assay in order to construct a glucose-measuring device, so may this micro-device. Karube (1988) has thus reported a Teflon-membrane covered device (see Fig. 5.25), which polarized at -1.1 V gave a current response which was directly proportional to oxygen, and could be employed in enzyme-linked assay.

Glutamate, which is produced in large amounts as a seasoning agent for the food industry, can be assayed indirectly by this oxygen sensor. Glutamate oxidase was immobilized on a cellulose triacetate membrane and mounted on the Teflon membrane of the oxygen electrode. The consumption of oxygen by the enzyme reaction is proportional to substrate glutamate, and produced a drop in the electrode current due to oxygen. A linear relationship was claimed for this electrode over the range 5–50 mM glutamate.

Weetall and Hotaling (1987/88) have devised a three-electrode system for glucose determination, produced by silk-screening graphite ink onto an acrylic-coated cardboard substrate. Using the enzyme glucose oxidase and the mediator 1,4-benzoquinone, they have employed the device in an enzyme-linked immunoassay, which although strictly a *solution assay*, uses an *in situ* 'immobilization' step, obviating the need to separate bound and unbound label in the immunoreaction.

Sample antigen (IgG) and enzyme-labelled antibody (anti-IgG–Enz) were incubated and then reacted with a second antibody, covalently coupled to magnetic particles. The 'magnetic complex' containing the enzyme label was accumulated on the electrode surface by placing the card over a permanent magnet. The bound enzyme label could then be estimated by addition of glucose and mediator.

This chapter thus demonstrates the breadth of materials and principles available to the imagination of the biosensor developer. It can be anticipated that future developments will include an increasing emphasis on micro-fabrication technology, and will employ both existing and emerging materials and methods.

References

Albery, W.J. and Bartlett, P.N. (1984). *J. Chem. Soc. Chem. Comm.*, pp. 234–236.

Albery, W.J., Bartlett, P.N. and Cass, A.E.G. (1987). *Phil. Trans. R. Soc. Lond. B.* **316**, 107.

Albery, W.J., Bartlett, P.N. and Craston, D.H. (1985). *J. Electroanal. Chem.* **194**, pp. 223–235.

Albery, W.J., Eddowes, M.J., Hill, H.A.O. and Hillman, A.R. (1981). *J. Amer. Chem. Soc.* **103**, pp. 3904–3910.

Allen, P.M., Hill, H.A.O. and Hillman, A.R. (1981). *J. Amer. Chem. Soc.* **103**, p. 3904.

Bartlett, P.N. and Whitaker, R.G. (1987a). *J. Electroanal. Chem.* **224**, pp. 37–48.

Bartlett, P.N. and Whitaker, R.G. (1987b). *J.C.S. Chem. Comm.* p. 1603.

Bessman, S.P., Thomas, L.J., Kojima, H., Sayler, D.F. and Layne, E.C. (1981). *Trans. Am. Soc. Artif. Intern. Organs* **27**, p. 7.

Blaedel, W.J. and Jenkins, R.A. (1976). *Anal. Chem.* **48**(8), p. 1240.

Brooks, S.L., Ashby, R.E., Turner, A.P.F., Calder, M.R. and Clarke, D.J. (1987/1988). *Biosensors* **3**, pp. 45–56.

Cardosi, M.F., Stanley, C.J. and Turner, A.P.F. (1986). *Proceedings 2nd International Meeting on Chemical Sensors*, Bordeaux, France.

Cass, A.E.G., Davis, G., Francis, G.D., Hill, H.A.O., Aston, W.J., Higgins, I.J., Plotkin, E.V., Scott, L D.L. and Turner, A.P.F. (1984) *Anal. Chem.* **56**, pp. 667–671.

Cenas, N.K. and Kulys, J.J. (1980). *Bioelectrochem. Bioenerg.* **8**, pp. 103–113.

Cenas, N.K., Pocius, A.K. and Kulys, J.J. (1984). *J. Electroanal. Chem.* **173**, p. 583.
Chao, S., Robbins, J.L. and Wrighton, M.S. (1983). *J. Amer. Chem. Soc.* **105**, p. 181.
Churchouse, S.J., Battersby, C.M., Mullen, W.H. and Vadgama, P.M. (1986). *Biosensors* **2**, p. 325.
Clark, L.C. (1973). 'Oxygen Supply', in Kessler *et al.* (eds) Urban and Schwarzenberg, p. 120.
Clark, Jr., L.C. and Clark, E.W. (1973). *Adv. Exp. Biol. Med.* **37A**, p. 127.
Comtat, M., Galy, M., Goulas, P. and Souppe, J. (1988). *Anal. Chim. Acta* **208**, p. 295.
Davies, P., Green, M.J. and Hill, H.A.O. (1986). *Enzyme Microb. Technol.* **8**, p. 349.
Delgani, Y. and Heller, A. (1987). *J. Phys. Chem.* **91**, p. 1285.
D'Costa, E.J., Turner, A.P.F., Higgins, I.J., Duine, J.A. and Dokter, P. (1984). *Gen. Microbiol. Quart.* **11**(1), p. M11.
D'Costa, E.J., Higgins, I.J. and Turner, A.P.F. (1986). *Biosensors* **2**, pp. 71–89.
Dhesi, R., Cotton, T.M. and Timkovich, R. (1983). *J. Electroanal. Chem.* **154**, p. 129.
Diaz, A.F., Castillo, J.I., Logan, J.A. and Lee, W.-Y. (1981). *J. Electroanal. Chem.* **129**, p. 115.
DiCosmo, R., Wong, C.-H., Daniels, L. and Whitesides, G.M. (1981). *J. Org. Chem.* **46**, p. 4622.
diGleria, K., Hill, H.A.O., McNeil, C.J. and Green, M.J. (1986). *Anal. Chem.* **58**, p. 1203.
Duine, J.A. and Frank, J. (1981). *Trends Biochem. Sci.* **6**, p. 278.
Eddowes, M.J. and Hill, H.A.O. (1977). *J. Chem. Soc. Chem. Comm.*, p. 71.
Enfors, S.-O. (1981). *Enzyme Microb. Technol.* **3**, pp. 29–32.
Foulds, N.C. and Lowe, C.R. (1986). *J. Chem. Soc. Faraday Trans. I* **82**, p. 1259.
Foulds, N.C. and Lowe, C.R. (1988) *Anal. Chem.* **60**, pp. 2473–2478.
Gorton, L.G. and Johansson, G. (1980). *J. Electroanal. Chem.* **113**, p. 151.
Gorton, L.G., Jaegfeldt, H.A., Torstensson, A.B.C. and Johansson, G.R. (1984). *US Patent 4,490,464*.
Gyss, C. and Bourdillon, C. (1987). *Anal. Chem.* **59**, pp. 2350–2355.
Higgins, I.J. and Hill, H.A.O. (1985). *Essays Biochem.* **21**, p. 119.
Huck, H., Schelter-Graf, A., Danzer, J., Kirch, P. and Schmidt, H. (1984). *Analyst* **109**, p. 147.
Ianniello, R.M., Lindsay, T.J. and Yacynych, A.M. (1982). *Anal. Chem.* **54**, pp. 1098–1101.
Ikariyama, Y., Furuki, M. and Aizawa, M. (1985). *Anal. Chem.* **57**, p. 496.
Jaegfeldt, H.A., Torstensson, A.B.C., Gorton, L.G.O. and Johansson, G. (1981). *Anal. Chem.* **53**, pp. 1979–1982.
Karube, I. (1988). *Biotech 2*, Ed. Hollenberg, C.P. and Sahm, H., pp. 37–49. Gustav Fischer, Stuttgart, N.Y.
Karube, I. and Sode, K. (1988). *Analytical Uses of Immobilised Biological Compounds for Detection, Medical and Industrial Uses.* NATO ASI Series, D. Reidel Publishing Company, p. 115.
Kirstein, L., Kirstein, D., Kühn, M., Scheller, F., Chojnacki, A., Dekowski, B. and Sens, H. (1983). DDR Patent Application WPC 12Q/254 523 3.
Kirstein, L., Kirstein, D. and Scheller, F. (1985). *Biosensors* **1**, p. 117.
Kulys, J.J. and Cenas, N.K. (1983). *Biochim. Biophys. Acta* **744**, pp. 57–63.
Kulys, J.J. and Razumas, V.J. (1983). *Biocatalysis in Electrochemistry of Organic Compounds.* Mokslas, Vilnius.
Kulys, J.J. and Samalius, A.S. (1983). *Bioelectr. B* **10**(4), p. 385.
Kulys, J.J., Samalius, A.S. and Svirmickas, G.J.S. (1980). *FEBS Lett.* **114**, pp. 7–15.
Leddy, J. and Bard, A.J. (1985). *J. Electroanal. Chem.* **189**, p. 203.
Lenhard, J.R. and Murray, R.W. (1978). *J. Am. Chem. Soc.* **100**, p. 7871.

Mascini, M. (1988). *Analytical Uses of Biological Compounds for Detection, Medical and Industrial Uses*. NATO ASI Series, pp. 153–167. D. Reidel Publishing Company.
Mell, L.D. and Maloy, J.T. (1975). *Anal. Chem.* **47**(2), pp. 299–307.
Mell, L.D. and Maloy, J.T. (1976). *Anal. Chem.* **48**(11), pp. 1597–1601.
Nabi Rahni, M.A., Guilbault, G.G. and deOlivera, N.G. (1986). *Anal. Chem.* **58**, p. 523.
Narasimhan, K. and Wingard, Jr., L.B. (1985). *Appl. Biochem. Biotech.* **11**, p. 221.
Narasimhan, K. and Wingard, Jr., L.B. (1986). *Anal. Chem.* **58**, p. 2984.
Powell, M.F., Wu, J.C. and Bruice, T.C. (1984). *J. Am. Chem. Soc.* **106**, p. 3850.
Razumas, K.J., Jasaitis, J.J. and Kulys, J.J. (1984). *Bioelectr. B* **12**(3–4), p. 297.
Scheller, F., Kirstein, D., Kirstein, L., Schubert, F., Wollenberger, U., Olsson, B., Gordon, L. and Johansson, G. (1987). *Phil. Trans. R. Soc. Lond.* **B316**, p. 85.
Schubert, F., Kirstein, D., Schroder, K.L., and Scheller, F.W. (1985). *Anal. Chim. Acta* **169**, p. 391.
Shichiri, M., Yamasaki, Y., Hakui, N. and Abe, H. (1982). *Lancet* **2**, p. 1129.
Shichiri, M., Hakui, N., Yamasaki, Y. and Abe, H. (1984). *Diabetes* **33**, p. 1200.
Shinohara, H., Chiba, T. and Aizawa, M. (1988). *Sensors and Actuators* **13**, pp. 79–86.
Silver, I.A. (1976). *Ion and Enzyme Electrodes in Biology and Medicine*, pp. 189–192. Eds Kessler, M. *et al.* Urban and Schwarzenberg.
Smith, V.J. (1987). *Anal. Chem.* **59**, p. 2256.
Takatsu, I. and Moriizumi, T. (1987). *Sensors and Actuators*, pp. 309–317.
Trinder, P. (1969). *J. Clin. Pathol.* **22**, p. 246.
Tse, P.H.S. and Gough, D.A. (1987). *Anal. Chem.* **59**, pp. 2339–2344.
Turner, A.P.F., Aston, W.J., Davis, W.J., Higgins, I.J., Hill, H.A.O. and Colby, J. (1984). *Microbial Gas Metabolism*, pp. 161, 170. Ed. Poole, R.K. and Dow, D.S. New York, Academic Press.
Turner, A.P.F., Hendry, S.P. and Cardosi, M.F. (1987). *Biotech '87* **1**(3), p. 125. Pinner, Online Publications.
Umana, M. and Waller, J. (1986). *Anal. Chem.* **58**, p. 2979.
Uosaki, K. and Hill, H.A.O. (1981). *J. Electroanal. Chem.* **122**, p. 321.
Updike, S.J. and Hicks, G.P. (1967). *Nature* **214**, pp. 986–988.
Varfolomeev, S.D. and Berezin, I.V. (1978). *J. Mol. Catalysis* **4**, pp. 387–399.
Weber, S.G. and Purdy, W.C. (1979). *Analyt. Lett.* **12**, p. 1.
Weetall, H.H. and Hotaling, T. (1987/1988). *Biosensors* **3**, pp. 57–63.
Wingard, Jr., L.B. (1982). *Bioelectrochem. Bioenerg.* **9**, p. 307.
Wingard, Jr., L.B. (1983). *Biotech '83*, p. 613. Pinner, Online Publications.
Wollenberger, U., Schubert, F., Scheller, F., Danielsson, B. and Mosbach, K. (1987). *Anal. Lett.* **20**(5), p. 657.
YonHin, B.F.Y. and Lowe, C.R. (1987). *Anal. Chem.* **59**, p. 2111.

Chapter 9

Potentiometric-type Biosensors

Potentiometric Enzyme Electrodes

In the same way that amperometric biosensors may be devised around some electroactive product of an enzyme catalysis (e.g. O_2 or H_2O_2 for glucose/glucose oxidase, see Chapter 8), potentiometric biosensors are centred on ion-selective electrodes which can directly or indirectly monitor a change in ion concentration which is related to the enzyme reaction,

$$\mathrm{E+S \rightleftharpoons ES \rightarrow E+P}$$

As with the amperometric-linked enzyme electrodes, both equilibrium and kinetic methods can be used. Kinetic measurements are, however, somewhat complicated by the fact that the measured potential is logarithmically related to the concentration of the measured species. Kinetic measurements can be simplified by the following considerations (Guilbault *et al.*, 1969). The measured potential is given by the Nernst equation (Chapter 3),

$$E = \text{constant} + RT/nF \ln a$$

giving

$$\frac{\mathrm{d}E}{\mathrm{d}t} = \frac{RT}{nF}\left(\frac{1}{a}\right)\frac{\mathrm{d}a}{\mathrm{d}t}$$

Assuming that in the initial stages of the reaction the term $(1/a)$ is relatively constant, the change in electrode potential with time is directly proportional to $\mathrm{d}a/\mathrm{d}t$ or the reaction rate. In general, however, behaviour is non-linear. Brady and Carr (1980) have shown that the potentiometric electrodes display a linear response to bulk substrate concentration only up to about one-tenth of the value of K_M (Michaelis constant), and that this linear range is essentially independent of the amount of immobilized enzyme. In order to establish this model of steady-state behaviour, they superimposed diffusional considerations, in the immobilized enzyme layer (Fick's second law), on Michaelis–Menten enzyme kinetics (see Chapter 2),

$$\frac{d[S]}{dt}=\frac{k_2[E_o][S]}{(K_M+[S])}$$

giving

$$\frac{d[S]}{dt}=\frac{D_S d^2[S]}{dx^2}-\frac{k_2[E_o][S]}{(K_M+[S])}$$

where x is the distance across the membrane. In steady-state operation, $d[S]/dt \rightarrow 0$, and comparing the dimensionless values for the dependent ([S]) and independent (x) variables:

$$[\bar{S}]=\frac{[S]}{[S_o]}$$

where $[S_o]$ = bulk concentration,

$$\bar{x}=\frac{x}{d}$$

where d = total membrane thickness,

$$\bar{K}_M=\frac{K_M}{[S_o]}$$

$$\frac{D_S([S_o])^2}{d^2}\,\frac{d^2[\bar{S}]}{d\bar{x}^2}=\frac{k_2[E_o][\bar{S}][S_o]}{K_M[S_o]+[S][S_o]}$$

$$\frac{d^2[\bar{S}]}{d\bar{x}^2}=\frac{d^2 k_2[E_o][\bar{S}]}{D_S[S_o](K_M+[S])}$$

$$=\frac{k_2[E_o]d^2}{D_S K_M}\times\frac{K_M[S]}{(K_M+[S])}$$

where the term

$$\frac{k_2[E_o]d^2}{D_S K_M}=\alpha d^2$$

and is known as the dimensionless *diffusion modulus* ('enzyme loading factor' $=\alpha$).

The system response will therefore depend on the factor K_M. Because of the influence of the enzyme kinetics on the thermodynamics of the Nernst potential, the 'Nernst' slope does not have the same significance. This model only considers

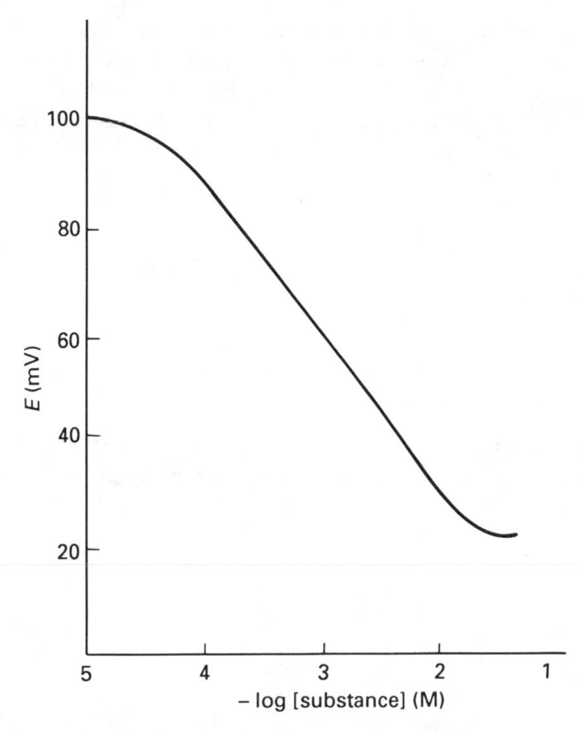

Fig. 9.1 Typical response curve for a potentiometric enzyme sensor.

simple enzyme kinetics. Extension of the model to include considerations of buffer capacity or activation or inhibition of the enzyme reaction will be made later.

The general shape of the typical response curve is shown in Fig. 9.1, with a linear portion generally in the range 10^{-4}–10^{-2} M, where deviations at low concentrations ($<10^{-4}$ M) indicate the detection limit of the potentiometric electrode, and at high concentrations (dependent on the K_M value) are due to saturation of the enzyme.

Other parameters also affect the absolute responses although do not generally alter the linear range. For example, enzyme loading can change the slope of the response. The slope increases with increase in enzyme loading until a level is reached which is no longer limiting (Fig. 9.2). pH is another important factor, since most enzymes have a pH range of maximum activity and on either side of this domain activity falls off rapidly. The pH optimum of an immobilized enzyme may not be the same as the soluble enzyme. In fact the pH profile alters depending on the carrier used. For example, bound to a negatively charged matrix, the pH optimum shifts to higher pH, while on a positively charged support, the converse is true. It can occur that the pH requirements for the base-sensing electrode are not compatible with the optimum pH for maximum activity of the electrode, in which case a compromise of pH conditions must be made!

The activity of the enzyme influences the response time of the electrode and this

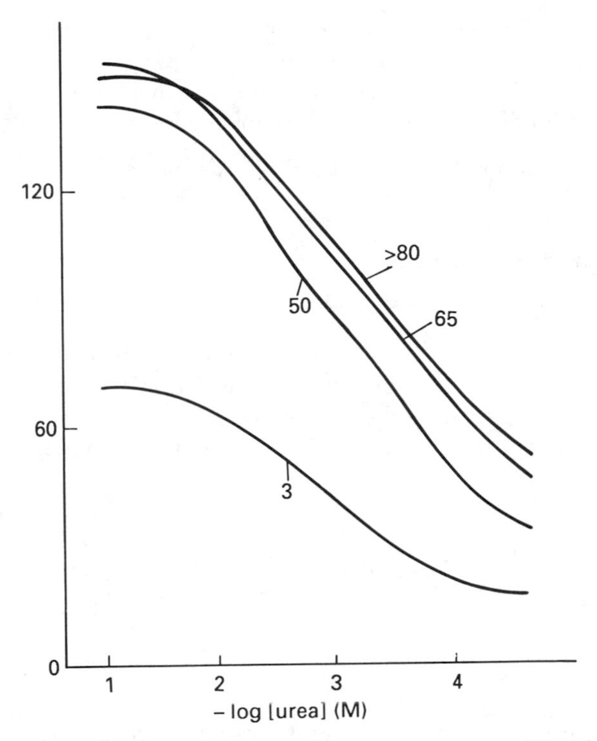

Fig. 9.2 Effect of enzyme loading on the response curve. Numbers on the curves refer to the approximate enzyme loading in units urease/cm^3 gel. (From data in Guilbault and Montalvo, 1970.)

is also affected by the thickness and permeability of the enzyme layer. Optimum response times are achieved with thin highly permeable layers using as high an enzyme activity as possible. In practice however more durable, thicker membranes, with a slower response time, are generally employed in order to prolong the lifetime of the sensor.

There are various configurations in which enzymes can be used with potentiometric electrodes for analytical purposes. *The base device for a potentiometric biosensor is generally the ion-selective electrode.* These devices have been discussed in Chapter 3.

In several cases, any one of many base sensors could be employed, so that any particular analyte could be targeted indirectly into a number of systems. Stability of the final biosensor is particularly dependent on an efficient enzyme immobilization procedure, since any slight deterioration will cause significant drift in the signal—as will any non-specific surface reaction where the measurement is concerned with changes in the accumulation of charge at the electrode surface.

Various immobilization methods have been applied with varying degrees of success for different systems:

(i) enzyme–buffer solution retained by dialysis membrane
(ii) physical entrapment of enzyme in synthetic matrix
(iii) glutaraldehyde crosslinking of enzyme and inert protein
(iv) chemical binding of enzyme to organic or inorganic support.

These will be encountered in examples discussed for each base sensor.

pH-linked Enzyme Electrodes

SUITABLE ENZYME SYSTEMS

The pH glass electrode was the first ion-selective electrode used in enzyme-linked analysis. Many enzyme reactions involve a change in hydrogen ion concentration in the vicinity of the enzyme-substrate interaction.

Hydrolysis of the β-lactam ring of penicillins, for example, is catalysed by penicillinase,

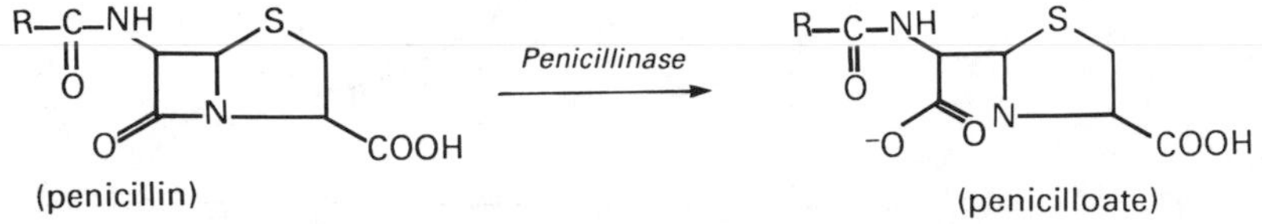

R–C(=O)–NH, S, N, O, COOH (penicillin) —*Penicillinase*→ R–C(=O)–NH, S, N, $^{-}$O, O, COOH (penicilloate)

causing an increase in the total concentration of hydrogen ion.

An electrode based on a pH probe coated with immobilized penicillinase can be employed to monitor penicillin. The probe shows near ideal behaviour with a response of 52 mV/decade over the range 5×10^{-2} to 10^{-4} M (Nilsson *et al.*, 1973).

An apparent improvement on this construction was reported by Cullen *et al.* (1974), who used penicillinase adsorbed onto a scintered glass disc, affixed to a pH electrode. The response for this electrode (56–58 mV/decade) was linear for the range 10^{-5}–3×10^{-3} M.

Similarly, a glucose assay can be based on the pH change concerned with the production of gluconic acid:

$$\text{glucose} + O_2 \xrightarrow{\textit{glucose oxidase}} \text{gluconic acid} + H_2O_2$$

With the glucose oxidase retained in solution behind a cellophane membrane, or entrapped in a polyacrylamide gel, at a pH glass electrode, the response to glucose was nearly linear (log scale) in the range 10^{-3} to 10^{-4} mol l^{-1} (Nilsson *et al.*, 1973).

At the other end of the pH scale, the same construction could be employed in conjunction with the enzyme urease for the determination of urea:

$$\text{Urea} \xrightarrow{\textit{urease}} 2NH_4^+ + HCO_3^-$$

As with the equivalent glucose electrode, the response was very dependent on the buffering capacity of the sample, but gave a relatively slow response to urea in the 5×10^{-5}–5×10^{-3} M range.

OPTIMIZATION OF IMMOBILIZATION TECHNIQUE

As already indicated, the method of immobilization make a significant contribution to operation of the sensor. Ideally the enzyme would be directly attached to the sensor's surface, without the need of retention by a membrane, and in a matrix that does not present a diffusional barrier.

Tor and Freeman (1986) have adapted an immobilization method, developed for the entrapment of whole cells, to give a gel matrix where enzyme is retained physically but where stability is much improved.

The method involves the crosslinking of linear chains of polyacrylamide hydrazine with dialdehydes such as glyoxal, in the presence of enzyme. It has been observed that polyacrylamide has a tendency to adhere strongly to glass surfaces. It would therefore appear to provide a suitable matrix for enzyme immobilization on a glass pH electrode. Moreover, the gels formed by this crosslinking procedure are highly porous and are therefore likely to provide little or no diffusion barrier.

In fact, Tor and Freeman found that a more robust immobilization matrix was formed by the crosslinking of copolymers of acrylamide and methacrylamide, since the polyacrylamide–hydrazide matrix resulted in a thick layer which cracked during crosslinking. The authors found that the best recipe was a copolymer acrylamide–methacrylamide hydrazide with the relative proportions 0.7:0.3. This produced a thin transparent durable membrane.

When penicillinase was immobilized by this method the linear response range was apparently extended, compared with previous reports, to 4×10^{-5}–10^{-3} M, and while storage stability data showed a 25% decrease in activity in 70 days (Fig. 9.3), the electrode calibration remained linear throughout the whole period.

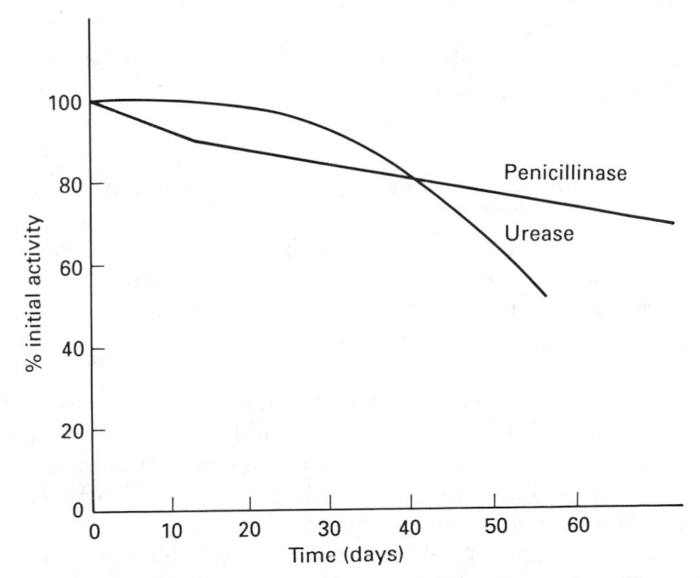

Fig. 9.3 Storage stability for urea and penicillin electrodes. (From data in Tor and Freeman, 1986.)

Urease too could be immobilized by this method to provide a sensor for urea, in this instance to give a concentration range comparable with previous reports (3×10^{-4}–2×10^{-5} M). This electrode showed a somewhat increased loss of activity of the order of 50% in 2 months, but here too the calibration curves remained linear.

It is interesting to note the apparently unchanged activity of this electrode for the first 20 days, followed by a significant loss of activity (see Fig. 9.3) and compare this with the gradual loss shown by the penicillin electrode from day 1. This demonstrates the response of different enzymes to identical immobilization procedures and identifies *the importance of tailoring the construction of the enzyme sensor to optimize the activity of the particular components under the required operation conditions.*

A particularly successful enzyme electrode employing this polyacrylamide gel immobilization procedure contained the enzyme acetylcholinesterase. Acetylcholine could be determined in the range 2×10^{-5}–10^{-3} M with a response time of <3 min. Remarkably, the data from storage stability experiments showed *no* loss of activity across the entire linear range in a 6-month period. This is obviously a great improvement on the 3–4 weeks previously obtained with a gelatin–glutaraldehyde immobilization procedure (Durand *et al.*, 1978).

Ammonia-linked Enzyme Electrodes

UREA

Urea, which has already been described in conjunction with a pH electrode, is an example of a target analyte that may be successfully linked with a number of different ion-selective electrodes:

$$\text{Urea} \xrightarrow{urease} 2NH_4^+ + HCO_3^-$$
$$NH_4^+ + OH^- \rightleftharpoons \boxed{NH_3} + H_2O$$
$$HCO_3^- + H^+ \rightleftharpoons \boxed{CO_2} + H_2O$$

Linking with an ammonia electrode is particularly favoured due to the high specificity.

An integral part of the ammonia electrode is the gas permeable polypropylene or Teflon membrane. With the urease chemically attached directly to this membrane, Guilbault and Mascini (1977) were able to produce an urea sensor with a range 5×10^{-5}–10^{-2} M, capable of a sample turnover rate of 3 min for up to 1000 samples.

The specificity of this system was demonstrated by Anfalt *et al.* (1973) who showed that even at high concentrations of alkali metal ion, the urea assay was unaffected in an electrode constructed from a commercial ammonia electrode with urease immobilized directly on the surface by glutaraldehyde crosslinking.

Aspartame, the artificial sweetener/sugar substitute, can be involved in various enzyme-catalysed conversions. Guilbault *et al.* (1988) have used the selective enzyme L-aspartase to direct aspartame estimation towards an ammonia-sensitive electrode:

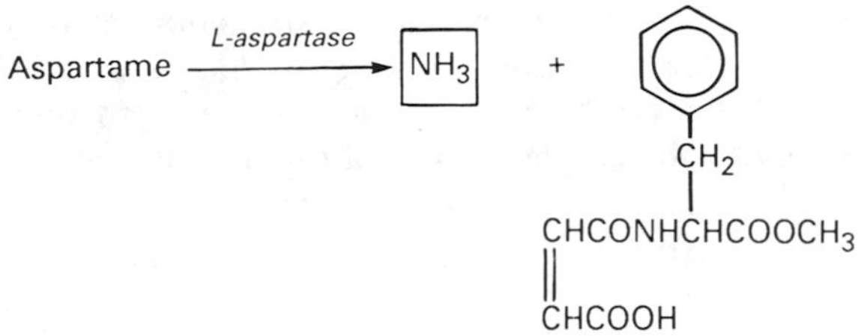

$$\text{Aspartame} \xrightarrow{\textit{L-aspartase}} \boxed{NH_3} + \text{C}_6\text{H}_5\text{-}CH_2\text{-}\overset{|}{\text{CHCONHCHCOOCH}_3} \;(\text{CHCONHCHCOOCH}_3 = \text{CHCOOH})$$

With the enzyme immobilized at an ammonia selective electrode by glutaraldehyde crosslinking, an enzyme probe could be produced with a −30 mV/decade response in the 1–10 mM range. This electrode was stable for a period of about 8 days.

Joseph (1985) has employed the alternative antimony micro-pH electrode to measure ammonia, assembled in above a 10 μm diameter capillary air gap in contact with the sample. The probe was adapted to be specific for urea, by immobilizing urease in a gel at the capillary tip by glutaraldehyde crosslinking. Typical response times were of the order of 30–45 s with a 60 s recovery time. As with any of the probes based on acidic or basic gases, the magnitude of the response depends on the buffering capacity of the sample, and in this case the equilibrium,

$$NH_3 + H_2O \rightleftharpoons NH_4^+ + OH^-$$

which defines the proportion of ammonium ions that are converted to ammonia. For this particular example, where both CO_2 and NH_3 could be involved, low pH samples alter the HCO_3^-/CO_2 equilibrium in favour of CO_2 and cause interference by this species.

CREATININE

A first attempt at the development of a creatinine sensor based on an ammonia electrode was made by Rechnitz and co-workers (Thompson and Rechnitz, 1974; Meyerhoff and Rechnitz 1976), with creatinase retained between the ammonia permeable membrane of an ammonia electrode and an outer cellophane membrane. The resulting device had a detection limit (5×10^{-4} M) which was too high to be suitable for clinical assay of serum samples.

However, immobilization of creatinase directly onto the polypropylene membrane of an ammonia electrode by glutaraldehyde crosslinking gave a creatinine electrode which was stable for 8 months and >200 assays and had a detection limit of 8×10^{-6} M (Guilbault *et al.*, 1980).

L-PHENYLALANINE

L-Phenylalanine can be selected with a very high degree of specificity by the enzyme phenylalanine ammonia-lyase in the reaction:

$$\text{L-phenylalanine} \xrightarrow{\textit{phenylalanine ammonia-lyase}} \textit{trans}\text{-cinnamate} + NH_3$$

This analyte is therefore an obvious target for an ammonia electrode-based sensor. The response was slow for this system (Hsiung *et al.*, 1977) with a very limited linear range of 10^{-4}–6×10^{-4}. Its major feature must therefore be its high specificity, due solely to the characteristics of the enzyme employed!

ADENOSINE

Adenosine assay may also be linked to an ammonia gas-sensing electrode through the enzyme adenosine deaminase.

$$\text{adenosine} + H_2O \xrightarrow{\textit{adenosine deaminase}} \text{inosine} + NH_3$$

Bradley and Rechnitz (1985) prepared an adenosine electrode with adenosine deaminase immobilized by glutaraldehyde crosslinking at an ammonia electrode. They attempted to characterize the electrode in terms of the enzyme kinetics, finding that a low level of enzyme activity was required in order for the enzyme kinetics to be rate limiting. At high loadings of immobilized enzyme, the response became limited by diffusion of ammonia through the glutaraldehyde layer. Similar correlations could be made about the effect of membrane thickness. The authors concluded that fast responses and enzyme control were compatible with thin, permeable, low-activity membranes, but that the dynamic properties of less fragile electrodes, prepared for analytical long-term stability, were not usually kinetically limited but determined by their 'geometrical parameters'.

INTERFERENCE

One of the major advantages with using an ammonia electrode as the base sensor is its insensitivity to other solution species. This does not however preclude interference from free ammonia. The normal range of ammonia concentration in blood serum is 40–80 μg/100 cm^3, but may rise up to 400 μg/100 cm^3. This clearly indicates that in order to obtain a reliable accurate analyte determination in an ammonia sensor-based assay, precautions must be taken to eliminate free ammonia from the sample.

Various methods have been suggested to achieve the elimination of this ammonia error. Among these, an enzymatic removal of ammonium ion with glutamate has been tried,

$$\text{NADH} + NH_4^+ + \alpha\text{-ketoglutarate} \xrightarrow{\textit{glutamate dehydrogenase}} \text{glutamic acid} + \text{NAD}^+$$

but with the enzyme immobilized, so as not to contaminate the sample, the process is very slow and the viscosity of the sample must be reduced (by undesirable dilution) in order to facilitate mixing and speed up the reaction.

In instances where the signal due to free ammonia is sufficiently below the saturation level for the sensor, the best method of removing the error is to make a differential measurement between the enzyme membrane electrode and a 'blank' enzyme free ammonia electrode.

Carbon Dioxide-linked Enzyme Electrodes

OXALATE

The carbon dioxide electrode may also form the base sensor for urea determination, but one of the most common carbon dioxide-linked bioassays is that for oxalate. The determination of this analyte in urine is particularly critical in patients with forms of hyperoxaluria.

Various enzyme-linked schemes have been considered for oxalate determination. Oxalate decarboxylase catalyses the reaction,

$$\text{oxalate} \xrightarrow{\textit{oxalate decarboxylase}} CO_2 + \text{formate}$$

but the sulphate and phosphate levels usually present in urine have been found to inhibit the enzyme.

Oxalate oxidase, on the other hand, catalyses the reaction

$$\text{oxalate} + O_2 \xrightarrow{\textit{oxalate oxidase}} 2CO_2 + H_2O_2$$

but is inhibited by several anions and cations. Furthermore, it has not been available commercially but must be isolated.

Kobos and Ramsey (1980) developed a carbon dioxide-based sensor where the enzyme, oxalate decarboxylase, was immobilized at the electrode surface by (a) entrapment by a dialysis membrane or (b) glutaraldehyde crosslinking. The electrodes prepared by method (b) showed a considerably broader optimal pH range (pH 2.5–4.5 versus pH 2.5–3.2 for (a)), but both systems had a limit of detection for oxalate of the order of 4×10^{-5} M.

The dialysis membrane trapped enzyme electrode also exhibited considerably poorer decay characteristics than the crosslinked enzyme. Indeed, whereas the former electrode showed a decreased response after 4 days, the latter system, stored with hydroquinone, showed no decrease after a month of daily use. Hydroquinone is reported to increase the enzyme activity of oxalate decarboxylase by up to 100%, if employed at a concentration *sufficiently low not to cause inhibition.*

Of particular importance with urine samples is the likely interference by phosphate or sulphate. While this was apparent for the dialysis membrane electrode, the glutaraldehyde crosslinked membrane showed no selectivity for these inhibitors, and calibration curves in the presence and absence of inhibitors were identical.

IMMUNOASSAY

The carbon dioxide electrode has also been employed in immunoassay where the enzyme label reaction can be linked to CO_2 production. Keating and Rechnitz (1985), for example, have proposed an assay for digoxin where digoxin immobilized on polystyrene beads competes with sample digoxin for peroxidase-labelled antibody (Fig. 9.4). After an incubation period, the beads are removed from the sample solution and the complexed enzyme label estimated by the

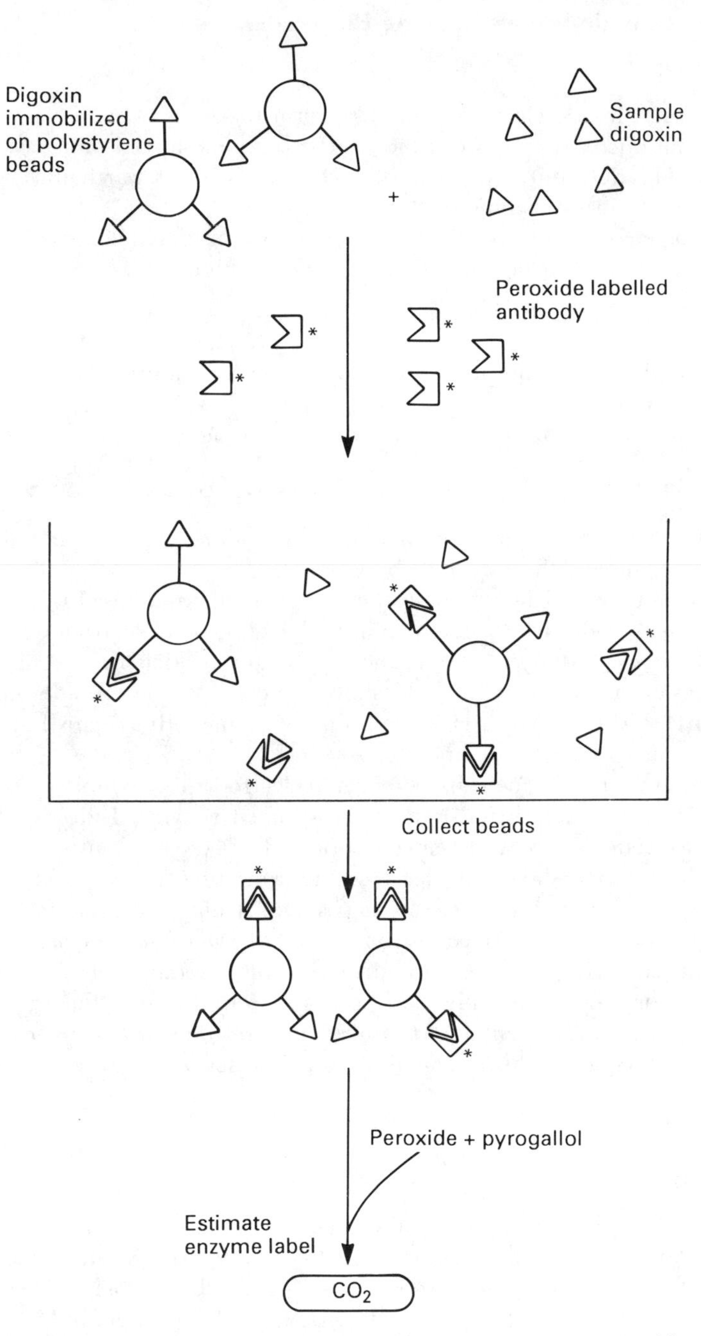

Fig. 9.4 Enzyme-linked immunoassay of digoxin via carbon dioxide electrode.

addition of the enzyme's substrates (peroxide and pyrogallol) and the measurement of CO_2:

$$H_2O_2 + \text{pyrogallol} \xrightarrow{\text{peroxidase}} CO_2$$

The rate of production of CO_2 in a competitive immunoassay such as this will be inversely proportional to the concentration of digoxin in the sample (more sample digoxin results in more digoxin–antidigoxin/peroxidase and less bead/digoxin–antidigoxin/peroxidase). This assay has a sensitivity down to picomolar concentrations, and while it is not assembled into a biosensor device to give a reagentless assay, it has the components suitable for biosensor exploitation.

Iodide-linked Enzyme Electrodes

An amperometric method for iodine titration is described in Chapter 8 as a means of assaying the peroxide produced in the glucose oxidase-catalysed oxidation of glucose:

$$\text{glucose} + O_2 \xrightarrow{\text{glucose oxidase}} \text{gluconic acid} + H_2O_2$$
$$H_2O_2 + 2I^- + 2H^+ \xrightarrow{\text{peroxidase}} 2H_2O + I_2$$

This assay may similarly be based on a potentiometric iodide membrane sensor.

Nagy *et al.* (1973) found, however, that although the sensor is very sensitive, interference by such reagents as ascorbic acid, tyrosine or other reducing agents, required that blood samples should be pretreated.

The 'iodide system' could in principle be employed whereever a change in peroxide concentration can be related to the concentration of target analyte.

L-Amino oxidase and peroxidase co-immobilized in a polyacrylamide gel, over the surface of an iodide electrode, gives a sensor for L-amino acids (Guilbault and Nagy, 1973):

$$\text{L-phenylalanine} \xrightarrow{\text{L-amino acid oxidase}} H_2O_2$$
$$H_2O_2 + 2H^+ + 2I^- \xrightarrow{\text{peroxidase}} I_2 + 2H_2O$$

However, as with many other iodide-linked sensors, determination of L-amino acids is more accurately performed in conjunction with one of the other potentiometric base sensors. Indeed, reports of the iodide-based sensor include problems of interference and specificity that are less prominent in the use of other systems.

Nevertheless, Boitieux *et al.* (1979, 1981) have reported enzyme-linked immunoassay based on a potentiometric iodide electrode, using a peroxidase enzyme label. With anti-hepatitis B surface antigen immobilized to a proteic gelatin membrane, and mounted over an iodide-sensitive electrode, hepatitis B surface antigen could be assayed with a sensitivity limit near $0.5\,\mu g\,l^{-1}$, comparing favourably with other enzyme immunoassay techniques.

Anti-oestradiol-17β immobilized to a gelatin membrane on the surface of an iodide electrode produced a similar sensor for oestradiol-17β. A competitive assay

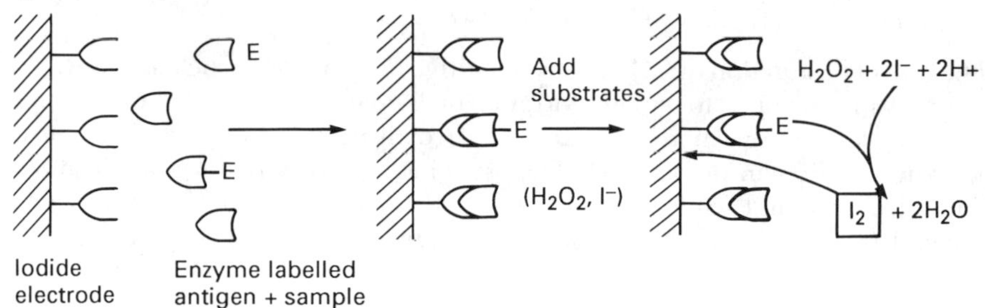

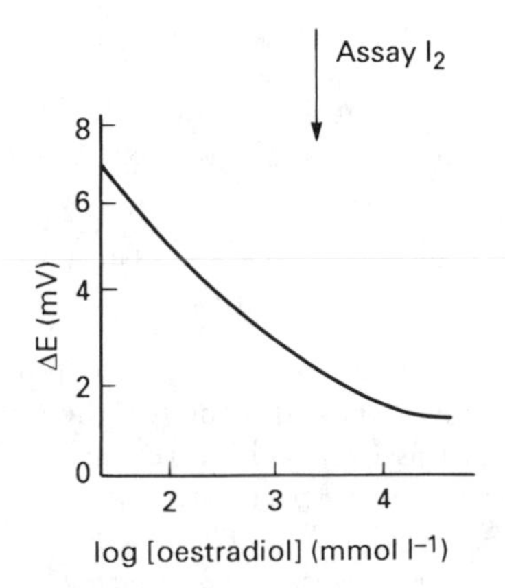

Fig. 9.5 Determination of 17β-oestradiol via an antibody-modified iodide electrode (Boitieux *et al.*, 1981).

was performed between enzyme-labelled antigen and sample antigen (Fig. 9.5) and the authors recorded a satisfactory determination of oestradiol-17β in the range 50 pmol l^{-1} to 10 nmol l^{-1}.

Ag^+/S^{2-} Electrodes

The Ag^+/S^{2-} electrode makes the determination of cysteine by direct potentiometry possible. In the presence of a thiol (e.g. cysteine) the reaction occurs:

$$Ag^+ + xR\text{–}S^{y-} \rightarrow Ag(R\text{–}S)_x^{(1-xy)}$$

Morf *et al.* (1974) predicted a -118 mV/decade response anticipating a 1:2 silver: cysteine complex, which compared favourably with the -108 mV/decade obtained. The limit of detection for this electrode was 2×10^{-4} M, but this non-enzymic determination is not selective only for cysteine but will respond similarly to any other thiols present.

An alternative approach (Guilbault *et al.*, 1972) is to employ the enzyme β-cyanoalanine synthase linked to a sulphide electrode,

$$CN^- + \text{cysteine} \xrightarrow{\beta\text{-cyanoalanine synthase}} HS^- + \beta\text{-cyanoalanine}$$

although this system was in fact developed for estimation of the enzyme, rather than its substrate. Unfortunately, since CN^- interferes with the electrode response, the 'mixed' calibration procedure that must be employed would reduce the ease of use for such a potentiometric device.

Many such examples of potentiometric-linked assays exist in the research literature. Their exploitation and development as biosensors does not necessarily result from these encouraging reports, since performance in 'real sampling environments' must be assessed together with many other factors in considering their probable success.

Analyte-selective Electrodes

Many of the problems already encountered with ion-selective electrode-linked biosensors, such as a free-ammonia error for the ammonia-linked assays, or a reducing-agent error for the iodide device, would be eliminated if the assay was not made indirectly in this fashion, via a 'third party' which might also be present independently in the sample, but was a direct estimation due to the potentiometric response caused by the enzyme-catalysed reaction with target analyte.

Unlike the ion-selective electrodes which follow an ion exchange at the membrane–solution interface, these electrodes usually involve probes of a noble metal and they exhibit Nernstian behaviour due to a change in the ratio of oxidized and reduced species at the electrode surface. As such they are often referred to as *potentiometric redox electrodes*. However, for biosensors based on this principle, it is not always clear as to the nature of the redox reaction.

POTENTIOMETRIC L-AMINO ACID ELECTRODE

A response was observed by Ianniello and Yacynych (1981), for a potentiometric sensor for L-amino acids. L-Amino acid oxidase (LAAO), which catalyses the reaction,

$$\underset{\displaystyle NH_3^+}{\text{R-CH-COO}^-} + H_2O + O_2 \xrightarrow{LAAO} \text{R}-\underset{\displaystyle \| \atop O}{\text{C}}-\text{COO}^- + NH_4^+ + H_2O_2$$

was covalently attached to a graphite electrode, modified by cyanuric chloride. The potentiometric response of this electrode for various amino acids was consistent with expected behaviour for the LAAO system—i.e. L-phenylalanine > L-leucine > L-methionine—with no response towards L-cysteine or D-phenylalanine. The electrode reaction is centred around the irreversible enzyme/O_2/H_2O_2 redox system:

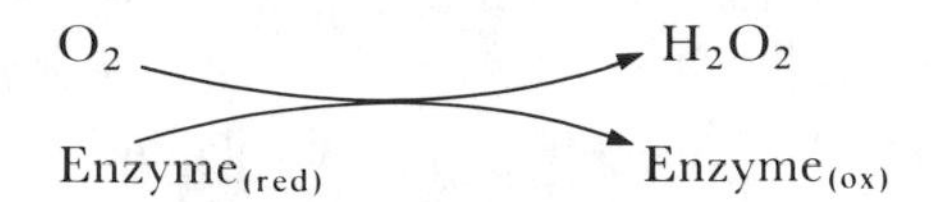

Since the O_2/H_2O_2 reaction is irreversible, it would not be expected to cause a Nernstian response. This can be confirmed at an 'inert' platinum electrode for various concentrations of peroxide: the potentiometric response is only of the order of -5 mV/decade change in peroxide concentration. This would suggest that the O_2/H_2O_2 couple does not initiate a simple oxidation or reduction response in the electrode potential.

By comparison at an untreated graphite electrode the response to H_2O_2 was -64 mV/decade while a cyanuric chloride-modified graphite electrode surface gave an equivalent response ($+79$ mV/decade) but of opposite slope. These findings would suggest a mechanism of interaction between the hydrogen peroxide and the electrode surface involving both oxidation and reduction of the hydrogen peroxide.

Clearly, the response of 40 mV/decade for L-phenylalanine cannot be attributed to a single redox couple, but rather to an interaction of enzyme–solution–surface species at an electrode whose surface pretreatment can drastically alter the results.

POTENTIOMETRIC GLUCOSE ELECTRODE

Similarly, Wingard and co-workers (Wingard *et al.*, 1980, 1983, 1984; Castner and Wingard, 1984; Procter *et al.*, 1985) observed that a glucose oxidase–catalase platinum electrode gave a potentiometric response to the presence of glucose. Subsequently they immobilized glucose oxidase with or without catalase by different techniques to platinum, gold or porous graphite. The enzyme(s) were trapped in a polyacrylamide gel, immobilized by glutaraldehyde crosslinking of the protein or else coupled covalently to an oxidized surface via an aminosilane and a glutaraldehyde bridge. The electrodes gave a linear response for glucose in the range 50–400 mg/100 ml and were thus suitable for measurements within the clinical blood-glucose level range of 90–120 mg/100 ml. Similar linearity was obtained whether the base electrode was graphite or platinum, but whereas the potential versus log-concentration slope was positive for graphite, for platinum it was negative (Fig. 9.6). Clearly this indicates once more that different surface reactions are involved at the two electrode materials and demonstrates that although the redox potential should be independent of the electrode material, the measured potential is influenced by surface reactions. *It is hardly surprising therefore that reports of measurements obtained via redox electrodes require such detailed information concerning the history and conditioning of the electrode surface.*

In the example cited above, the potentiometric response is *generated* by the glucose oxidase-catalysed oxidation of glucose.

On platinum the change in potential of the electrode with respect to glucose

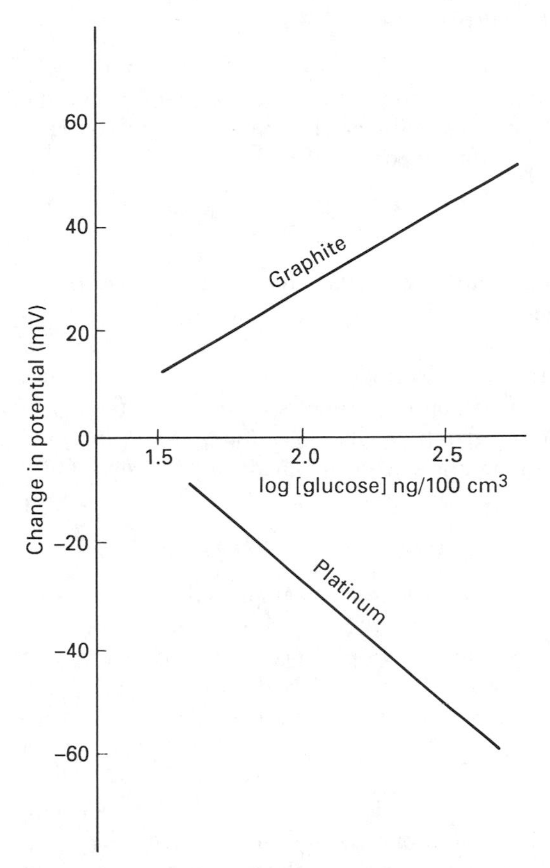

Fig. 9.6 Response of potentiometric redox electrodes (Pt and graphite), coated with crosslinked glucose oxidase, to oxygenated glucose solutions at pH 7.4. (Data from Wingard *et al.*, 1984.)

concentration produced a negative slope (−40 mV/decade). With the overall enzyme-catalysed reaction being

$$\text{glucose} + O_2 \xrightarrow{\text{glucose oxidase}} \text{gluconic acid} + H_2O_2$$

it would be anticipated that, as with the L-amino acid electrode, a major contribution to the electrode processes would come from the oxidation or reduction of peroxide: A slope of −40 mV/decade recorded for increasing glucose concentration is indicative of slightly less than a two-electron transfer and probably results from several reactions occurring together (both oxidations and reductions). In aqueous solutions, the formation of platinum oxide is well known. The mechanism proposed involves reversible attachment of hydroxyl groups at the surface

$$Pt + 2H_2O \rightarrow Pt(OH)_2 + 2H^+ + 2e^-$$

followed by an irreversible rearrangement, allowing penetration of the –OH groups into the platinum lattice. It follows that the formation of $Pt(OH)_2$ at the electrode surface will be influenced by the presence of H^+ ions, so that an interaction has been imagined:

$$H_2O_2 \rightarrow O_2 + 2H^+ + 2e^-$$
$$Pt(OH)_2 + 2H^+ + 2e^- \rightarrow Pt + 2H_2O$$

Almost certainly, however, other reactions also contribute to the overall potential difference. The formation of platinum oxide for example, is thought to begin with the oxidation of the –OH groups on the surface of the electrode.

At a porous graphite electrode, the positive slope of 30 mV/decade may be indicative of the oxidation of peroxide in conjunction with hydroquinone and aldehyde functional groups on the electrode, that can undergo two-electron oxidations to quinone and carboxylic acid, respectively.

POTENTIOMETRIC LACTATE ELECTRODE

A different type of redox reaction was employed in the construction of a lactate electrode (Shinbo *et al.*, 1979), where a redox couple was attached to potentiometrically inert electrode surface to produce a redox-sensitive electrode.

Assays of lactate have often been based on its enzyme-catalysed oxidation of lactate in the presence of an electron acceptors such as hexacyanoferrate(III),

$$\text{lactate} + 2Fe(CN)_6^{3-} \xrightarrow{\text{lactate dehydrogenase}} \text{pyruvate} + 2Fe(CN)_6^{4-} + 2H^+$$

and the lactate concentration related to the concentration ratio of hexcyanoferrate(III)/(II), followed by a redox electrode.

The sensor membrane for this redox electrode is prepared from PVC containing dibutylferrocene (PVC–Fc), and the complete enzyme electrode constructed by immobilizing the enzyme in a gelatin layer on the redox electrode surface. The potential difference across the PVC–Fc membrane will be generated by the ferrocyanide : ferricyanide ratio. An S-shaped response to lactate is recorded for this sensor, which possesses a narrow linear range (10^{-4}–10^{-3} M) due to the low K_M value for the enzyme (1.2×10^{-3} M).

Antibody–antigen Electrodes

SELECTIVE MEMBRANE POTENTIALS

Ion-selective electrodes measure the potential that develops across a membrane that is selective for a particular ion. In principle, membranes could be designed to respond selectively to many other analytes although transport of the species through the membrane may be too slow to be useful for larger species.

The development of a potential across a membrane to which a biorecognition molecule has been attached has long been used to monitor biological activity.

Measurements have been made for both enzyme–substrate systems and antibody–antigen reactions. In the former case, the change in the membrane characteristics reflects the utilization of the sample substrate by the membrane-bound enzyme. This has been used, for example, to show the effect of added benzoylarginine to a trypsin membrane or a mixture of pyruvate and β-diphosphoyridin nucleotide to a membrane exposed to lactic acid dehydrogenase (del Castillo *et al.*, 1966). However, as described in other sections, many different approaches have been made to enzyme-linked sensors and so it is perhaps more interesting to investigate this technique for bioaffinity reactions, where identification of a physicochemical parameter that can be monitored is less easily made. Indeed, it is often difficult to attribute any accessible parameter to the equilibrium complexation of a bioaffinity molecule (BaM) and its complement (Comp),

$$\text{BaM} + \text{Comp} \rightleftharpoons \text{BaMComp}$$

since changes are likely to be associated with charge distribution or 'molecular size' rather than oxidation state or H^+ concentration etc., as in enzyme-catalysed reactions.

ANTI-DIGOXIN–DIGOXIN

Existing ion-selective electrodes can be adapted in order to perform such an analysis. For example, Keating and Rechnitz (1984) have described a potentiometric ionophore modulation immunoassay (PIMIA) for antibodies, based on an antigen modification of a potassium ionophore. Applied to an assay of antibodies to digoxin, the antigen digoxin was coupled to one of the potassium ionophores (*cis*-dibenzo-18-crown-6 or benzo-15-crown-5). The resulting conjugate, immobilized in a PVC membrane on a conventional ion-selective electrode provided a recognition surface for the digoxin antibody. Introduction of sample antibody results in binding to the bound antigen at the membrane–solution interface, causing a change in the membrane potential. Obviously the main disadvantage of such a system is the dual recognition function of the ionophore–antigen membrane. The assay must be performed at a constant background K^+ concentration, entailing extensive pretreatment of the sample to remove any interfering ions.

ANTI-hCG–hCG

A more tailored immuno-surface for a potentiometric electrode has been proposed for the anti-hCG–hCG system. Anti-human choriogonadotropin (anti-hCG) may be attached to a cyanobromide-treated titanium oxide electrode. This antibody electrode is reported to respond to solution hCG with a positive shift in potential (Yamamoto *et al.*, 1980). The reaction between antibody and antigen depends on the affinity constant for the reaction

$$\text{anti-hCG} + \text{hCG} \overset{k_a}{\rightleftharpoons} \text{anti-hCG–hCG}$$

so that the rate of reaction will be proportional to concentration of sample hCG.

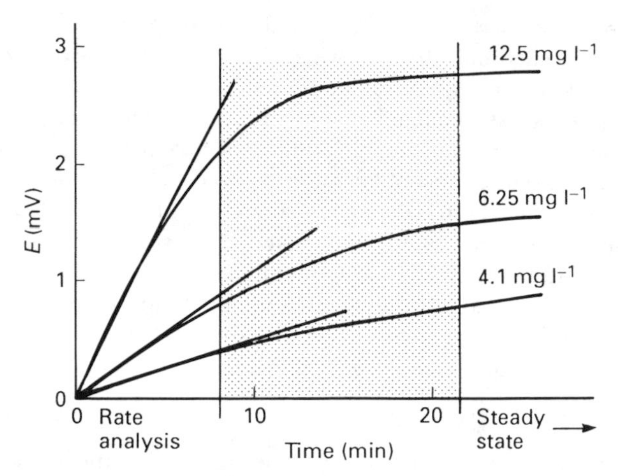

Fig. 9.7 Response curves for hCG at an anti-hCG electrode. (From data in Yamamoto *et al.*, 1980.)

Processing the assay data as a rate of reaction rather than a steady-state value can achieve some improvement in signal quality and assay time. As seen in Fig. 9.7, a steady-state measurement for this hCG electrode cannot be made until the reaction is complete ($\sim$20 min), whereas the rate of reaction is estimated on the straight-line portion of the response curve ($<$6 min). Slope analysis, where a straight line can be drawn through a noisy signal, recorded over a period of some minutes, can also give a better signal-to-noise ratio than a single-point estimation on a noisy 'plateau'.

The detection limit for this system was estimated to be about 0.1 μg/ml and here it is proposed that the source of the potential shift must be associated with changes in charge distribution or activity at the electrode–solution interface, during antibody–antigen complexing, rather than the ion-exchange process which can be assigned to the previous example.

WASSERMANN ANTIBODY

An antigen-binding membrane specific for the Wassermann antibody, used for non-treponemal serology tests for syphilis, was prepared by casting a lipid antigen, of cardiolipin, phosphatidylcholine and cholesterol, in triacetylcellulose (Aizawa *et al.*, 1977). Cardiolipin shows a net negative charge due to phosphate groups, which is reflected by the change in membrane potential with antigen concentration. After immunological reaction with antibody containing serum, the membrane potential changed to reach steady-state within 20 min. The potential shift was consistent with a decrease in negatively charged groups, which was originally considered to result from antigen–antibody interaction at the membrane–solution interface, causing a change in charge density. However, Collins and Janata (1982) showed that this Wassermann antibody-sensitive membrane would also respond to changes in concentration of various inorganic ions and suggested that the immunological response that had been recorded was entirely a secondary effect due to the ion-exchange properties of the membrane.

It is frequently necessary to exercise some caution, therefore, in interpreting the nature and origin of the potentiometric response recorded from these 'biorecognition membranes'.

BIOAFFINITY MEMBRANE FOR RIBOFLAVIN ASSAY

The principle behind this measurement seems very similar to that involved in the assays described above, except here the reaction of interest is an indirect one: a *competitive* binding between an affinity species with its membrane-bound complement and a solution analogue.

Riboflavin binding protein (RBP) is selective for riboflavin, showing reversible binding with an affinity constant of about 1.3×10^{-9} M. This is some orders of magnitude greater than for flavin analogues such as acriflavin (1.8×10^{-7} M) or flavin adenine dinucleotide, FAD (1.4×10^{-5} M), so that RBP bound to one of these analogues will be displaced by riboflavin to form a more stable complex.

This competition is the basis for an affinity membrane developed by Yao and Rechnitz (1987). A membrane with acriflavin bound to *both* sides was prepared by glutaraldehyde coupling to an acetylcellulose–octadecylamine membrane:

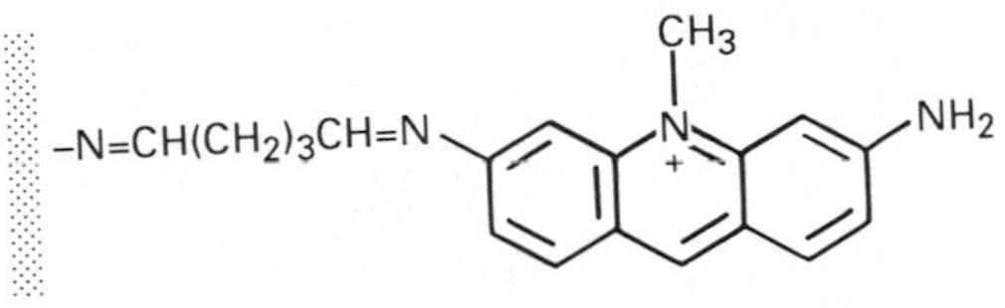

and an FAD-bound membrane was prepared by coupling through the ribityl moiety activated with cyanogen bromide:

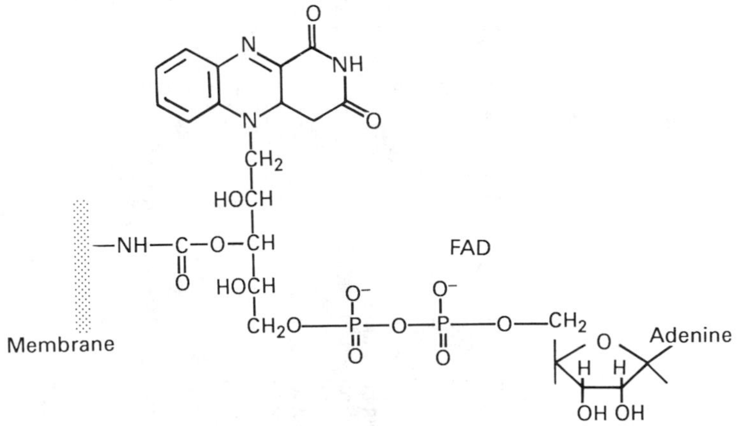

These membranes are 'activated' at pH 7 by reaction with RBP to give the RBP–flavin analogue complex at both surfaces of the membrane (Fig. 9.8). Since

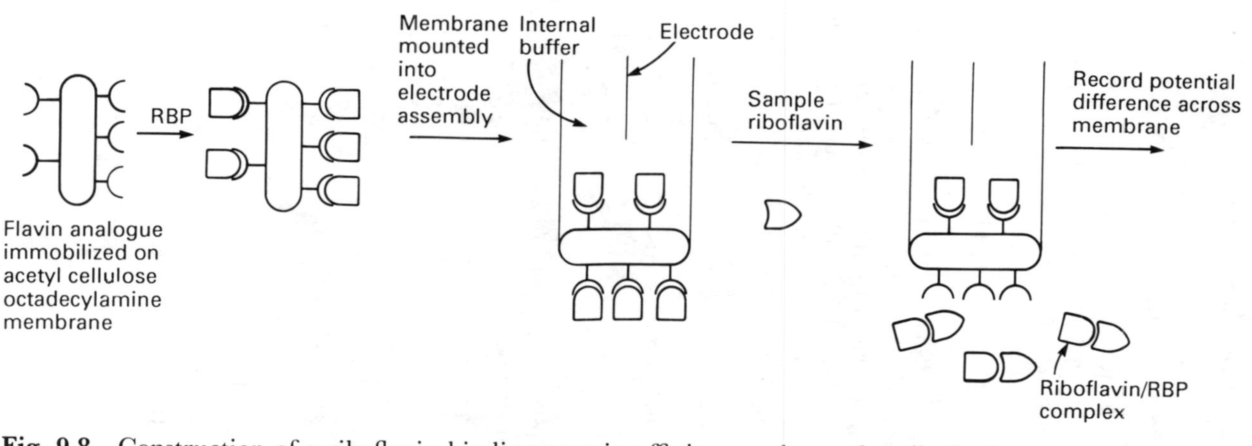

Fig. 9.8 Construction of a riboflavin binding protein affinity membrane for riboflavin assay.

the isoelectric point of RBP is 4.6, it is negatively charged at the working pH. This property means that when RBP is displaced from the membrane surface complex by flavin, the change in the charge at the surface is accompanied by a change in membrane potential. The direction of this change will depend on the charge characteristics of the base flavin analogue. Since FAD has a negatively charged site, for example, the removal of RBP produces a negative shift in membrane potential, while breaking of the RBP–acriflavin complex gives a positive change.

Both of these membranes respond due to a competition for RBP between the membrane-bound flavin analogue and riboflavin itself. The bioaffinity constant for the displacement of the membrane-bound complex by riboflavin can be derived:

$$\text{RBP} + \text{riboflavin} \rightleftharpoons \text{RBP–riboflavin} \qquad (K_d = 1.3 \times 10^{-9}\ \text{M})$$
$$\text{RBP} + \text{acriflavin} \rightleftharpoons \text{RBP–acriflavin} \qquad (K_d = 1.8 \times 10^{-7}\ \text{M})$$
$$\text{RBP} + \text{FAD} \rightleftharpoons \text{RBP–FAD} \qquad (K_d = 1.4 \times 10^{-5}\ \text{M})$$

(1) RBP–acriflavin + riboflavin ⇌ RBP–riboflavin + acriflavin

with $K_d = (1.3 \times 10^{-9})/(1.8 \times 10^{-7}) = 7.2 \times 10^{-3}$ M, and

(2) RBP–FAD + riboflavin ⇌ RBP–riboflavin + FAD

with $K_d = (1.3 \times 10^{-9})/(1.4 \times 10^{-5}) = 9.3 \times 10^{-5}$ M.

These values show that the equilibrium in (2) lies further to the right than in (1), and this is reflected in the relative response times for the two membranes: 12 min for (1) as opposed to 6–8 min in (2).

This type of 'competition assay' is most efficient when the bioaffinity constants that describe the two competing complexes are separated by several orders of magnitude.

In this form, the main requirement for this system seems to be that one of the complexing agents involved in the equilibrium should have a net charge, so that the charge density at one of the membrane solution interfaces (external) varies with respect to the other interface (internal) during the reaction (see Fig. 9.8).

STATE OF IMMUNE RESPONSE

Blood levels for the pathogenic microbe *Candida albicans* provide an important indicator of immune response. *Candida albicans* is found under healthy conditions, but the levels rise when immunological function is impaired. Since *C. albicans* has a net negative charge, its interaction with antibody bound at a membrane surface could considerably alter the membrane potential.

Anti-*C. albicans* was bound to a cellulose triacetate membrane and the difference in membrane potential recorded against concentration of *C. albicans*. The range of detection was 10^4–5×10^5 cells cm^{-3} (Matsuoka *et al.*, 1985). Response to other organisms observed in blood was minimal, except when *Saccharomyces cereviciae* was applied to the system. These organisms are sufficiently closely related to expect them to exhibit common antigenic structures. It would not be inconceivable, therefore, that the low-purity antiserum employed to make the membrane showed some cross-reactivity.

In view of earlier interpretations of the response of the Wassermann antibody, however, the origin of this antibody–antigen initiated change in membrane potential may also be a secondary indirect one.

Applications to FET Devices

The advantages of deploying ISFET devices in place of ISEs (see Chapter 4) are also applicable to enzyme-based FETs (ENFET) and immuno-based FETs (IMFET). However, as yet, there are only a few reports of the application of these devices to biological systems.

THE BIOFET

In the enzyme FET, the enzyme is immobilized in a gel over the ion-selective membrane of the base FET sensor (Fig. 9.9), so that the principle behind the union is the same as that already encountered for ISE-linked enzyme sensors. In practice, a dual gate ISFET is usually employed (Fig. 9.10), where one of the

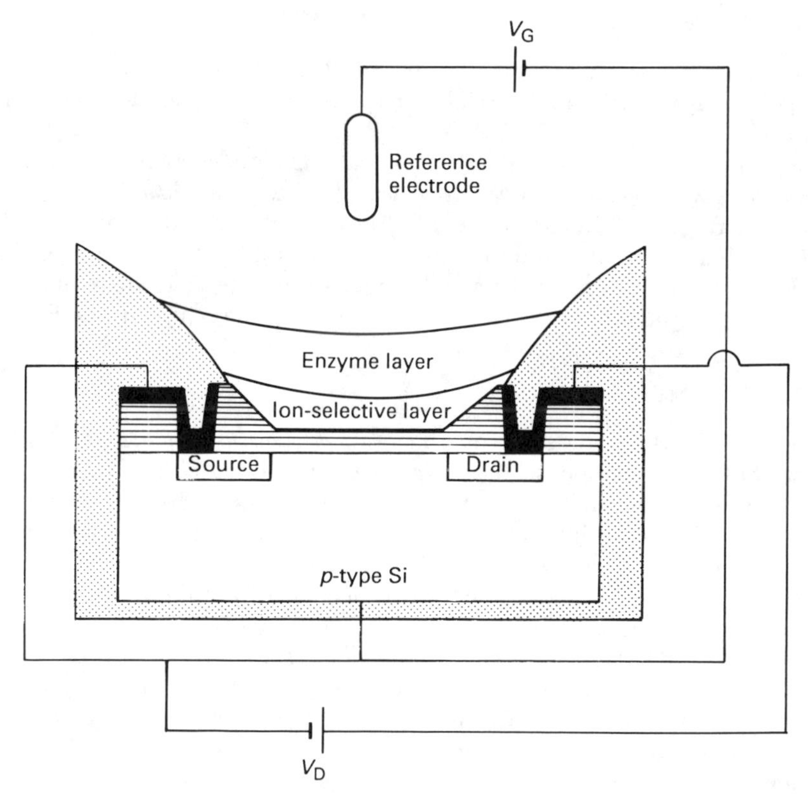

Fig. 9.9 Schematic diagram of an ENFET.

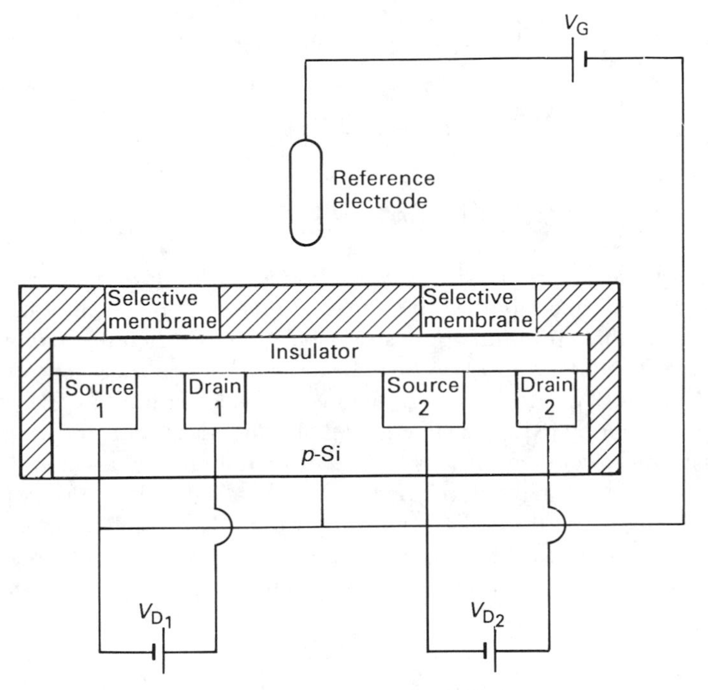

Fig. 9.10 Schematic diagram of a dual-gate ENFET.

FETs can act as a reference system and is assembled in the same way as the sample FET, but contains a blank enzyme-free gel membrane. This arrangement allows some automatic compensation of fluctuations in solution pH, temperature, etc.

ENFETs could be based on any of the FET counterparts of the ISEs already discussed in this chapter, but the most widely used base FET device is the pHFET. In unbuffered solution, the response of an H^+-sensing enzyme device has been characterized. However, many physiological samples, such as blood, have a relatively high buffering capacity. Eddowes (1987) has considered the enzyme-linked assay of substrate, resulting in a net change in H^+ concentration, where the buffering capacity of the analyte contributes to the response. In a buffer solution the protonation equilibrium occurs with buffer anion B^-:

$$H^+ + B^- \rightleftharpoons HB \qquad K_b = \frac{[B^-][H^+]}{[BH]}$$

Concentration profiles for the participating species across the enzyme layer, and in particular the H^+ concentration at the pH-sensitive surface have been derived by consideration of the relevant differential equations, describing the diffusional mass-transport coupled reactions.

$$D_s \frac{d^2[S]}{dx^2} + \frac{k_2[E][S]}{K_M + [S]} = 0$$

$$D_{H^+}\frac{d^2[H^+]}{dx^2}+\frac{k_2[E][S]}{K_M+[S]}-k_b[B^-][H^+]+k_{-b}[BH]=0$$

$$D_{BH}\frac{d^2[BH]}{dx^2}+k_b[B^-][H^+]+k_{-b}[BH] \qquad =0$$

$$D_B\frac{d^2[B^-]}{dx^2}+k_b[B^-][H^+]+k_{-b}[BH] \qquad =0$$

Solution of these equations with suitable boundary limits gives a model for the response of an enzyme-modified pH sensitive device in buffer solution. Eddowes' model predicts that optimum response will be obtained when the enzyme kinetics are sufficiently rapid for the response to be under mass-transport control, and that the usable range of the device will be defined at the lower limit by $[B^-]>[S]$ and at the upper limit by $[S]>K_M$. The model also suggests that changes in bulk pH cannot be simply compensated by a direct differential measurement.

Caras *et al.* (1985) have developed a theoretical model for the immobilized enzyme which extends the earlier treatment of $d[S]/dt$ to account for the feedback influence of product (in the case of a pHFET, H^+) on the enzyme reaction and thus the enzyme kinetics. In general where products or secondary substrates of the

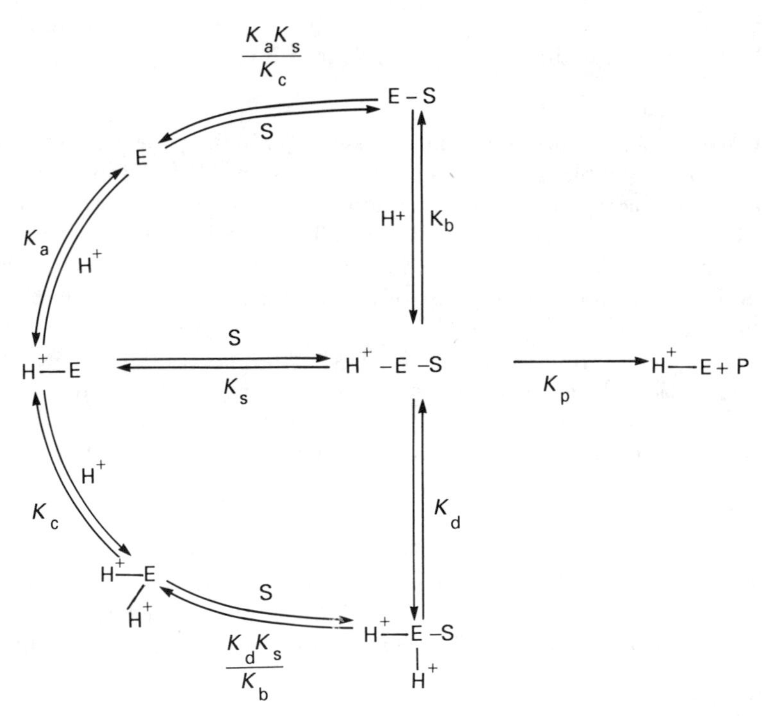

Fig. 9.11 pH-dependent model for the enzyme reaction. (After Caras *et al.*, 1985.)

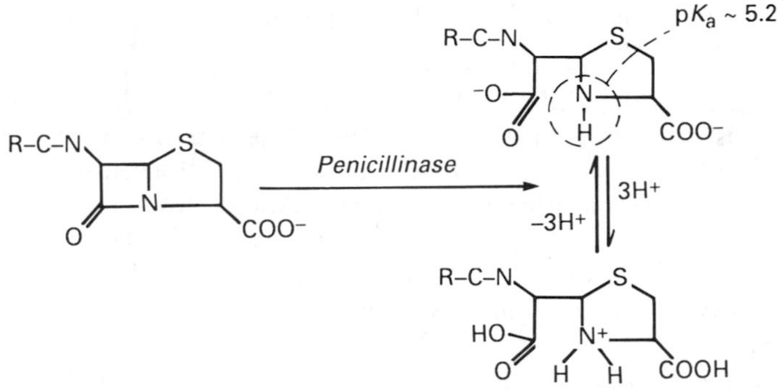

Fig. 9.12 pH dependency of the action of penicillinase on its substrate, penicillin.

enzyme reaction influence the enzyme kinetics, they will generally alter the enzyme electrode response, so they should be included in the model (Fig. 9.11). For example, under the conditions indicated in the scheme, the Caras *et al.* model evaluates the diffusion modulus (see p. 217) as:

$$\Phi^2 = d^2\left[\frac{k_2[\mathrm{E_o}]}{K_S D_S\left[1+\dfrac{[\mathrm{H}^+]}{K_b}+\dfrac{K_a}{[\mathrm{H}^+]}\right]}\right]$$

PENICILLINASE pH FET

In a penicillin-sensitive FET utilizing penicillinase (Caras and Janata, 1985) the enzyme was covalently bound to a polyacrylamide gel matrix deposited on a bare Si_3N_4 gate, which had been pretreated with BSA–glutaraldehyde in order to ensure good adhesion of the enzyme polymer. The effect of H^+ on the response can be effectively modelled by consideration of enzyme kinetics in the presence of H^+. However in this instance, the existence of a group with $pK_a \sim 5.2$ (Fig. 9.12) means that at pH values below this the group will be protonated, thus removing free H^+. This results in a narrower dynamic range, since under conditions of high proton production (i.e. high substrate concentration), the R′R″NH group will be protonated, the H^+ will not be involved in the schemes proposed, and the experiment will deviate from the model.

GLUCOSE OXIDASE pHFET

A glucose-sensitive FET is prepared by covalently attaching the enzyme glucose oxidase to polyacrylamide gel in a manner similar to that above (Caras *et al.*, 1985). The response of this ENFET to glucose followed a theoretical model if the treatment described above was extended to include the influence of both O_2 and H^+ (H_2O_2 is removed 'instantaneously' by the presence of catalase).

$$\text{glucose} + O_2 \xrightarrow{\text{glucose oxidase}} H^+ + \text{gluconate} + H_2O_2$$

Hanazato *et al.* (1987) investigated the possibility of preparing a glucose FET entirely by photolithographic techniques. A glucose oxidase–photopolymer membrane was developed on the pre-silanized gate region of a FET, using the water-soluble photopolymer polyvinylpyrrolodone (PVP) sensitized with 2,5-bis(4′-azido-2′-sulphobenzal)cyclopentanone (BASC). However, the photopolymerized membrane peeled away from the surface and good adhesion had to be induced by the introduction of additional chemical steps. Glucose oxidase and BSA in the polymer was crosslinked with glutaraldehyde to give a well-adhering polymer layer.

The resulting enzyme membrane FET, deployed in the differential mode against a bare gate FET, responded to glucose up to 3 mM, and under conditions of *oxygen* saturation, the linear range could be considerably extended. The output of the device decayed over a period of 15 days, to about 50% of the initial response. In fact the response of the enzyme FET increased with increase in the content of glucose oxidase, but at high enzyme loadings a uniform film could not be formed on the FET.

UREASE pHFET

Van der Schoot and Bergveld (1987, 1988) have proposed a system which is independent of the buffer capacity, by the integration of an ENFET with a pH actuator electrode, which will control the pH inside the immobilized enzyme layer. (Compare with the oxygen-stabilized system in Chapter 8.) Since the sensor effectively operates at constant pH, the term *chemostatic enzyme sensor* has been proposed. The base sensor is surrounded by a noble-metal electrode which can be used as either anode or cathode, to generate OH^- or H^+ by the electrolysis of water, and thus control the pH in the vicinity of the gate. Changes in pH at the ENFET are measured with respect to a reference FET, and the pH response due to the urease-catalysed decomposition of urea, related to the current flow required to restore the pH at the pH actuator electrode.

GLUCOSE OXIDASE–UREASE DUAL FUNCTION pHFET

Nakamoto *et al.* (1988) have reported a dual-function enzyme FET device prepared using photolithographic techniques selectively to expose the gate areas of two pH-ISFETs on the same chip, and allow the consecutive immobilization of glucose oxidase and urease, respectively, on the two gate areas by glutaraldehyde crosslinking (Fig. 9.13). Differential measurements of urea and glucose were made at the respective gates against a common pH-ISFET. Glucose showed a linear response range up to 90 mg/dl and urea over a similar range, but with a much smaller differential output voltage. No cross-reactivity was recorded between the two enzyme FETs, demonstrating the accuracy of the photolithographic patterning technique.

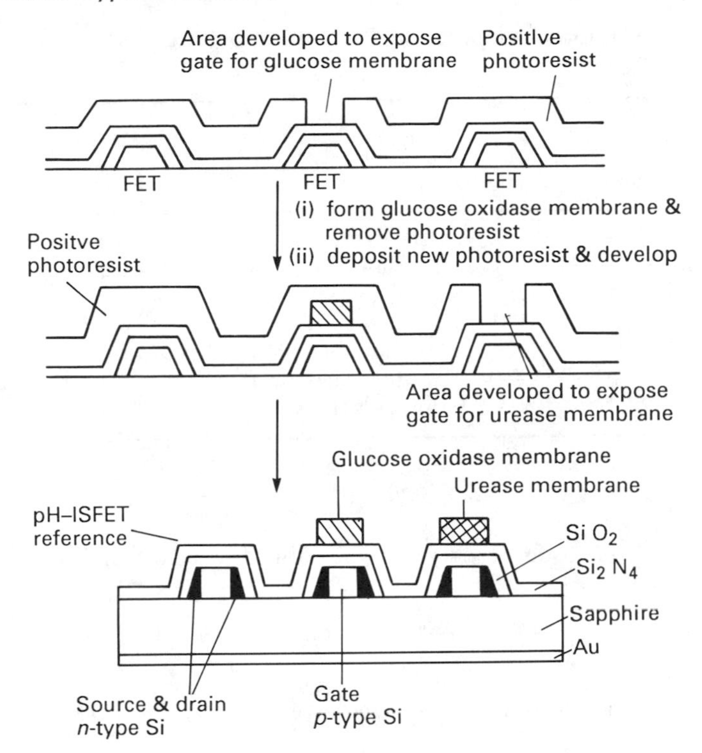

Fig. 9.13 Construction of multiple-gate ENFETs by photolithographic techniques.

PdMOSFET

Use of the gas FETs (see Chapter 4) has been investigated in conjunction with biological molecules. For the hydrogen-sensitive FETs this is centred on the hydrogenase–dehydrogenase (HDH) system,

$$H^+ + NADH \xrightarrow{HDH} \boxed{H_2} + NAD^+$$

so that links with the NAD-dependent enzymes could be envisaged. For example, it has been employed in an NADH regeneration system involving alanine dehydrogenase (Danielsson *et al.*, 1982, 1988) co-immobilized with hydrogenase,

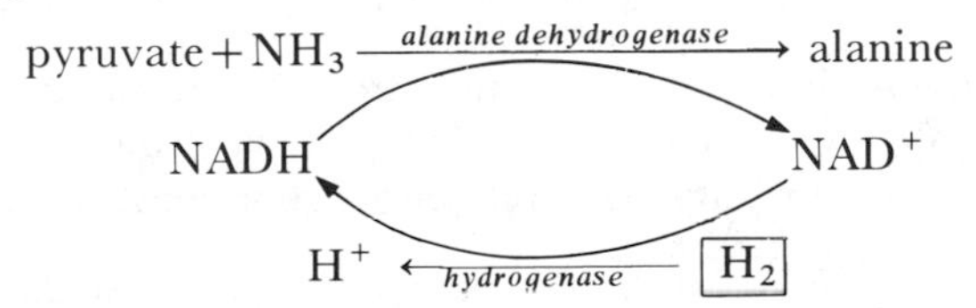

so that many combinations of this sort should be possible, since it has been shown

Table 9.1 Ammonia-sensitive FET-linked substrate assay. (Winquist *et al.*, 1984, 1985)

Enzyme-linked reaction		*(μM)*	*Response to 10 μM (mV)*
Urea $\xrightarrow{\text{urease}}$	$CO_2 + 2NH_3$	0.2–40	16
L-Asparagine $\xrightarrow{\text{asparaginase}}$	Asparate + NH_3	0.2–40	8
L-Aspartate $\xrightarrow{\text{aspartase}}$	Fumarate + NH_3		8
L-Glutamate + NAD^+ $\xrightarrow{\text{glutamate dehydrogenase}}$	α-Oxoglutarate + NH_3 + NADH		8
Creatinine $\xrightarrow{\text{creatinine iminohydrolase}}$	*N*-methylhydantoin + NH_3	0.2–30	8
Adenosine $\xrightarrow{\text{adenosine deaminase}}$	Inosine + NH_3		8

that immobilized hydrogenase is relatively stable. Indeed ethanol can be included in the sequence,

$$C_2H_5OH + NAD^+ \xrightarrow{\text{alcohol dehydrogenase}} CH_3CHO + NADH + H^+ \xrightarrow{\text{hydrogenase}} NAD^+ + \boxed{H_2}$$

so that it could be assayed via hydrogen evolution at a PdMOSFET structure.

AMMONIA-SENSITIVE FET

Several combinations of substrate–enzyme that would be relevant here have been discussed earlier in this chapter. In 'flow-through' designs, with the ammonia-sensitive FET downstream of a column containing immobilized enzyme, many of these combinations have been tested (Winquist *et al.*, 1984, 1986). The activity of the column was high enough to ensure complete substrate conversion and, as would be predicted from the equations for the respective enzyme reactions with substrate, the FET response to added substrate was about twice the magnitude for urea, where 2 mol of ammonia are produced per mole of substrate, compared with the other enzyme–substrate combinations (Table 9.1).

However, modification of the FET device to bring the enzyme 'column' directly in contact with the ammonia-sensing element, would complete this transition to ENFET. The possibility has been tested for urea (Winquist *et al.*, 1985). The IrMOS device, covered with a Teflon gas-permeable membrane, was modified with urease by immobilizing it directly onto the membrane by crosslinking with glutaraldehyde. In a dual-gate system, urea could be determined in an endogenous ammonia concentration, against a reference probe constructed with inactivated enzyme.

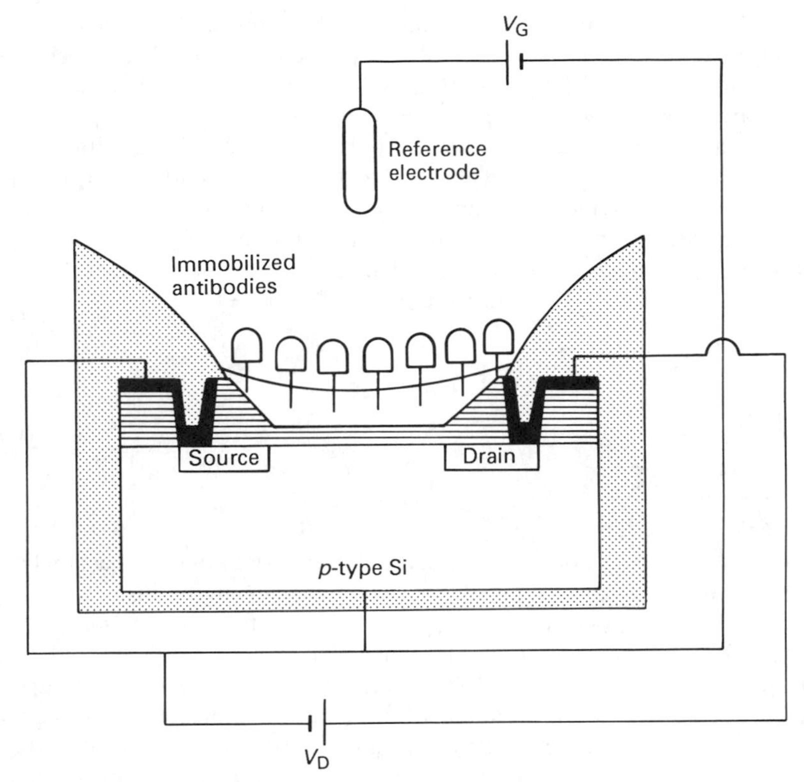

Fig. 9.14 Schematic diagra of an ImmunoFET.

Immuno-sensitive FETs

The problem of performing an unlabelled immunoassay is one that would seem ideally suited to solution by a FET-type device. A direct-reading immunoprobe must be based on a surface reaction (Fig. 9.14), since the selective permeation of the large molecules involved, in a manner akin to that involved in ion-selective membranes, is unlikely. In order to realize such a probe, therefore, would require a direct reading of interfacial charge density. The immunoreaction can be described by the equilibrium,

$$\mathrm{Ab} + \mathrm{Ag} \rightleftharpoons \mathrm{AbAg}$$

characterized by the equilibrium constant,

$$K = [\mathrm{AbAg}]/[\mathrm{Ab}][\mathrm{Ag}]$$

If the antibody is covalently attached to the surface of a non-ionic substrate, it becomes part of the double layer at the substrate–solution interface and will form specific binding sites for antigen within the double layer. These antibodies are polyelectrolytes whose net charge (polarity and magnitude) will depend on the surrounding environment, most notably the pH. On binding the antigen (which may be charged or neutral) the resulting charge of the complex will cause a change in the charge distribution in the double layer.

A FET can be considered as a device that responds to changes in interfacial charge distribution (see Chapter 4), due to the series combination of C_{dl}, the capacitance of the double layer and C_o, the capacitance of the insulator causing the charge in the inversion layer of the semiconductor of Q_i. When adsorption takes place at the insulator interface, the charge Q_{ads} is transferred to the interface, across the double layer. The effect on Q_i will therefore be:

$$Q_i = \frac{Q_{ads}}{C_{dl}} \times \frac{C_o C_{dl}}{(C_o + C_{dl})}$$

so that,

$$\frac{Q_i}{Q_{ads}} = \frac{C_o}{(C_o + C_{dl})}$$

This predicts that the amount of charge which is mirrored in the inversion layer of the FET due to the adsorption of a charged species at the insulator–solution interface will be only some fraction (Q_i/Q_{ads}) of the total charge adsorbed (Q_{ads}).

In view of these considerations, Janata and co-workers (Janata and Huber, 1980; Janata and Blackburn, 1984; Blackburn, 1987) have estimated the theoretical sensitivity for an immuno-FET model. Assuming that the Langmuir adsorption isotherm can be applied and that the binding equilibrium between antibody and antigen can be described by the equilibrium constant K, then the interfacial potential can be described by

$$\phi_{sol-mem} = \frac{Q_i}{C_o} = \frac{Q_{ads} C_o}{C_o(C_o + C_{dl})} = \frac{zF}{(C_o + C_{dl})} \times \frac{K[Ag][B]}{1 + [Ag]}$$

where z is the ionic charge of the antigen and [B] is the total surface concentration of binding sites. By assigning various typical values, this equation can be solved for $\phi_{sol-mem}$ for a detection limit of 10 mV assuming that each antibody occupies $10\,nm^2$ and that K (the equilibrium constant) has values between 10^5 and 10^9 (see Chapter 2). A detection limit of $\geqslant 10^{-12}$ M is calculated, which as can be seen from earlier discussions of detection targets (see Chapter 1) is well within the requirements for an immunoprobe.

In order to apply this treatment to an IMFET, however, the membrane–solution interface must be ideally polarized, i.e. must act as a perfect capacitor, so that charge cannot cross the interface. In practice, there is always a leakage current associated with this interface, modelled by the introduction of a resistance R in parallel with the capacitance (C_{dl}), so that the leakage current will decay in an exponential manner, characterized by the time constant RC_{dl}. Janata and Blackburn (1984) calculate that a charge-transfer resistance $> 10^7\,\Omega$ to attain a usable time constant > 100 s would be required.

Immuno-FET Responses

The antigen-binding membrane developed by Aizawa *et al.* (1977), specific for the Wassermann antibody would seem ideal for application to an IMFET. Applied to the gate region of a CHEMFET, the device responded to additions of the antibody. Two mechanisms for a CHEMFET response have been considered:

(1) the migration of ions into an ion-selective membrane, causing a change in the charge distribution through the membrane; and
(2) the adsorption of charged species at the interface.

The dimensions of the protein molecules involved in immunoassay would restrict their involvement in an ion-exchange type process in the membrane, so that the immunoresponse would be of the second type. The response of the membrane to small inorganic ions present and a significant ion-exchange current density (10^{-5}–10^{-6} A cm^{-2}), led Collins and Janata (1982) to propose that the immunoresponse was due to the effect of adsorbed protein coupling to a mixed membrane potential. The primary response, therefore, was to the presence of small inorganic ions which was affected by the presence of protein adsorbed on the membrane surface.

In view of these findings, other 'IMFETs', which have been reported to show a direct response to the adsorption of protein in the gate region of a CHEMFET device, can probably be interpreted as an indirect secondary phenomenon due to the coupling of the adsorption with the ion-exchange process. This assumption is consistent with the results of Janata and Huber (1980) for various FET structures directed at the measurement of interfacial charge, particularly since the charge-transfer resistance was considerably less than the desired $10^7\ \Omega$. Indeed, in the absence of an ideally polarized surface, the direct potentiometric immunosensor cannot be realistically devised and it remains for highly specific secondary effects to be imaginatively coupled to the immunoreaction.

References

Aizawa, M., Kato, S. and Suzuki, S. (1977). *J. Membrane Sci.* **2**, p. 125.
Anfalt, T., Granelli, A. and Jagner, D. (1973). *Anal. Lett.* **6**, p. 969.
Blackburn, G.F. (1987). *Biosensors*, eds, Turner, A.P.F., Karube, I. and Wilson, G.S. Oxford University Press, pp. 481–530.
Boitieux, J.-L., Desmet, G. and Thomas, D. (1979). *Clin. Chem.* **25**(2), p. 318.
Boitieux, J.-L., Lemay, C., Desmet, G. and Thomas, D. (1981). *Clin. Chem. Acta* **113**, p. 175.
Bradley, C.R. and Rechnitz, G.A. (1985). *Anal. Chem.* **57**, pp. 1401–1404.
Brady, J.E. and Carr, P.W. (1980). *Anal. Chem.* **52**, p. 977.
Caras, S.D. and Janata, J. (1985). *Anal. Chem.* **57**, p. 1924.
Caras, S.C., Petelenz, D. and Janata, J. (1985). *Anal. Chem.* **57**, p. 1920.
Castner, J.F. and Wingard, L.B. (1984). *Anal. Chem.* **56**, pp. 2891–2896.
Collins, S. and Janata, J. (1982). *Anal. Chim. Acta* **136**, p. 93.

Cullen, L.F., Rusling, J.F., Schleifer, A. and Papariello, G.J. (1974). *Anal. Chem.* **46**, p. 1955.

Danielsson, B., Winquist, F., Malpote, J.Y. and Mosbach, K. (1982). *Biotechnol. Lett.* **4**, p. 673.

Danielsson, B., Mosbach, K., Winquist, F. and Lundström, I. (1988). *Sensors and Actuators* **13**, pp. 139–146.

del Castillo, J., Rodriguez, A., Romero, C.A. and Sanchez, V. (1966). *Science* **153**, p. 185.

Durand, P., David, A. and Thomas, D. (1978). *Biochim. Biophys. Acta* **527**, p. 277.

Eddowes, M.J. (1987). *Sensors and Actuators* **11**, p. 265.

Guilbault, G.G. and Montalvo, J.G. (1970). *J. Am. Chem. Soc.* **92**, p. 2533.

Guilbault, G.G. and Nagy, G. (1973). *Anal. Lett.* **6**, p. 301.

Guilbault, G.G. and Mascini, M. (1977). *Anal. Chem.* **49**, p. 795.

Guilbault, G.G., Smith, R.K. and Montalvo, J.G. (1969). *Anal. Chem.* **41**, p. 600.

Guilbault, G.G., Gutknecht, W.F., Kuan, S.S. and Cochran, R. (1972). *Anal. Biochem.* **46**, p. 200.

Guilbault, G.G., Chen, S.P. and Kuan, S.S. (1980). *Anal. Lett.* **13**(B18), p. 1607.

Guilbault, G.G., Lubrano, G.J., Kauffmann, J.-M. and Patriarche, G.J. (1988). *Anal. Chim. Acta* **206**, p. 369.

Hanazato, Y., Nakako, M., Maeda, M. and Shiono, S. (1987). *Anal. Chim. Acta* **199**, pp. 87–96.

Hsiung, C.P., Kuan, S.S. and Guilbault, G.G. (1977). *Anal. Chim. Acta* **90**, p. 45.

Ianniello, R.M. and Yacynych, A.M. (1981). *Anal. Chim. Acta* **131**, p. 123.

Janata, J. and Huber, R.J. (1980). *Ion Selective Electrodes in Analytical Chemistry*, Vol. 2, Ed. Freiser, H. New York, Plenum Press, pp. 107–174.

Janata, J. and Blackburn, G.F. (1984). *Ann. N. Y. Acad. Sci.* **428**, pp. 286–292.

Joseph, J.P. (1985). *Anal. Chim. Acta* **169**, p. 249.

Keating, M.Y. and Rechnitz, G.A. (1984). *Anal. Chem.* **56**, pp. 801–806.

Kobos, R.K. and Ramsey, T.A. (1980). *Anal. Chim. Acta* **121**, p. 111.

Matsuoka, H., Tamiya, E. and Karube, I. (1985). *Anal. Chem.* **57**, p. 1998.

Meyerhoff, M. and Rechnitz, G.A. (1976). *Anal. Chem.* **85**(2), pp. 277–285.

Morf, W.E., Kahr, G. and Simon, W. (1974). *Anal. Chem.* **46**, p. 1538.

Nagy, G., von Storp, H. and Guilbault, G. (1973). *Anal. Chim. Acta* **66**, p. 443.

Nakamoto, S., Ito, N., Kuriyama, T. and Kimura, J. (1988). *Sensors and Actuators* **13**, pp. 165–172.

Nilsson, H., Akerlund, A. and Mosbach, K. (1973). *Biochim. Biophys. Acta* **320**, p. 529.

Procter, A., Castner, J.F., Wingard, L.B. and Hercules, D.M. (1985). *Anal. Chem.* **57**, pp. 1644–1649.

Shinbo, T., Sugiura, M. and Kamo, N. (1979). *Anal. Chem.* **51**, p. 100.

Thompson, H. and Rechnitz, G.A. (1974). *Anal. Chem.* **46**(2), pp. 246–249.

Tor, R. and Freeman, A. (1986). *Anal. Chem.* **58**, p. 1042.

van der Schoot, B.H. and Bergveld, P. (1987). *Anal. Chim. Acta* **199**, pp. 157–160.

van der Schoot, B.H. and Bergveld, P. (1988). *Analytical Uses of Immobilised Biological Compounds for Detection, Medical and Industrial Uses*, NATO ASI Series, p. 195. D. Reidel Publishing Company.

Wingard, L.B., Wolfson, S.K., Lui, C.C., Yao, S.J., Schiller, J.G. and Drash, A.L. (1980). *Enzyme Eng.* **5**, p. 197.

Wingard, L.B., Cantin, L.A. and Castner, J.F. (1983). *Biochim. Biophys. Acta* **748**, p. 21.

Wingard, L.B., Castner, J.F., Yao, S.J., Wolfson, S.K., Drash, A.L. and Lui, C.C. (1984). *Appl. Biochem. Biotechnol.* **9**, p. 95.

Winquist, F., Spetz, A., Lundström, I. and Danielsson, B. (1984). *Anal. Chim. Acta* **163**, p. 143.

Winquist, F., Lundström, I. and Danielsson, B. (1985). *Sensors and Actuators* **8**, p. 91.

Winquist, F., Lundström, I. and Danielsson, B. (1986). *Anal. Chem.* **58**, p. 145.

Yamamoto, N.Y., Nagasawa, Y., Shuto, S., Tsubomura, H., Sawai, M. and Okumura, H. (1980). *Clin. Chem.* **26**(11), pp. 1569–1572.

Yao, T. and Rechnitz, G.A. (1987). *Anal. Chem.* **59**, p. 2115.

Chapter 10

Evolving Optical Biosensors

Indicator Labelled Bioassay

As discussed in earlier chapters, the majority of traditional bioassays are based on an optical technique. Many of these assays are suitable for conversion to 'solid-state' sensor devices through the use of an optical wave-guide transducer in the extrinsic or intrinsic mode (see Chapter 6).

It is possible to follow an evolutionary pathway for the optical bioassay from a solution reaction with added reagents through a reagentless extrinsic device and onto the exploitation of the intrinsic properties of waveguides (see Chapter 6). Devices in all categories of this development can be expected to find application.

Solid-phase Absorption Label Sensors

As was seen in Chapter 6, the initial requirement in transferring a solution spectrophotometric indicator assay to a waveguide sensor, was the immobilization of the indicator in an optical modulator. In the case of indicator-linked bioassay there will be the additional requirement of immobilization of the biorecognition molecule. While any of the established solution assays are potentially able to be converted to solid phase, in practice only a limited selection are suitable. In the first instance the criteria that need to be considered are

(1) choice of immobiliaztion support;
(2) immobilization of the indicator with retention of activity in the desired range;
(3) immobilization of the biorecognition molecule with retention of activity;

Table 10.1 LED light sources

Light source	λ_{max} (nm)
Blue	455–465
Yellow	580–590
Green	560–570
Red	635–695
Near infra-red	820
Infra-red	930–950
Detectors	
Photodiode	560 (460–750)
	750
	800
	850
	900 (350–1150)
Phototransistor	940

(4) cell geometry; and
(5) choice of source and detector components.

One of the aims of a biosensor is to create a portable device. This criterion would therefore require miniature low-power components such as LED light sources and photodiode detectors. The use of these components in a pH probe has already been demonstrated in Chapter 6. However, although the spectral range covered by LEDs is constantly increasing, their employment would place some degree of restriction on the wavelengths available for indicator excitation (Table 10.1). Nevertheless, the supporting electronics using these components can be exceedingly simple where only basic signal calibration is required (Fig. 10.1).

The optical properties of the immobilization matrix will depend on the chosen cell geometry. Optical sensors for pH, O_2, etc. have been described (see Chapter 6) where the indicator is covalently immobilized to polyacrylamide-coated microspheres and the scattered light detected. An ammonia sensor relied on light being reflected from an indicator-coated membrane. In these examples, the support serves as an optically inert reflecting phase.

ALBUMIN

Goldfinch and Lowe (1980) have reported an immobilization matrix which is optically clear, and is placed in the light path between source and detector. The Doumas test for albumin (Doumas *et al.*, 1971) is based on the formation of a complex between the triphenylmethane dye, bromocresol green and albumin in a solution buffered at about pH 3.8. On binding albumin the dye changes from yellow to blue, so that the spectral properties can be exploited in an assay for albumin.

The glutathione-conjugate of bromocresol green (Fig. 10.2) was covalently

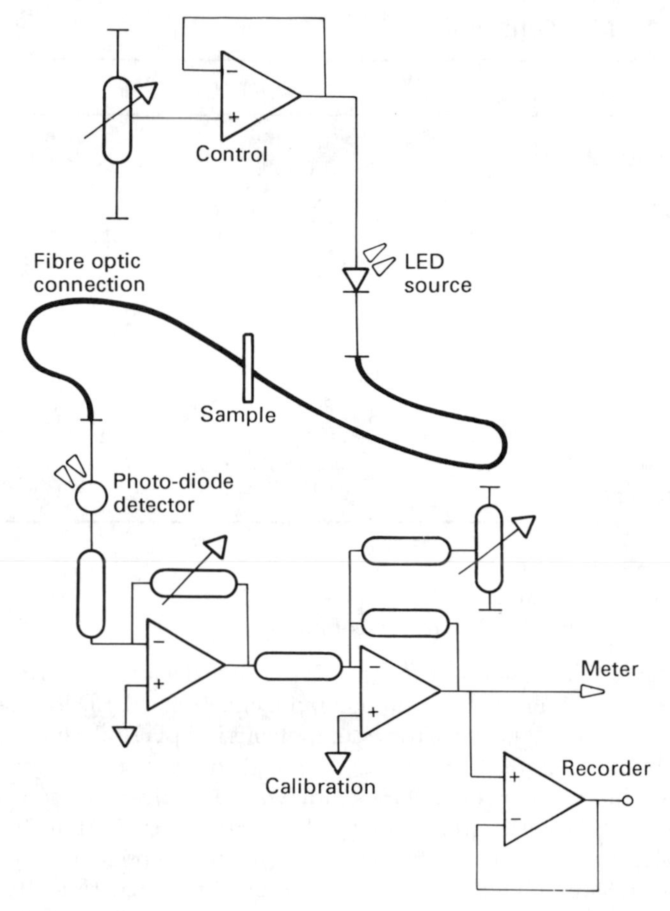

Fig. 10.1 Basic control circuit for optical assay employing LED light source and photodiode detector.

attached to a CNBr-activated optically clear cellulose membrane. The dyed indicator membrane was mounted as an integral part of a flow-through cell in the light path between a red LED source (λ_{max} 630–633 nm) and a silicon photodiode detector. The spectral properties of this indicator are particularly suitable for employment in this way since the albumin complex and free dye are well separated on the wavelength axis, and λ_{max} for the LED corresponds well with the λ_{max} for the albumin complex (Fig. 10.3). This system realizes a low-cost reagentless assay for albumin, linear in the 5–35 mg/ml range. Since the free dye can be regenerated from the albumin–dye complex, the assay cell is also potentially re-usable.

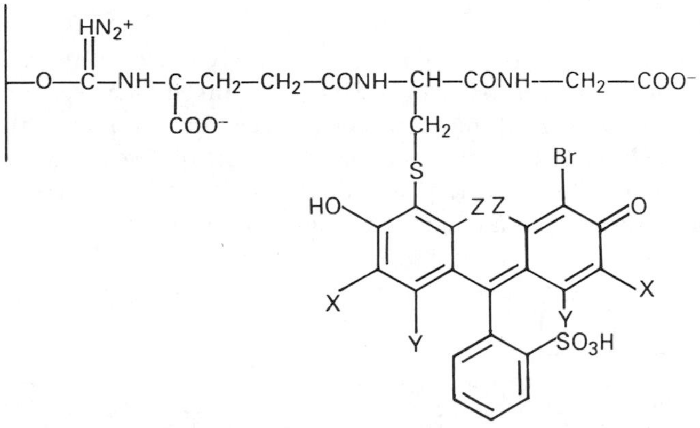

Fig. 10.2 Dye-modified cellulose membrane, glutathione dye conjugates: bromocresol green, X = Br, Y = CH_3, Z = H. Bromothymol blue, X = $-CH(CH_3)_2$, Y = H, Z = CH_3.

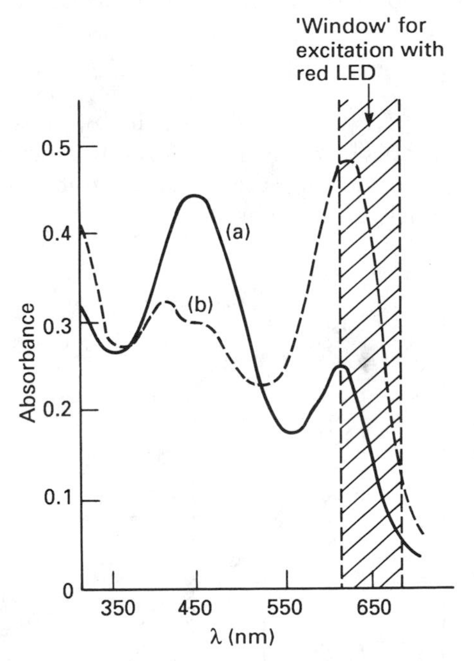

Fig. 10.3 Absorption spectrum for the bromocresol green–glutathione conjugate.

PENICILLIN

These triphenylmethane dyes are in fact primarily pH indicators, and as such could form the basis of a pH probe. Similarly, linked to a secondary reaction involving a change in $[H^+]$ they could be targeted towards many other analytes. This technique has already been seen for other types of base sensor, most notably the pH electrode. The enzyme penicillinase, for example, reacts with its substrate penicillin to give penicilloic acid and thus an increase in H^+:

RCONH … COOH (penicillin) —*Penicillinase*→ H^+ + RCONH … ^-O … N H … COOH (penicilloate)

An enzyme-indicator membrane has been prepared by covalently attaching the enzyme to the glutathione-conjugated bromocresol green-dyed cellulose membrane, via a carbodiimide reaction (Goldfinch and Lowe, 1984). The resulting membrane responds to penicillin or penicillin analogues in the 0–10 mM range. The half-life for this sensor is estimated to be in excess of one year.

UREA

At the other end of the pH spectrum, the reaction of the enzyme urease on its substrate urea results in a net increase in pH. An optical membrane tuned to urea can be constructed by attaching urease to a bromothymol blue-modified membrane, by glutaraldehyde crosslinking.

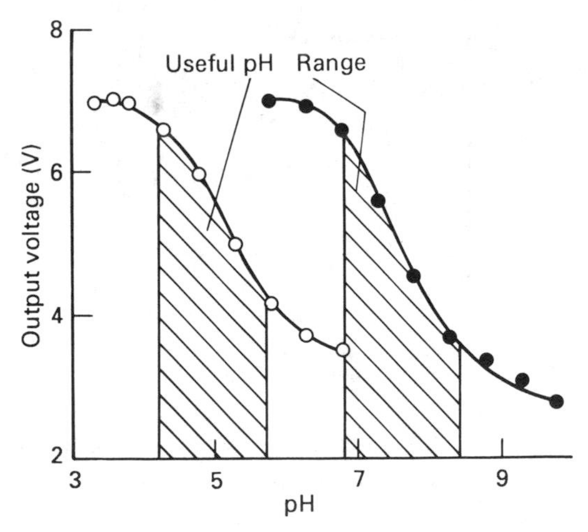

Fig. 10.4 pH response of bromocresol green (○) and bromothymol blue (●) membranes. (Goldfinch and Lowe, 1984).

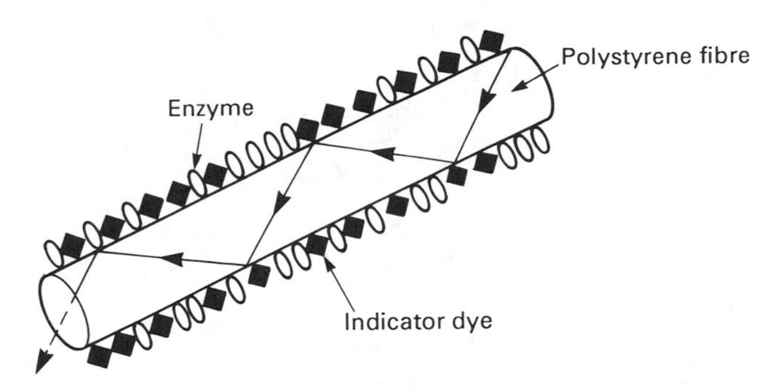

Fig. 10.5 Model of an intrinsic optical enzyme sensor based on a polystyrene fibre.

These two indicator dyes have pK_a values separated by 2 pH units in the immobilized form (Fig. 10.4). Bromocresol green (pK_a 5.1) has a working pH range of 3.3–6.8, whereas bromothymol blue (pK_a 7.5) changes from yellow to blue over the range pH 6–10. Optimization of the enzyme–indicator system relies on the correct choice of pH indicator to achieve maximum activity.

INTRINSIC MODEL

To develop an intrinsic optical modulator with these indicator and biochemical components requires an immobilization matrix with optical waveguide properties. Bromocresol green has been immobilized on a polystyrene fibre together with adsorbed penicillinase (Hall, unpublished results). Here the assay reagents are directly attached to the optical transducer, and the fibre responds to absorption of light propagating along the waveguide, by dye molecules immobilized within the evanescent field (Fig. 10.5).

p-NITROPHENYLPHOSPHATE

Arnold (1985) reported a feasibility study for an enzyme-based fibre-optic sensor, incorporating an immobilization matrix with light-scattering properties. These features obviously require a different modulator design to those employing an optically 'clear' enzyme matrix. The enzyme was covalently attached to the inner layer of a double nylon membrane, covering the common end of a bifurcated fibre bundle (Fig. 10.6). The enzyme in this model system was alkaline phosphatase, directed at the model analyte *p*-nitrophenylphosphate:

$$O_2N^{+}\!-\!C_6H_4\!-\!O\!-\!P(O^-)(=O)\!-\!O^- + HOH \xrightarrow[\text{phosphatase}]{\text{Alkaline}} {}^-O_2N^{+}\!=\!C_6H_4\!=\!O + O^-\!-\!P(OH)(=O)\!-\!OH$$

(*p*-nitrophenoxide)
(λ_{max} 404 nm)

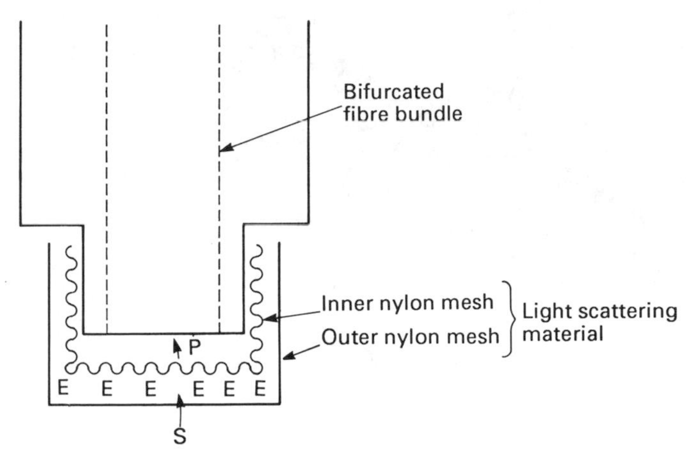

Fig. 10.6 Arnold design for a model enzyme sensor, using the enzyme alkaline phosphatase.

The product of this enzyme-catalysed reaction acts as its own indicator since it has a large molar extinction coefficient at 404 nm, and so no additional optical indicator is required. This obviously restricts the applicability of this model to systems that can be linked to alkaline phosphatase (consider for example, the amplification system involving alkaline phosphatase, see Chapter 8).

Chemiluminescent Labels

As was seen in Chapter 6, the most widely used chemiluminescent reagent is the cyclic hydrazide, luminol, employing peroxidase-catalysed hydrogen peroxide as the oxidizing system. Since so many oxidase enzymes produce hydrogen peroxide this technique has been linked to a number of different analytes.

Freeman and Seitz (1978) have evaluated a fibre-optic sensor for hydrogen peroxide based on the luminol reaction, by immobilizing peroxidase in a polyacrylamide gel containing luminol at the end of a fibre optic. This design of biosensor should illustrate one of the particular characteristics of an optical sensor. The luminescence is detected *in situ*, that is *without diffusion* of the luminescent species to the probe surface being involved prior to its detection, and it is thus independent of the thickness of the membrane under conditions of $[S] \ll K_M$. However, since in this instance the enzyme probe reaction was second order with respect to hydrogen peroxide, the rate-limiting step is the first-order mass transfer of the substrate to the probe surface. As peroxide is consumed by the probe, it is not a passive measurement so this combination leads to depletion of the substrate immediately next to the enzyme layer (Fig. 10.7). At high peroxide concentrations this can therefore cause significant differences between stirred and unstirred samples. Since this transduction system requires no light source, Aizawa *et al.*

(1984) have attached an acrylamide enzyme gel directly to a photodiode detector. The biophotodiode responded to sample hydrogen peroxide in the range 1–10 mM when luminol was added to the sample solution. The photocurrent reached a maximum in 2 min.

In linking such a chemiluminescent based probe to hydrogen peroxide-producing reactions such as

$$\text{glucose} + O_2 \xrightarrow{\textit{glucose oxidase}} \text{gluconic acid} + H_2O_2$$

the kinetics of the signal-producing step with respect to other reactions involved in the assay system are critical to the linearity of the concentration response. For the biophotodiode a near linear response was reported for the range 0.1–1.5 M glucose.

Bioluminescence

Ford and DeLuca (1981) have developed an assay method for testosterone and androsterone linked to the bacterial luciferase system via NADH (see Chapter 6), with the enzymes covalently linked to Sepharose 4B:

$$\text{androsterone} + NAD^+ \xrightarrow{\textit{hydroxysteroid dehydrogenase}} \text{5-androstane 3,17-dione} + NADH$$

$$\text{testosterone} + NAD^+ \xrightarrow{\textit{hydroxysteroid dehydrogenase}} \text{4-androstene 3,17-dione} + NADH$$

Testosterone was detected in the 0.8 pmol to 1 nmol range and androsterone in the 0.3 pmol to 2 nmol range, with a reproducibility of ± 1–3%. This is not a sensor since the detector (transducer) is not an integral part of the device, and in practice

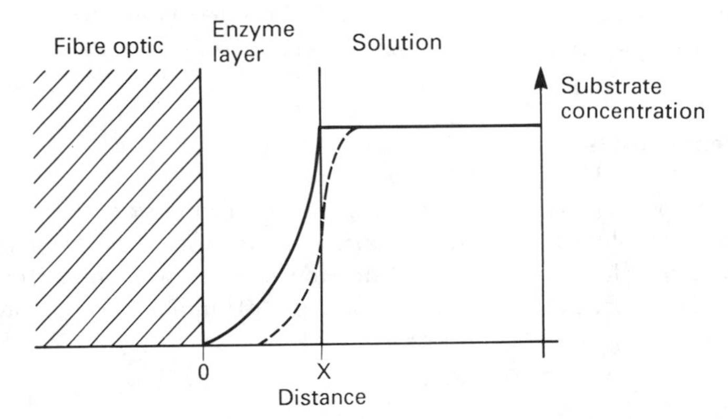

Fig. 10.7 Concentration profiles for substrate concentration in the layers next to the optical probe, in a chemiluminescent sensor where substrate is consumed (——) without depletion of surface substrate and (----) when the reaction is sufficiently rapid to deplete the surface concentration of substrate.

Table 10.2 Reactions involving immobilized bacterial luciferase

Analyte	*Coupling reactions**	*Range*
ATP	—	$0.12–2\times10^3$ pmol
NADH/NAD(P)H	—	1 pmol–50 nmol/10 pmol–200 nmol
Androsterone (A), testosterone (T)	$A(T)\text{–}OH + NAD^+ \xrightleftharpoons{HDH} A(T){=}O + NADH$	0.5 pmol–1 nmol
Ethanol (E)	$E + NAD^+ \xrightarrow{ADH}$ acetaldehyde $+ NADH$	0.01–10 pmol
D-Glucose (G)	$(G) + ATP \xrightleftharpoons{HK}$ G-6-phosphate(G-6-P) $+ ADP$	2–100 pmol
G-6-Phosphate	$G\text{-}6\text{-}P + NADP^+ \rightleftharpoons$ G-6-P-gluconolactone $+ NADPH$	1 pmol–20 nmol
L-Lactate (L)	$L + NAD^+ \xrightleftharpoons{LDH}$ pyruvate $+ NADH$	2 fmol–1 pmol
L-Malate (M)	$L\text{–}M + NAD^+ \xrightleftharpoons{MDH}$ oxaloacetate $+ NADH$	

*Enzymes: HDH = hydroxysteroid dehydrogenase; ADH = alcohol dehydrogenase; HK = hexokinase; LDH = lactate dehydrogenase; MDH = malate dehydrogenase.

the Sepharose 4B is stirred in the sample solution. A closer approach to a biosensor has been made with the luciferase and oxidoreductase co-immobilized onto acrylamine-coated glass beads glued to a glass rod but this method is not reported to give as good a precision as the Sepharose 4B beads. The activity and stability of luciferase immobilized on various supports has been studied by Brovko *et al.* (1980) and Ugarova *et al.* (1982). In principle, any of the immobilization matrices would show potential for application in an optical cell very similar in design to that reported by Peterson *et al.* for a pH probe (see Chapter 6). They have already been tested for many analytes targeted by NAD-dependent oxido-reductases, linked into the bacterial luciferase system (Table 10.2).

Fluorescent Labels

The use of fluorescent labels in the development of solid-state sensors has been more widely employed than the colorimetric counterparts—probably since fluorescent labels have already been developed for a wide range of solution bioassay where either a 'reactive' indicator (e.g. a pH indicator) or a non-reactive 'tag' (e.g. antibody label) is required.

ESTERASES

A similar design to that described by Arnold was employed by Wolfbeis (1986) in a probe for the kinetic determination of enzyme activities. In this instance, however, the immobilized species were the substrates for cholinesterase and related carboxylesterases and the enzyme took the role of analyte. The fluorescent indicator, 1-hydroxypyrene-3,6,8-trisulphonate (HPTS), was immobilized on an ion-exchange membrane and then acylated to produce the desired ester. Since source and detection wavelengths were different, the membrane could be mounted over the end of a single 100 μm fibre which was divided at an optical coupler towards source and detector.

^-O_3S OR ^-O_3S SO_3^- —*eterase*→ ^-O_3S OH ^-O_3S SO_3^- ⇌ (pK_a 7.0) ^-O_3S O^- ^-O_3S SO_3^-

(substrate) (HPTS) (HPTS phenolate ion)

Enzyme activity was measured by the fluorescence at 520 nm, following excitation of the phenolate anion at 460 nm. The determination was limited by the rates of enzymic versus non-enzymic hydrolysis, which varies with different enzyme–ester combinations.

PENICILLIN

The change in $[H^+]$ which accompanies the enzyme-catalysed conversion of penicillin to penicilloic acid has been linked to the fluorescent pH indicator fluorescein isothiocyanate (FITC) (Fuh *et al.*, 1988). This is the same configuration of sensor as the Peterson pH probe, with the enzyme penicillinase immobilized on the indicator-modified glass bead in a thin film, by glutaraldehyde crosslinking. Working in the linear pH range for the indicator, and at an optimum pH for enzyme activity (pH 6.8), the sensor would respond to penicillin up to 10 mM with a detection limit of 0.1 mM. A slight decrease in the activity of the probe was noticed, amounting to 5% in a 5-day period, but this is consistent with usual findings involving glutaraldehyde crosslinked enzymes.

CONCANAVALIN A BINDING PROTEIN

A competitive binding sensor based on fluorescence, more akin to an immunoassay, has been developed for glucose, which depends on the numerical aperture of the optical fibre and the fact that the light source can be focused to some extent, so that components can be immobilized outside the 'field of view' of the fibre.

Concanavalin A (ConA), a binding protein for glucose and other sugars of similar molecular weight, is immobilized by glutaraldehyde crosslinking on the inner surface of a hollow dialysis fibre, outside the 'field of view' as demonstrated in Fig. 10.8 (Schultz *et al.*, 1982). Fluorescein-labelled dextran is introduced into

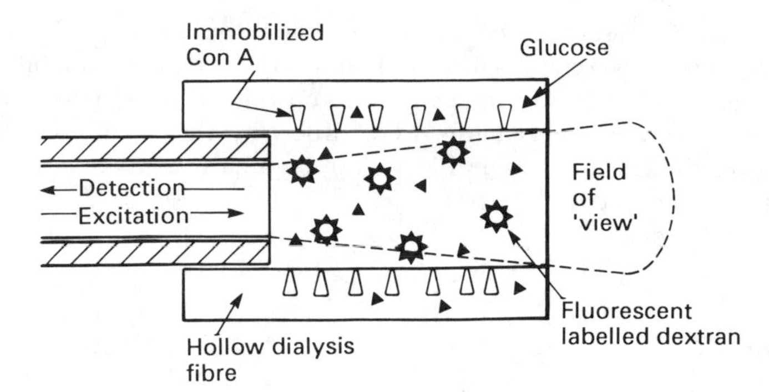

Fig. 10.8 Competitive optical immunoassay probe for glucose, involving the glucose binding protein concanavalin A (ConA) and fluorescent-labelled dextran.

the optical modulator cell as the competing sugar ligand. Increasing the glucose concentration displaces the labelled dextran from the immobilized concanavalin A, causing it to diffuse freely into the volume illuminated by the fibre. This in turn leads to a glucose concentration-related increase in fluorescence. In a later version of this sensor reproducibility was enhanced by improving the ConA immobilization efficiency and optimizing the concentration of ConA required for maximum response (Mansouri and Schultz, 1984; Srinivasan *et al.*, 1986). It was found that this was best achieved by covalently attaching the ConA via a glutaraldehyde spacer to the inside of the oxidized cellulose fibre. By performing the coupling with a low concentration of glutaraldehyde also present on the other side of the dialysis fibre, the amount of ConA bound could be optimized. The resultant sensor responded to plasma glucose in the range 0.5–4 mg/ml with a drift of 15% in 15 days and *in vitro* measurements in blood suggested that the sensor could measure to an accuracy of ± 0.13 mg/ml.

The response of this sensor to any particular sugar will depend on the relative affinity ratios for the two equilibria concerned (compare biotin affinity sensor, Chapter 9).

$$A + Bp \rightleftharpoons A:Bp; \quad K_a = \frac{[A:Bp]}{[A][Bp]}$$

$$a^* + Bp \rightleftharpoons a^*:Bp; \quad K_{a*} = \frac{[a^*:Bp]}{[a^*][Bp]}$$

where a^* is labelled analyte analogue, A is analyte and Bp the common binding protein. Total analyte analogue will therefore be given by

$$[a^*]_t = [a^*] + [a^*:Bp]$$

total analyte by

$$[A]_t = [A] + [A:Bp]$$

and total binding protein by

$$[Bp]_t = [Bp] + [A:Bp] + [a^*:Bp]$$

$$= [Bp] + K_a[A][Bp] + [a^*]_t - [a^*]$$

$$= \frac{[a^*:Bp]}{[a^*]K_a^*} + \frac{K_a[A][a^*:Bp]}{[a^*]K_a^*} + [a^*]_t - [a^*]$$

so that

$$[a^*] = [a^*:Bp]\left[\frac{1+K_a[A]}{K_a^*[a^*]}\right] + [a^*]_t - [Bp]_t$$

$$\frac{[a^*]}{[a^*]_t} = \left[\frac{[a^*]_t - [a^*]}{[a^*]_t}\right]\left[\frac{1+K_a[A]}{K_a^*[a^*]}\right] + 1 - \frac{[Bp]_t}{[a^*]_t}$$

$$\frac{[a^*]}{[a^*]_t} = \left[1 - \frac{[a^*]}{[a^*]_t}\right]\left[\frac{1+K_a[A]}{K_a^*[a^*]}\right] + 1 - \frac{[Bp]_t}{[a^*]_t}$$

showing that the normalized response for $[a^*]/[a^*]_t$ is dependent on two groups of parameters: $[Bp]_t$ and $(1+K_a[A])/(K_a^*[a^*]$. (Similar conclusions can be drawn about thc ratio $[A]/[A]_t$.) More detailed examination of these interdependencies shows that $[Bp]_t > [a^*]_t$ is required for efficient operation, so there is not an excessive amount of unbound a^*, but if $[Bp]_t/[a^*]_t$ is too large then the signal due to analyte-induced release of labelled analogue (a^*) is small. The best concentrations of $[Bp]_t$ and $[a^*]$ will depend on the relative values for K_a and K_a^*.

Instead of these extrinsic configurations, an intrinsic optical transducer can be devised with the fluorescent label immobilized directly on the waveguide, and coupled with light propagating in the evanescent field.

Immunological Sensors

There is a considerable literature concerning the use of total internal reflection fluorescence to examine the non-specific interaction of proteins with surfaces in different states of hydrophobicity and composition. Newby *et al.* (1984), for example, have shown that rhodamine-labelled IgG adsorbed on the remote sensor tip of an optical fibre could be detected due to fluorescence coupling into the fibre via the evanescent wave.

The use of the technique for specific interactions at biorecognition surfaces also attracts attention for immunoassay applications. The original work (Kronick and Little, 1973, 1975) employed the haptens, morphine and phenylarsonic acid, immobilized at a quartz waveguide as the hapten–albumin conjugate. Fluorescein-labelled antibody was reacted with the immobilized hapten, so that the fluorescent label was within the evanescent field. On addition of sample

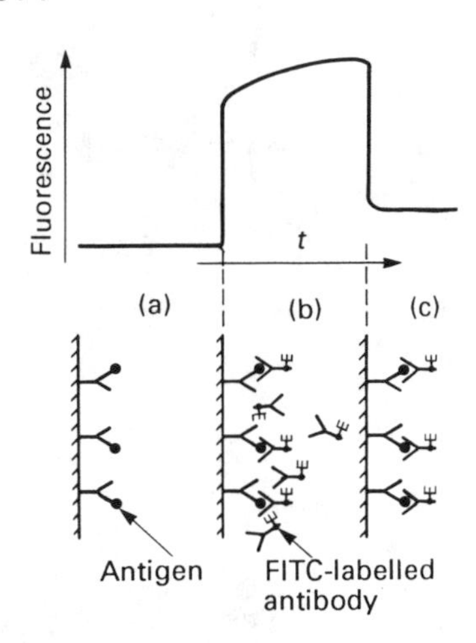

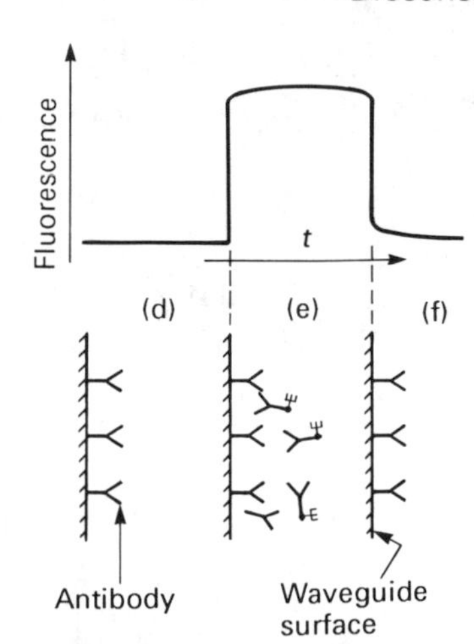

Fig. 10.9 Sandwich immunoassay using fluorescent-labelled antibody. Comparison of the signal due to (a)–(c) specific and non-specific interaction of the labelled antibody with the probe, and (d)–(f) non specific signal due to the labelled antibody.

solution, hapten-labelled antibody was displaced from the surface-immobilized hapten, thus removing fluorescent label from the evanescent field at a rate proportional to the concentration of sample hapten. A detection limit for morphine of 0.2 μmol/l was claimed.

Sutherland *et al.* (1984) investigated two sandwich immunoassay systems with anti-IgG immobilized on quartz slides (Fig. 10.9a) or fibres. Anti-IgG was immobilized on the waveguide by glutaraldehyde linkage at an activated surface. The immobilized antibody was incubated with sample antigen and then with a second fluorescein-labelled antibody (Fig. 10.9b). The binding of the labelled antibody within the evanescent field was related to concentration of sample antigen. With many of these immuno-equilibria the problem of non-specific surface interactions can be significant. In this instance, the baseline or non-specific signal due to the label, was provided by the response of labelled antibody in the absence of antigen. As can be observed from the diagrammatic representation of the sensor response in Fig. 10.9(d), this optical signal is in fact instantaneous and does not depend on the kinetics of the binding equilibria between antibody and antigen. The 'zero' baseline could therefore be estimated on the time axis by curve analysis, thus obviating the need for a sample blank.

Hirschfeld (1984) designed a disposable capillary-fill immunosensor, with the capillary volume defining a reproducible sample size. Incorporating an intrinsic optical transducer format, this concept was adopted in a capillary-fill immunosensor (Smith, 1986; Badley *et al.*, 1987), consisting of two glass slides (Fig. 10.10) separated by a narrow capillary space ($< 100\,\mu$m). The lower slide functions as a waveguide, and has antibody immobilized to its upper surface.

Fluorescein-labelled antigen is retained on the lower surface of the upper slide in a soluble matrix. On addition of a solution of sample antigen, the labelled antigen is released into the sample solution and diffuses across the gap to compete for the immobilized antibody sites on the waveguide (Fig. 10.10b):

$$\mathrm{Ab}+\mathrm{Ag}+\mathrm{Ag}^{*} \rightleftharpoons \mathrm{Ab}:\mathrm{Ag}+\mathrm{Ab}:\mathrm{Ag}^{*}$$

At equilibrium the amount of antibody-bound labelled antigen (Ab : Ag*) or free labelled antigen (Ag*) can be related to the concentration of sample antigen. Alternatively, the ratio Ab : Ag*/Ag* would give a normalized response which compensated for sampling fluctuations.

Obviously, the diffusion process will dominate for reaction times greater than that required to establish the immunological equilibrium. With a 10 nM concentration of protein, for example, with forward and reverse rate constants given by $k_1 \sim 10^5\ \mathrm{M}^{-1}\,\mathrm{s}^{-1}$ and $k_2 \sim 10^{-4}\ \mathrm{s}^{-1}$ respectively, the time for equilibrium to be established is usually found to exceed 900 s.

Since the time for diffusion across the capillary gap is given by

$$t = d^2/D$$

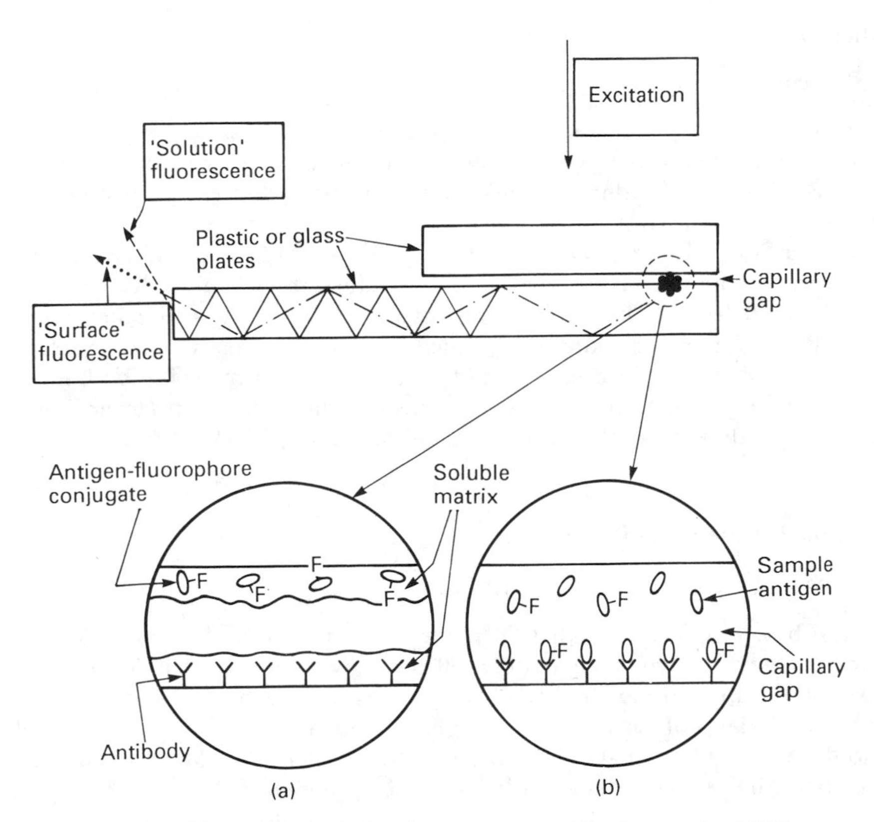

Fig. 10.10 Capillary-fill sandwich immunoassay (Badley *et al.*, 1987).

where d is the gap and D the diffusion constant for the antigen, then diffusion will dominate when t is greater than the equilibrium time. With an equilibrium time of this order and $D \sim 10^{-11}\ m^2\ s^{-1}$, that means for gaps >0.09 mm, or conversely for a 0.1 mm gap, diffusion will dominate for concentrations >9 nM. It can be seen that narrower gaps would give measurements independent of diffusion for much higher concentrations of protein.

In this device, the fluorescent label is excited with a photographic flash, and the fluorescence collected by the waveguide and detected by photodiodes at the flat perpendicular end of the guide. The fluorescence from the solution labelled antigen couples into the guide by refraction at angles below the critical angle (see Chapter 6), while the fluorescence from the antibody–antigen complex at the interface between the guide and the solution, couples into the guide via the evanescent field (see Fig. 10.10). Emission from the end of the guide, measured with respect to angle (ϕ) will therefore give a normalized estimate of the sample antigen.

Fluorescence due to the solution label, which has coupled into the guide by refraction, will propagate along the guide and emerge at angles

$$\cos\phi < n_1 \cos\theta_c$$

where θ_c is the critical angle, i.e.

$$\cos\phi < \sqrt{(n_1^2 - n_2^2)}$$

The device was reported initially with a human immunoglobulin G–antibody model, but further tests are reported to have resulted in reproducible results for a number of analytes within clinically useful precision and range (Badley *et al.*, 1987).

The authors point out that an advantage of these waveguide sandwiches is that they can be manufactured using the mass production technology for producing liquid crystal displays. The capillary fill feature of this device means that sampling and filling is automatic, taking a reproducible sample volume. The 'cell' geometry also has implications for other types of biosensor, and its exploitation has been described for electrochemically based biosensors, where film electrodes are laid on one plate of the sandwich, within the capillary space (Birch, 1987).

External Reflection Techniques

BREWSTER ANGLE MEASUREMENTS

It was shown in Chapter 6 that for plane (TM) polarized light incident on an interface between two phases of different refractive indexes, the reflection, R_p, is zero for the Brewster angle, i.e. when $\tan^{-1}\theta = n_1/n_2$ where n_1 and n_2 are the refractive indexes of the media where light is incident and refracted, respectively, and θ is the incident angle. For a transparent incident phase and an absorbing substrate, the refractive index n_2 is a complex number, $\mathcal{N}_2$,

$$\mathcal{N}_2 = n_2 + ik_2 \text{ (where } k_2 \neq 0)$$

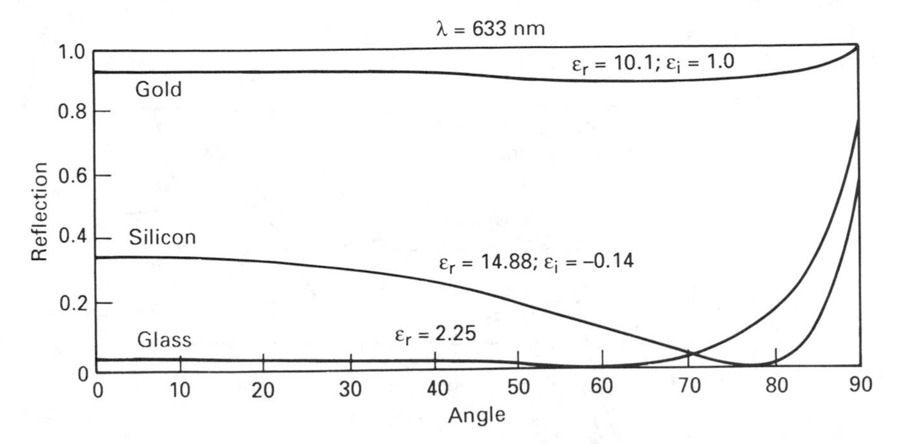

Fig. 10.11 The effect of the relative magnitude of the real and imaginary parts of the refractive index on the reflection of *p*-polarized light.

R_p shows a minimum at the so-called *pseudo-Brewster angle*. This minimum is sharp for highly polarizable materials with low absorption, i.e. a material with a larger real part and smaller imaginary component to the dielectric (Fig. 10.11). Polished silicon is such a substrate with a suitable refractive index. The angle θ at which a minimum in R_p is observed is very sensitive to the presence of overlayers on the substrate surface. The potential suitability of silicon for the immobilization of biomolecules suggests that biomolecular reactions may be resolved by this technique. The change in R_p at a given angle of incidence ($\approx$Brewster angle) can be related to the absolute thickness of the surface layer with a resolution which will depend on the instrumental resolution and the maximum reflectance change (ΔR) for a layer of thickness d:

$$\Delta d_{\min} = \left(\frac{\Delta R}{\Delta d}\right)^{-1} \Delta R_{\min}$$

Assuming the reflection measurement can be obtained with an accuracy $\Delta R_{\min} = 0.025\%$, then calculations made for a typical sensor model suggest a minimum thickness change ($\Delta d_{\min}$) of 0.05 nm is detectable.

Arwin and Lundström (1985) and Welin *et al.* (1984) have tested the technique for immunological reactions between antibody and antigen and suggested an upper detection limit for their feasibility model of 0.8 μg/cm^2, and a lower limit of 0.02–0.05 μg/cm^2, which corresponds to a surface thickness change of 0.2–0.4 nm. Welin *et al.* (1984) suggest that the technique would allow the detection of 1 μg/cm^3 protein in solution. An antigen such as human IgG, human serum albumin (HSA), bovine serum albumin (BSA) or γ-globulin was coated onto silicon surfaces, and the response to specific antibody recorded as a change in reflectivity. Cross reactivity between immobilized γ-globulin was also tested and found to be insignificant (Fig. 10.12). A potential weakness of such a method,

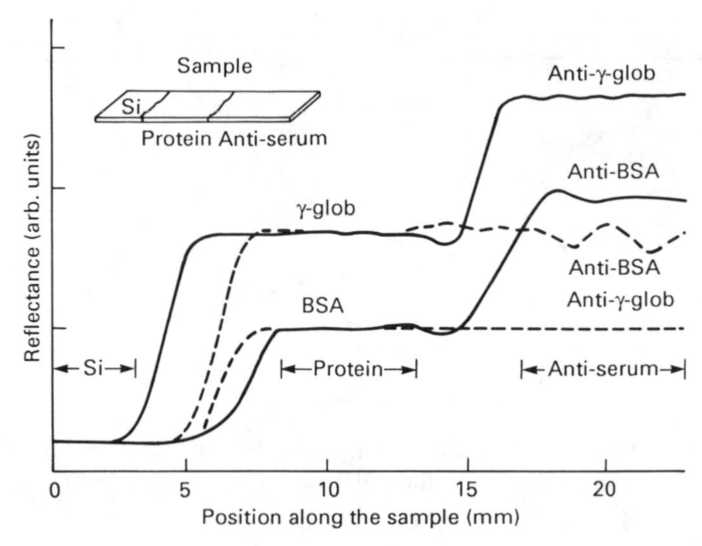

Fig. 10.12 Thickness profiles measured along an antigen-covered silicon sample incubated with antisera. (——) Samples exposed to specific antisera; (----) samples exposed to non-specific sera. (From Arwin and Lundström, 1985, with permission.)

which makes no distinction between specific and non-specific adsorption in the measurement, is however the interference from non-specific binding.

Similar observations were made by Stange *et al.* (1988) for silicon wafers coated with HSA, and their response to solution antibody. A minimum antibody concentration of $10\,\mu g/cm^3$ is reported here.

In these test models where the biorecognition antigen was coated onto the silicon from serum, and the response to specific and non-specific proteins tested in serum solutions, the residual adsorption was negligible, but in crude biological samples this could become a significant part of the signal. In devising the immobilized biorecognition surface, therefore, it is necessary to inhibit any non-specific interaction with the base surface which cannot be separated from the main signal.

This technique allows reactions to be monitored at a surface easily at high speed, using instrumentation which is considerably less complex than that required for ellipsometry. On the other hand, however, the method does not show the same accuracy as ellipsometry and yields only one gross property, the optical thickness $n_{\text{film}}d$. Particular care must therefore be exercised in interpreting changes in reflectance, which may be due to surface inhomogeneities.

INTERFERENCE ENHANCED REFLECTIVITY

The high attenuation of reflection behaviour from dielectric multilayers has been exploited to study the adsorption/desorption of ultra-thin organic layers (Laxhuber *et al.*, 1986).

The technique requires a non-absorbing dielectric (e.g. SiO_2) with dielectric constant $\varepsilon_1 = \varepsilon_{1,r}$, deposited to a thickness of about one-quarter of a wavelength on a dielectric $\varepsilon_0 = \varepsilon_{0,r} - i\varepsilon_{0,i}$ (e.g. silicon), which exhibits a highly reflecting surface. A layer deposited at the SiO_2–sample interface (dielectric constant $\varepsilon_2 = \varepsilon_{2,r} - i\varepsilon_{2,i}$) can be studied, since reflection from its surface modulates that emanating from the lower surfaces.

It is possible to predict the effect of layer characteristics on the attenuation of the measurement, and it can be seen that (Fig. 10.13) the most significant changes in reflectance, calculated as a function of SiO_2 thickness, occur for *s*-polarized light, and in the range 90–120 nm and 160–190 nm SiO_2, with a reflectance minimum at about 140 nm.

With this technique, scattering caused by surface inhomogeneities, which always leads to a decrease in reflectance, can be accounted for. A comparable measurement can be made in the region of decrease in reflectance with thickness (90–120 nm SiO_2) and increase in reflectance with thickness (160–190 nm SiO_2). The scattering error will be common to both measurements and can therefore be eliminated from the signal.

Other multiple dielectric layer sandwiches can also be devised, suitable for producing an analyte recognition interface (Hall and Duschl, forthcoming). The method is particularly sensitive to changes in the imaginary part of the dielectric constant of the overlayer, so that it is possible to devise many analyte-sensitive overlayers which undergo a transition from a coloured to a transparent (or *vice versa*) state on exposure to analyte.

It is also worth considering analyte-selective membranes in this technique. Ion-selective membranes and many bioselective membranes undergo a change in optical thickness in response to the analyte. Since the determination of absolute layer thickness by this method is at least twice as sensitive as that obtained by Brewster angle measurements, it may offer some advantages. Although interference enhanced reflectivity (IER) has only recently been exploited for sensor developments, it promises to become a most powerful tool.

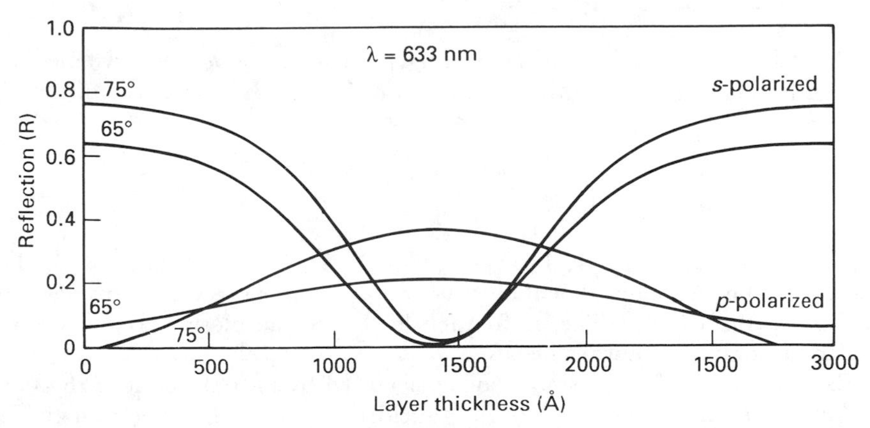

Fig. 10.13 Reflection of *s*- and *p*-polarized light at a surface: silicon–SiO_2. The effect of the thickness of the SiO_2 layer.

Surface Plasmon Resonance (SPR)

The wave treatment of Chapter 6 can be developed to consider the case of an oscillating electron density at the interface between two materials of different dielectric constants. Oscillations in the electron density of the valence electrons can be induced in the interior of a solid on excitation by electrons or light. Alternatively, the action of an exterior electric field on a plasma boundary causes discontinuities in the normal component (TM) of the electric field across the interface. The oscillating surface charges so produced move as a surface wave along the interface:

$$\sigma_{(x,t)} = \sigma \exp i(\mathbf{k}_x x - \omega t)$$

The oscillations couple with high-frequency electromagnetic fields extending into space (z-direction), so that the dependence of E_z on z is given by

$$E_z = \text{const} \exp i(\mathbf{k}_x x - \omega t) \exp i\mathbf{k}_z z$$

where the wave vector $\mathbf{k}_z$ is coupled to $\mathbf{k}_x$ according to,

$$\mathbf{k}_x^2 + \mathbf{k}_z^2 = \varepsilon(\omega/c)^2$$

for a medium with dielectric constant ε. Thus when $\mathbf{k}_x > \sqrt{\varepsilon}(\omega/c)$, then $\mathbf{k}_z$ is imaginary and the field decreases exponentially from the interface with no transport of energy (non-radiative). The wave is thus confined at the interface with $\mathbf{k}_x$ (the wave vector) in the propagation direction and $\mathbf{K}_1$ and $\mathbf{K}_2$ the wave vectors in the $+z$- and $-z$-directions,

$$\mathbf{K}_1 = [\mathbf{k}_x^2 - \varepsilon_1(\omega/c)^2]^{1/2}$$
$$\mathbf{K}_2 = [\mathbf{k}_x^2 - \varepsilon_2(\omega/c)^2]^{1/2}$$

The boundary conditions are satisfied only for p-polarized light (TM).

Since the tangential components of the electric and magnetic fields must be continuous across the interface,

$$H_1 = H_2;\ \mathbf{K}_1/\varepsilon_1 = \mathbf{K}_2/\varepsilon_2$$

so that for a surface plasmon wave existing at the interface between two media of dielectric constants ε_1 and ε_2, substitution of $\mathbf{K}_1$ and $\mathbf{K}_2$ yields the dispersion relation

$$\mathbf{k}_x = \left(\frac{\omega}{c}\right)\left(\frac{\varepsilon_1\varepsilon_2}{\varepsilon_1 + \varepsilon_2}\right)^{1/2}$$

and $\mathbf{k}_x^2 > (\omega/c)^2\varepsilon_1, (\omega/c)^2\varepsilon_2$; then $\varepsilon_1 < 0$ and $|\varepsilon_1| > \varepsilon_2$, or $\varepsilon_2 < 0$ and $|\varepsilon_2| > \varepsilon_1$. The most commonly employed media for the propagation of surface electromagnetic waves are the metals, since at frequencies below the plasma frequency, the dielectric constant is always negative.

Excitation of surface plasmon can be achieved by incident light of the same frequency and $\mathbf{k}_x$ component. For light incident at an interface between air and the plasmon material, with wave vector $\mathbf{k}$, then the x component of the vector $(\mathbf{k}_l)$

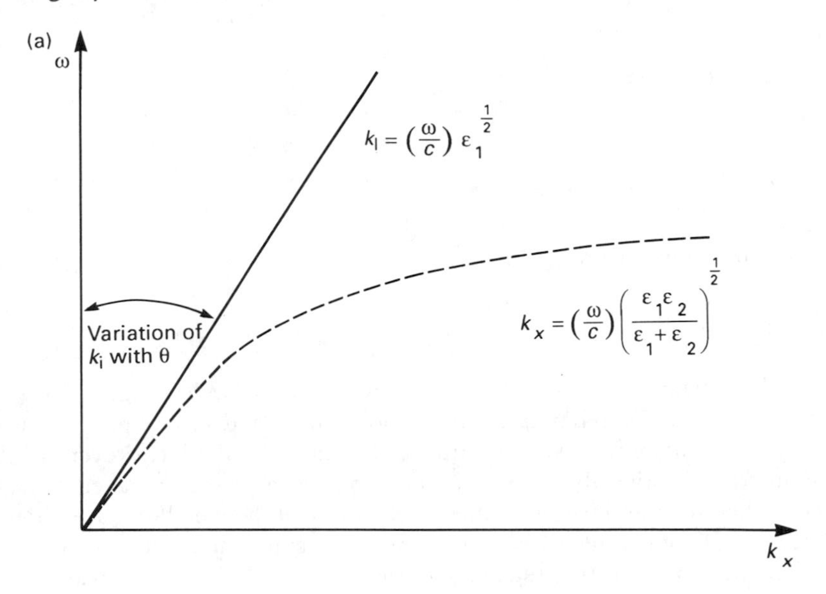

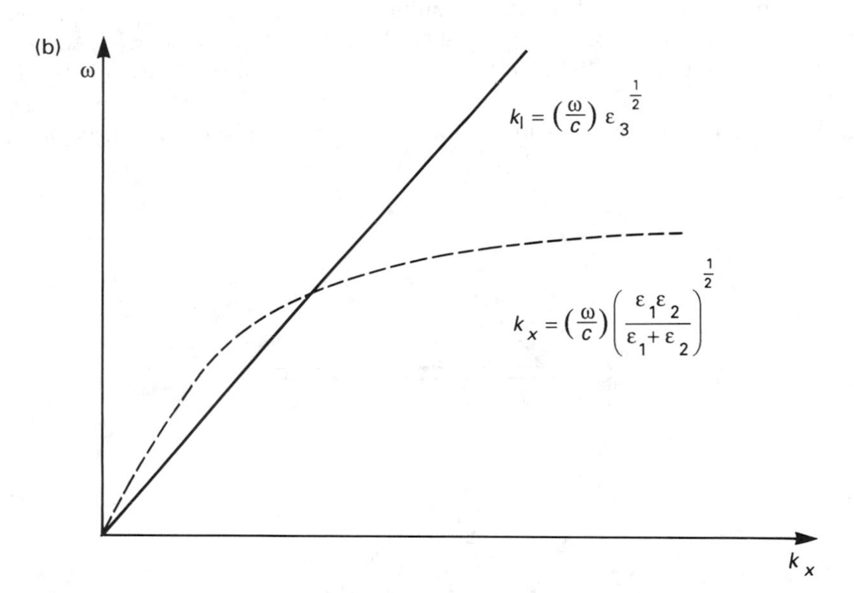

Fig. 10.14 (a) Dispersion relations for light in a medium with dielectric constant, ε_1, and for a surface plasmon wave at a metal film with dielectric constant ε_2, at the $\varepsilon_2/\varepsilon_1$ interface; (b) dispersion relations for light incident from a medium with dielectric constant, ε_3, and for a surface plasmon wave at a metal film with dielectric constant ε_2 at the $\varepsilon_2/\varepsilon_1$ interface. $\varepsilon_3 > \varepsilon_1$.

where,

$$\mathbf{k}_1 = |\mathbf{k}|\sin\theta$$

will be

$$\mathbf{k}_1 = \left(\frac{\omega}{c}\right)\varepsilon_1^{1/2}\sin\theta$$

having a maximum value

$$\mathbf{k}_1 = \left(\frac{\omega}{c}\right)\varepsilon_1^{1/2}$$

$\mathbf{k}_x$ therefore always remains larger than $\mathbf{k}_1$, and as can be seen in Fig. 10.14(a), the plots of the dispersion relations for the incident light and surface plasmon waves do not intersect at any point and excitation does not take place. However, if light is incident from another dielectric medium, such as a prism or waveguide (Fig. 10.15a), then the dielectric constants on each side of the metal film are different, i.e. $\varepsilon_3 > \varepsilon_1$. If total internal reflection of the incident light occurs at the **2/3** interface, then by adjusting the angle of incidence of the incoming wave, the $\mathbf{k}_x$ component of the wave vector will equal the wave vector of the surface plasmon wave at the **1/2** interface. If the film is sufficiently thin then the evanescent wave will couple media **1** and **3**. As can be seen in Fig. 10.14(b) the dispersion curves for the light incident at the **2/3** interface and the surface plasmon wave now show points of intersection. Excitation will take place at these frequencies and values of $\mathbf{k}_x$. Varying the angle of incidence will change the dispersion curve for the

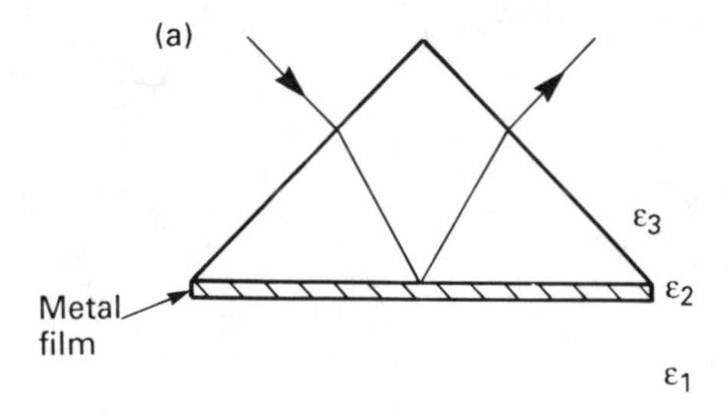

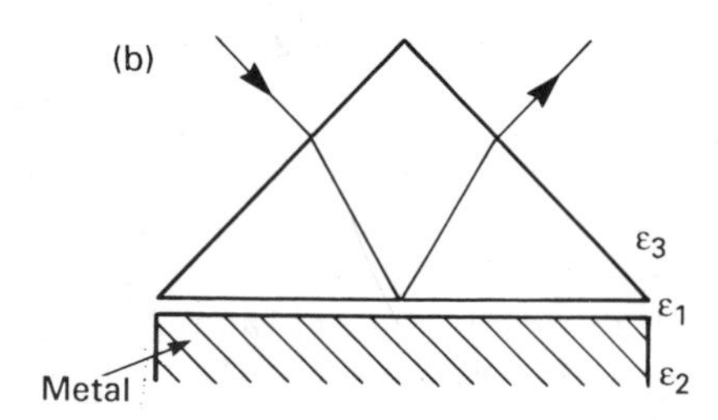

Fig. 10.15 Excitation configurations for surface plasmon resonance.

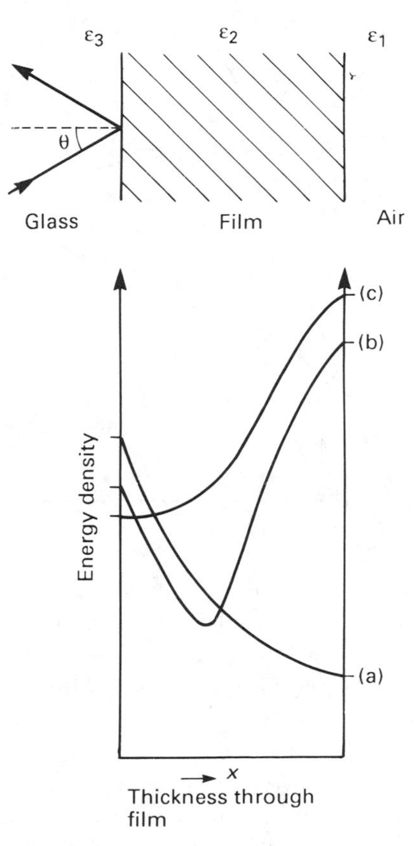

Fig. 10.16 Field energy distribution through thin metal film. Curve (a) exponential decay of field energy outside the resonance area; curve (b) at incident angles near to resonance, the field reaches a maximum at the boundary, but drops to a minimum within the plasma; curve (c) when resonance is induced the field energy rises across the plasma to a maximum at the boundary.

incident wave, thus changing the points of intersection and therefore the resonant frequencies. This excitation model is known as the *Kretschmann configuration* (Kretschmann, 1971). An alternative combination is found in the *Otto configuration* (Otto, 1968), but here the surface plasmon wave is excited across an air gap (Fig. 10.15b).

Normally, on total internal reflection the field energy decays exponentially in the x-direction from the **2/3** interface (Fig. 10.16a). When the dispersion relations of the incident light and the **1/2** surface plasmon wave cross and resonance is induced, then the spatial distribution of the energy density reaches a maximum at the **1/2** interface (Fig. 10.16c). The efficiency of this coupling will depend on the film thickness. The film must be thin enough so that the intensity of the exciting

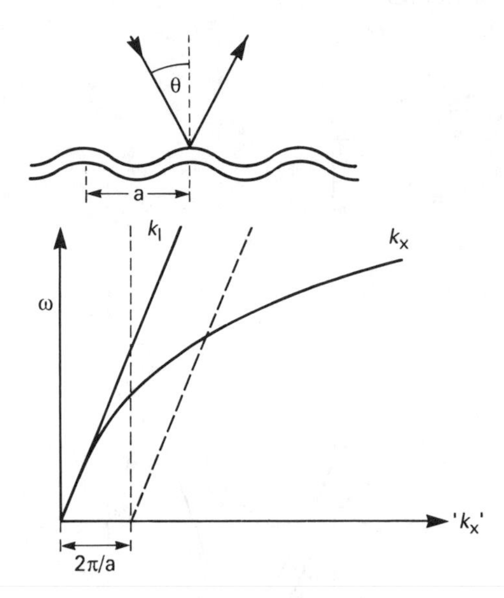

Fig. 10.17 Coupling of incident light and the surface plasmon mode at a sinusoidal metal grating of period a. Dispersion relations for light incident from a medium with dielectric constant, ε_1, with effect of roughness vector $2\pi/a$ and for a surface plasmon wave at a metal film with dielectric constant ε_2 at the $\varepsilon_2/\varepsilon_1$ interface.

field in the boundary does not become too weak. At incident angles very near to resonance, intermediate cases for the field energy exist with minima within the plasma (Fig. 10.16b) (Raether, 1980).

So far only light incident on a smooth plane surface has been considered. For a plane of incidence drawn perpendicular to the bulk metal, light incident on a rough surface will show both radiative and non-radiative modes for a given θ. A rough surface can be described by a spectre of sinusoidal profiles:

$$\sum_{i}^{n} \frac{h_i}{2} \sin(2\pi x/a_i)$$

where h is the grove depth. This serves as a source of pseudo-momentum which can be added to, or subtracted from, the light vector $\mathbf{k}_l$.

If the surface can be characterized by a single sinusoidal profile of period a, then the grove profile will be given by:

$$z = (h/2)\sin(2\pi x/a)$$

where h is the depth of the grating. When two surface plasmon modes are connected by a multiple of the grating vector, $2\pi/a$, then coupling between oppositely propagating modes can occur; $\mathbf{k}_x$ then becomes

$$\mathbf{k}_x = \frac{\omega}{c} \varepsilon_1^{1/2} \sin\theta + \frac{m2\pi}{a}$$

where $m = \pm 0, 1, 2 \ldots$ and the incident light can couple with the surface plasmon mode (Fig. 10.17) (Raether, 1977).

By appropriately selecting h it is possible to cause R_p to fall to zero at the resonance angle. The efficiency of absorption of light by a diffraction grating has been dealt with by Hutley and Maystre (1976). Expressing the energy of the diffracted wave as a function of h and θ,

$$\left| r \frac{\sin\theta - \alpha^z}{\sin\theta - \alpha^p} \right|^2$$

where α^z and α^p are numerically calculated complex numbers for different values of h, gives total absorption of the incident light when $\sin\theta = \alpha^z$. Conversely, when $h=0$ then $\alpha^z = \alpha^p$, and absorption is at a minimum (Fig. 10.18).

Experimentally, resonance is observed when coupling takes place between the incident light and the electromagnetic surface wave, as a minimum in the intensity of the reflected light. For a given frequency, a plot of reflectivity against angle of incidence produces a curve like that shown in Fig. 10.19. Since the dispersion relation for the surface plasmon resonance is defined by the dielectric constants ε_1 and ε_2, then the resonance condition will be very sensitive to variations in the refractive index immediately adjacent to the plasmon film, at the **1/2** interface. Adsorption or binding of molecules at this surface will be observed as a shift in the position of the reflectance minimum.

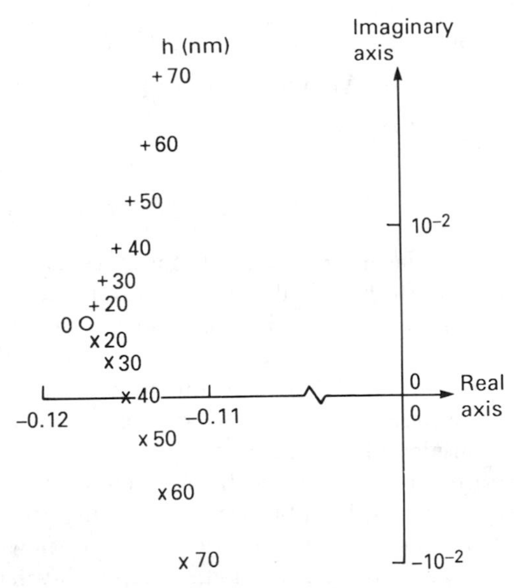

Fig. 10.18 Values of α^z and α^p corresponding to different values of h (depth of the grating), plotted on a complex plane. $(+) = \alpha^p$, $(\times) = \alpha^z$. (From Hutley and Maystre, 1976, with permission.)

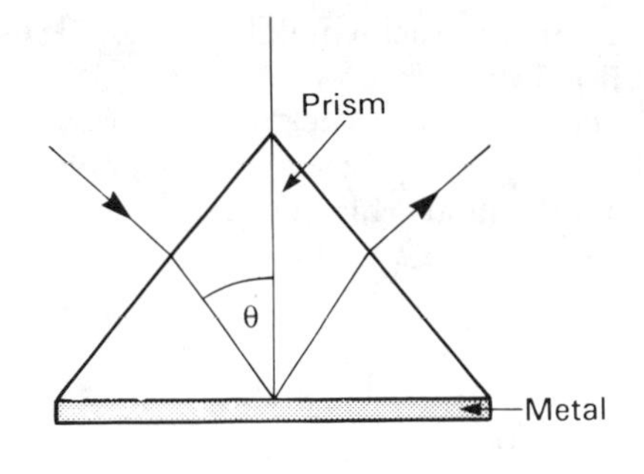

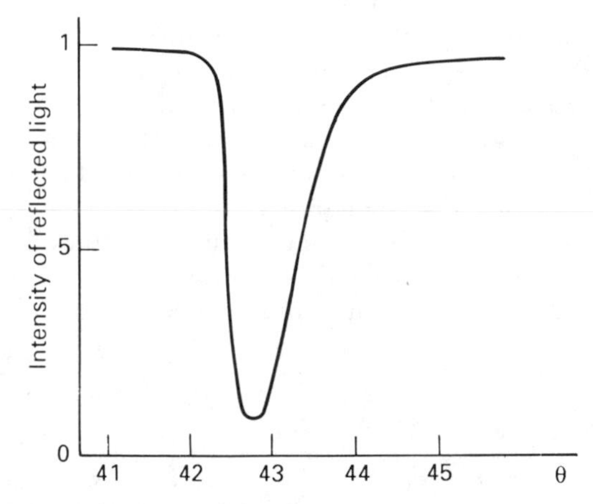

Fig. 10.19 Experimental observation of surface plasmon resonance.

APPLICATIONS

The application of surface plasmon resonance to the investigation of surface layers has been demonstrated for both optically inert and optically excited layers at a metal interface. In the former case, it has been shown that the technique can provide information about the layer structure and thickness, without the need for more complex ellipsometric equipment (Gordon and Swalen, 1977), while in the latter case optical excitation of the immobilized layer is possible via the surface plasmon wave with considerable enhancement.

For dyes with absorption bands in the visible frequency range the technique is a highly sensitive means of providing information of the molecular layer structure (Pockrand *et al.*, 1978), and Benner *et al.* (1979) have shown that the SPR-enhanced fluorescence, emitted from a rhodamine 6G layer on silver can be coupled to a new surface plasmon mode at the Stokes shifted wavelength. Similarly, a SPR-enhanced Raman spectrum can be recorded (Girlando *et al.*, 1980; Knoll *et al.*, 1982) of organic films as thin as a monolayer.

The propagation distance for the surface plasmon wave along the metal interface is related to the frequency. In the visible region it is only of the order of 10^{-4}, but in the infra-red, distances of 3 cm are realistic (Bhasin *et al.*, 1976). In contrast to the single-point geometry already considered, therefore, cells can be designed within the dimensions of the propagation path, which will allow interaction with surface species along the path of the surface plasmon wave (Fig. 10.20).

In view of the foregoing discussion on optical assay methods earlier in this chapter and also in Chapter 6, it would appear that surface plasmon resonance might be a technique which can offer some important improvements to many photometric assays. This is not only true for estimations performed via an optical label, but also for unlabelled determinations such as the direct monitoring of antibody–antigen interactions. In fact SPR-linked quantitative analyses have concentrated to date more on the development of these direct assays than on exploiting the SPR-enhanced optical properties of labels or even the analyte or immobilized recognition molecule themselves. This is however an area of great potential and interest, and we can anticipate SPR-linked optical assay techniques in many forms.

Nylander *et al.* (1982/83) have exploited what is normally an inherent problem of the anaesthetic gas halothane to devise a detection system. The silicon–glycol copolymer oils absorb halogenated hydrocarbons causing a change in the refractive index of the oil. Silicon oil was spun-coated from a 2% solution in trichlorocthane onto a thin silver film, which had been evaporated onto a glass slide. The slide was mounted in the Kretschmann configuration for SPR, downstream from an anaesthetic machine (Fig. 10.21). Absorption and desorption of the halothane into the oil occurred in the millisecond range, and the shift in the plasmon resonance angle was proportional to halothane concentration up to 2% and it was estimated that a limit of detection, better than 10 ppm, could be obtained.

The use of SPR has also been directed towards the challenge of unlabelled immunoassay (Liedberg *et al.*, 1983). The antigen human immunoglobulin (IgG) was adsorbed onto a silver film, deposited onto glass. Binding of sample antibody

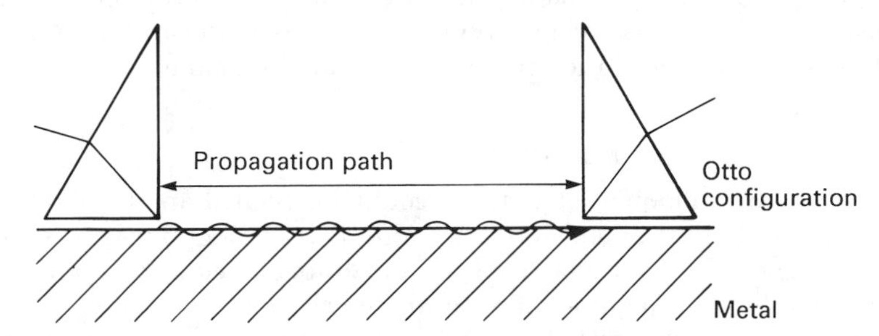

Fig. 10.20 Geometry for launching surface plasmon waves, which will propagate along the metal surface, and allow decoupling and detection along the propagation path.

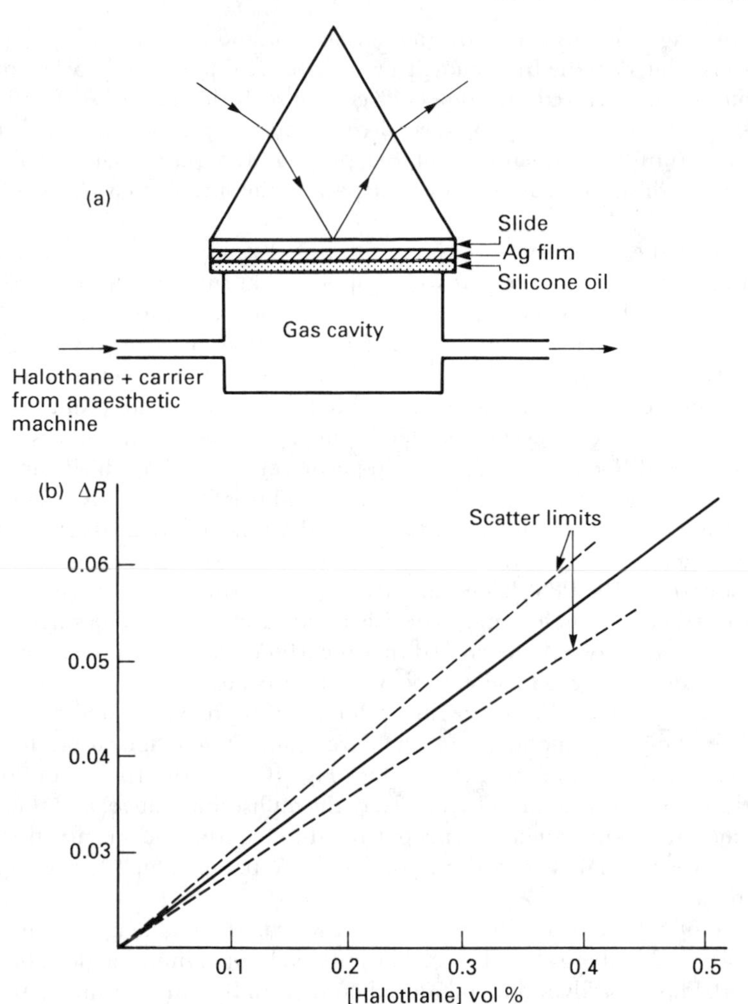

Fig. 10.21 (a) Detection of halothane with surface plasmon resonance; (b) dependence of ΔR on halothane concentration, showing linear relationship. Dashed lines incorporate scatter giving an estimated maximum error $<0.05\%$ halothane.

(anti-IgG) to the immobilized antigen could be related to the shift in the resonance angle. The maximum derivative ($d\theta/dt$) versus log[anti-IgG] showed a near linear portion between 20 and 200 μg/ml, although below 20 μg/ml this plot was insufficiently accurate to be of good analytical use.

Flanagan and Pantell (1984) have also examined the applicability of SPR to immunoassay. They have employed a similar configuration to examine the binding between human serum albumin (HSA) adsorbed onto an Ag plasma film, and anti-HSA, and compared the experimental curves with an approximated theoretical prediction for organized layers of different thicknesses. Correlation of theoretical and experimental results is likely to assist optimization of this technique as an assay method, and its development into a biosensor.

Cullen *et al.* (1987/88) have investigated immunocomplex formation at gold diffraction gratings. Both of the above immunological models were employed. In the case of human IgG the antigen was immobilized on the gold grating by adsorption, and for HSA the antibody (anti-HSA) was the immobilized species and the antigen the target analyte.

This feasibility study with diffraction gratings has confirmed this geometry as a sensitive SPR system for biosensing. The results are comparable with the Kretschmann configuration model previously discussed for both immunological models. A plot of angular shift versus log[protein] showed a comparable linear range to the equivalent plot of $d\theta/dt$ versus log[protein] proposed by Liedberg *et al.*, but Cullen *et al.* have not attempted to extrapolate the linearity to concentrations $<20\,\mu g/ml$ (Fig. 10.22).

A major concern with immunoassay techniques which do not employ a label is to distinguish between non-specific interactions of the target analyte with the sensor surface, and specific interaction between antibody and antigen. Surface plasmon resonance responds to changes in the dielectric constant at the plasmon

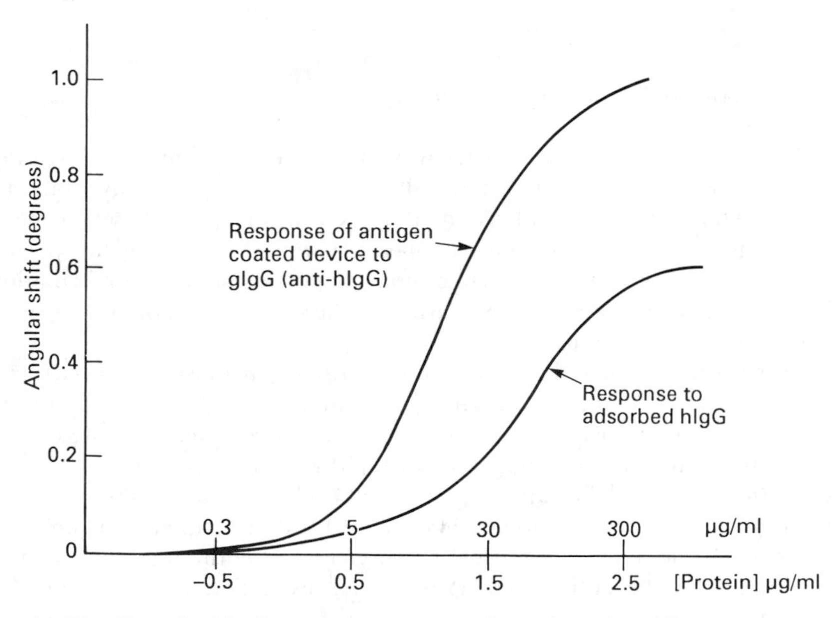

Fig. 10.22 Protein binding curves, as a surface plasmon resonance measurement of change in excitation angle at a given incident frequency. Recorded for human IgG adsorbed to a gold metal grating and the formation of an immunocomplex with goat IgG anti-human IgG at the human IgG-coated surface. gIgG = goat immunoglobulin G; hIgG = human immunoglobulin G. (Cullen *et al.*, 1988.)

interface, but it does not identify the cause of that change. In order to eliminate non-specific interactions, therefore, it is necessary to ensure that the biorecognition molecule is efficiently immobilized at the metal interface, completely 'blocking' the metal surface from interaction with other molecules.

The devices described above used the recognition protein immobilized by adsorption on the metal. This may not provide a satisfactory method for many applications, but Cullen *et al.* were able to show that, at least for their device, a non-specific protein, ovalbumin, gave no response on a surface treated with adsorbed anti-HSA, while on the surface with adsorbed human IgG, a response equivalent to only 7.4% of complementary protein was seen.

Other groups have addressed themselves to the problem of covalent attachment of the biorecognition molecule. Daniels *et al.* (1988) and Liley *et al.* (unpublished) have deposited a thin layer of oxides of silicon on the metal film, to which the biomolecule can be attached via a silane (see Chapter 1). Although the procedure described in these two studies is significantly different, the principle is similar. In the latter case the technique has been directed towards the development of a simple-to-use multiple DNA probe, while the former study is concerned with the model system biotin–avidin. For biotin, covalently bound at a gold film and recorded at a fixed angle, a significant correlation is reported between the rate of change of reflectivity and the concentration of avidin over the range 10^{-8}–10^{-7} M.

Surface-enhanced Raman Spectroscopy

The use of the surface electromagnetic wave to excite the optical properties of surface-immobilized species is a particularly sensitive technique. Assays could be constructed which employ an indirect method exciting an optical label, or made directly, targeting the inherent optical characteristics of the biological species. Raman spectroscopy is very sensitive to minor changes in molecular conformation (see Chapter 6), and enhanced by excitation by the surface plasmon wave it may prove a powerful bioanalytical method.

In fact, surface-enhanced Raman spectra (SERS) have been observed for a number of small neutral molecules and the technique has also been applied to observe interactions of biological molecules, for example compounds related to the catecholamine neurotransmitters (Lee *et al.*, 1988) and the heme chromophore in dilute solutions of myoglobin and cytochrome C (Cotton *et al.*, 1980).

With a constant miniaturization (see Chapter 6) and the frequent concomitant improvement in the quality of optical components, a continuing expansion of optical devices can be anticipated. Any optical excitation is 'fair game' for exploitation in a biosensor!

References

Aizawa, M., Ikariymoto, Y., Kumo, H. (1984). *Anal. Lett. B.* **17**(7), pp. 555–564.

Arnold, M.A. (1985). *Anal. Chem.* **57**(2), p. 565.

Arwin, H. and Lundström, I. (1985). *Anal. Biochem.* **145**, p. 106.

Badley, R.A., Drake, R.A.L., Shanks, I.A., Smith, A.M. and Stephenson, P.R. (1987). *Phil. Trans. R. Soc. Lond.* **B316**, pp. 143–160.

Bhasin, K., Bryan, D., Alexander, R.W. and Bell, R.J. (1976). *J. Chem. Phys.* **64**, pp. 5019–5025.

Benner, R.E., Dornhaus, R. and Chang, R.K. (1979). *Opt. Commun.* **30**, p. 145.

Birch, B.J. (1987). *Practical Chemical and Biological Sensors, RSC Analytical Division Meeting*, December, 1987.

Brovko, L.Y., Kost, N.V. and Ugarova, N.N. (1980). *Biochemistry SSR* **45**, p. 1199.

Cotton, T.M., Schultz, S.G. and van Duyne, R.P. (1980). *J. Am. Chem. Soc.* **102**, pp. 7960–7962.

Cullen, D.C., Brown, R.G.W. and Lowe, C.R. (1987/88). *Biosensors* **3**(4), pp. 211–216.

Daniels, P.B., Deacon, J.K., Eddowes, M.J. and Pedley, D.G. (1988). *Sensors and Actuators* **15**, pp. 11–18.

Doumas, B.T., Watson, W. and Biggs, H.G. (1971). *Clin. Chim. Acta* **31**, p. 87.

Flanagan, M.T. and Pantell, R.H. (1984). *Electron. Lett.* **20**(23), p. 968–970.

Ford, J. and DeLuca, M. (1981). *Anal. Biochem.* **110**, p. 43.

Freeman, T.M. and Seitz, W.R. (1978). *Anal. Chem.* **50**(9), p. 1242.

Fuh, M.S., Burgess, L.W. and Christian, G.O. (1988). *Anal. Chem.* **60**, p. 433.

Girlando, A., Philpott, M.R., Heitmann, D., Swalen, J.D. and Santo, R. (1980). *J. Chem. Phys.* **72**, p. 5187.

Goldfinch, M.J. and Lowe, C.R. (1980). *Anal. Biochem.* **109**, p. 216.

Goldfinch, M.J. and Lowe, C.R. (1984). *Anal. Biochem.* **138**, p. 430.

Gordon II, J.G. and Swalen, J.D. (1977). *Opt. Commun.* **22**, p. 374.

Hall, E.A.H. and Duschl, C. (forthcoming). *Textbook on Chemical Sensors*, eds. Worrell, W. and Weppner, W.

Hirschfeld, T. (1984). *IEEE Biomed.*, **31**(8), p. 582.

Hutley, M.C. and Maystre, D. (1976). *Opt. Commun.* **19**, p. 3.

Knoll, W., Philpott, M.R., Swalen, J.D. and Girlando, A. (1982). *J. Chem. Phys.* **77**, p. 2254.

Kretschmann, E. (1971). *Z. Phys.* **241**, p. 313.

Kronick, M.N. and Little, W.A. (1973). *Bull. Am. Phys. Soc.* **18**, p. 782.

Kronick, M.N. and Little, W.A. (1975). *J. Immunol. Meth.* **8**, p. 235.

Laxhuber, L.A., Rothenhäusler, B., Schneider, G. and Möhwald, H. (1986). *J. Appl. Phys. A* **39**, p. 173.

Lee, N.-S., Hsieh, Y.-Z., Paisley, R.F. and Morris, M.D. (1988). *Anal. Chem.* **60**, pp. 442–446.

Liedberg, B., Nylander, C. and Lundström, I. (1983). *Sensors and Actuators* **4**, p. 299.

Mansouri, S. and Schultz, J.S. (1984). *Bio/technology* **2**, p. 885.

Newby, K., Reichert, W.M., Andrade, J.D. and Benner, R.E. (1984). *Appl. Opt.* **23**(11), p. 1812.

Nylander, C., Liedberg, B. and Lind, T. (1982/83). *Sensors and Actuators* **3**, pp. 79–88.

Otto, A. (1968). *Z. Phys.* **216**, p. 398.

Pockrand, I., Swalen, J.D., Santo, R., Brillante, A. and Philpott, M.R. (1978). *J. Chem. Phys.* **69**, pp. 4001–4011.

Raether, H. (1977). *Phys. Thin Films* **9**, p. 145.

Raether, H. (1980). *Tracts in Modern Physics*, Vol. 88. Berlin, Springer-Verlag.
Schultz, J.S., Mansouri, S. and Goldstein, I.J. (1982). *Diabetes Care*, **5**, pp. 245–253.
Smith, A.M. (1986). *Biotech '86* **D9**. Pinner, Online Publications.
Srinivasan, K.R., Mansouri, S. and Schultz, J.S. (1986). *Biotech. Bioeng.* **28**, p. 233.
Stange, U., Groome, N., Tarassenko, L. and Hutchins, M.G. (1988). *Biomaterials* **9**, p. 58.
Sutherland, R.M., Dähne, C., Place, J.F. and Ringrose, A.R. (1984). *Clin. Chem.* **30**, p. 1533–1538.
Ugarova, N.N., Brovko, L.Y. and Kost, N.V. (1982). *Enzyme Microb. Technol.* **4**, p. 224.
Welin, S., Elwing, H., Arwin, H., Lundström, I. and Wikström, M. (1984). *Anal. Chim. Acta* **163**, p. 263.
Wolfbeis, O.S. (1986). *Anal. Chem.* **58**(13), p. 2875.

Chapter 11

Other Transducing Techniques

Conductance Methods

Many chemical reactions in solution involve a change in the ionic species. Associated with this change will be a net change in the conductivity of the reaction solution. Solution conductance measurements are non-specific and their widespread analytical use has therefore been restricted. Where the specificity does not play a significant role, however, conductance measurements are capable of extreme sensitivity.

Conductance (S) is the reciprocal of resistance (R), expressed in reciprocal ohms (mhos). The standard unit of conductance is specific conductance K, which is defined as the conductance of a 1 cm cube of liquid at a given temperature. The observed conductance of a solution depends inversely on the distance d, between the electrodes, and their area A:

$$S=\frac{1}{R}=\frac{AK}{d}$$

Electrolytic conductance measurements usually involve the determination of the resistance of the sample between two parallel electrodes (Fig. 11.1). Solution resistance is determined by the migration of all ions present, and is thus the result of *any* ionized species.

However, although this configuration is the one employed in the routine screening of bacteriology samples to give a microorganism count, the origin of the change in conductivity, which is related to microbial activity, is not precisely

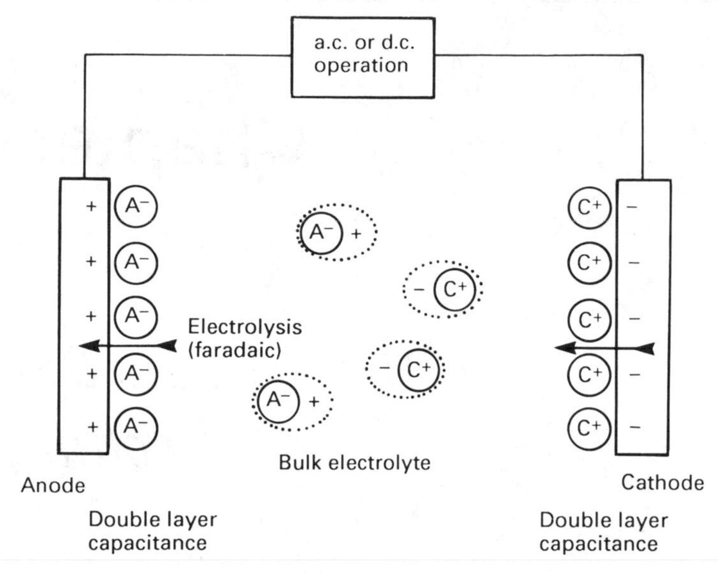

Fig. 11.1 Measurement of electrolytic conductance. Migration of ions from the bulk electrolyte and Faradaic reactions.

known. Nevertheless, examination of the morphology of conductance curves in particular growth media and in the presence and absence of inhibitors, can give some indication concerning the identity of the microorganism, so that the establishment of a library of organism growth patterns could become a powerful tool in the identification of microorganism cultures. Routine automated screening of large numbers of clinical blood samples is performed with commercially available conductivity monitoring systems, and a similar experimental 'set-up' has been investigated as a means of following and controlling ethanol fermentation (Miike *et al.*, 1984).

It is however the nature of a biosensor to induce selectivity by introducing a biospecific reaction surface and to measure the selective reaction with the target analyte at that surface. Frequent involvement of ionic species and thus conductance changes in the course of many of these biospecific reactions would suggest that the technique could be widely applicable.

In common with other types of biosensors, the biorecognition molecule must be immobilized at the transducer. In this instance, where changes in conductance are to be measured, that infers that it should be deposited between two ohmic conductors in a suitable matrix. This matrix may be an 'inert' system which allows ion mobility, or else the change in conductance may be more directly related to changes in the layer itself, in the form of a semiconductive film.

The use of conductivity to develop an enzyme-linked assay probably originates with the determination of urea in solution, due to the specific conductivity change induced by the addition of the enzyme urease to the sample solution (Chin and Kroontje, 1961; Hanss and Rey, 1971):

Table 11.1 The factors which cause conductance change

Enzyme	*Origin of conductance change*
Amidases	Generation of ionic groups
Dehydrogenases and decarboxylases	Separation of unlike charges
Esterases	Proton migration
Kinases	Change in the degree of association of ionic species (chelation)
Phosphatases and sulphatases	Change in the size of charge-carrying groups

$$O{=}C(NH_2)_2 + H_2O \xrightarrow{urease} CO_2 + 2NH_3$$

This technique was soon applied to various other enzyme systems with varying degrees of success (Table 11.1) (Lawrence, 1971; Lawrence and Moore, 1972) and multichannel systems were proposed.

A microelectronic conductance device which employs the samc tcchnique, but shows all the required characteristics of a biosensor has been described by Watson *et al.* (1987/88). The basic conductance transducer comprised two serpentined and interdigitated gold conductor tracks (Fig. 11.2), produced by photolithographic techniques (see Chapter 5), to provide ohmic contact with the deposited film.

This electrode geometry provides a very large ratio of electrode perimeter to electrode spacing. Since the measured conductance (S) can be described by the relationship between the interface area and the distance between the electrodes, it can be seen that a large A/d ratio will give the greatest attentuation of the conductance response.

Urease was immobilized over the interdigitated network in this device by glutaraldehyde crosslinking (see Chapter 1), and conductance monitored by an a.c. conductimetric monitoring technique, making a differential measurement between sample and reference cells. Both d.c. and a.c. modes could be employed, but a.c. is often favoured since it more easily compensates for double-layer charging, Faradaic processes and other phenomenon associated with d.c. current flow.

The differential measurement employed here accommodates non-specific variations in background conductivity of the sample, and the device is reported to respond to urea in both buffer and clinical serum samples in the range 0.1–10 mM urea. The apparent K_M for the immobilized enzyme was higher than in solution, as would be anticipated for the diffusional limitations of the immobilization matrix. The measurement parameter employed in this device is broadly applicable to a wide spectrum of enzyme-linked reactions, so that in principle it could be

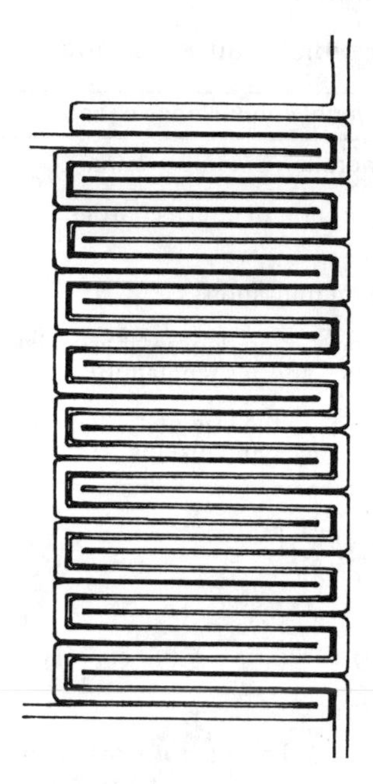

Fig. 11.2 Interdigitated electrodes to provide ohmic contact with an overlaying film, in which conductance changes are measured. The interface-area/inter-electrode distance ratio is large to attenuate the response.

developed for other target analytes. Glucose oxidase–glucose is potentially one such system for which it could be exploited. Malmros *et al.* (1987/88), however, have chosen to investigate a glucose sensor based on a conductivity change which results from an interaction with doped polyacetylene films, following the enzyme initiated sequence:

$$\text{glucose} + O_2 \xrightarrow{\text{glucose oxidase}} \text{gluconic acid} + H_2O_2$$

The conductivity of polyacetylene can be modified to span a large range by chemical doping with a small amount of iodine. These doped films also show a change in conductivity due to the presence of hydrogen peroxide. The latter response can be related to glucose in an enzyme-linked assay but it can be considerably enhanced by the inclusion of lactoperoxide (LPO)/KI in the assay buffer. The enzyme catalyses the peroxide oxidation of I^-

$$H_2O_2 + 2H^+ + 2I^- \xrightarrow{LPO} I_2 + 2H_2O$$

and a significant contribution to the enhanced conductivity change is likely to be due to the production of molecular iodine, which will be available for polymer doping.

A glucose-sensitive doped polyacetylene film was produced by adsorbing glucose oxidase onto a film which had been appropriately doped with iodine to give suitable 'base' conductivity. The change in conductivity due to glucose was measured as a function of time over a fixed time interval. The measurement was performed by the application of a rectangular wave potential of 500 mV, across the film between two electrodes and current sampling towards the end of a 100 μs pulse period. The pulse is sufficiently short not to induce Faradaic processes (see Chapter 5) and thus with this control regime the response was shown to be linear in the range 10–180 mg dl^{-1}

Many other films also show a change in conductivity on exposure to specific analytes. For example, polypyrrole has already been encountered in previous chapters as a film which could be electrochemically deposited, and whose conducting/insulating properties could be manipulated by growth conditions and doping. Thin films of polypyrrole deposited over interdigitated electrodes have been evaluated for their response to various gases. Sensitivity can be 'tuned' by altering the polymerization conditions, but no clear pattern for directing this sensitization has yet emerged. Miasik *et al.* (1986) have demonstrated a conductance-related response for polypyrrole exposed to NH_3 (decrease in S with $[NH_3]$) or NO_2 or H_2S (increase in S with concentration), but no response to several other gases such as H_2, CO, CO_2 or CH_4.

Like pyrrole, the metal phthalocyanines (MPc) are stable *p*-type organic semiconductors, whose central metal atom affects not only the conductivity of the crystal, but also its sensitivity to different gases or vapours. Several metal phthalocyanines are sensitive to NO_2 with PbPc showing high sensitivity and high conductance (Bott and Jones, 1984). A PbPc NO_2 conductance sensor is reported to retain stability in continuous operation for 6 months, and in conjunction with other gas sensors its application was demonstrated in a mining environment (Bott and Jones, 1985).

Alternatively, Blanc *et al.* (1988) have shown that poly(fluoro-aluminium)phthalocyanine, deposited over interdigitated electrodes on alumina or silica, shows large increases in conductivity due to NO_2, even at the 1 ppm level (Fig. 11.3a), in comparison with the response to oxygen which induced no noticeable effect at $< 10^4$ ppm (Fig. 11.3b).

Wohltjen *et al.* (1985) have developed a copper-complexed phthalocyanine (CuPc) to detect ammonia, with a detection limit < 0.5 ppm. The semiconducting film was laid down over an interdigitating electrode array, using a Langmuir–Blodgett film technique. In order to render the CuPc compatible with this deposition method, a peripherally substituted compound was employed, namely tetracumylphenoxyphthalocyanine (Fig. 11.4).

The resultant device is reported to show a fully reversible response to ammonia with a signal magnitude related to film thickness, and with no observable degradation in performance over a period of several weeks. The film was selective towards electron-donor vapours, with nitrogen dioxide also causing a response, but vapours such as benzene, carbon dioxide or cyanogen chloride gave a negligible effect, even at $> 10\,000$ ppm.

The measurement regime employed here utilized a d.c. conductance measurement with a bias potential < 1 V. The authors argue that the more complex a.c. technique does not guarantee stability and that ratiometric measurement made

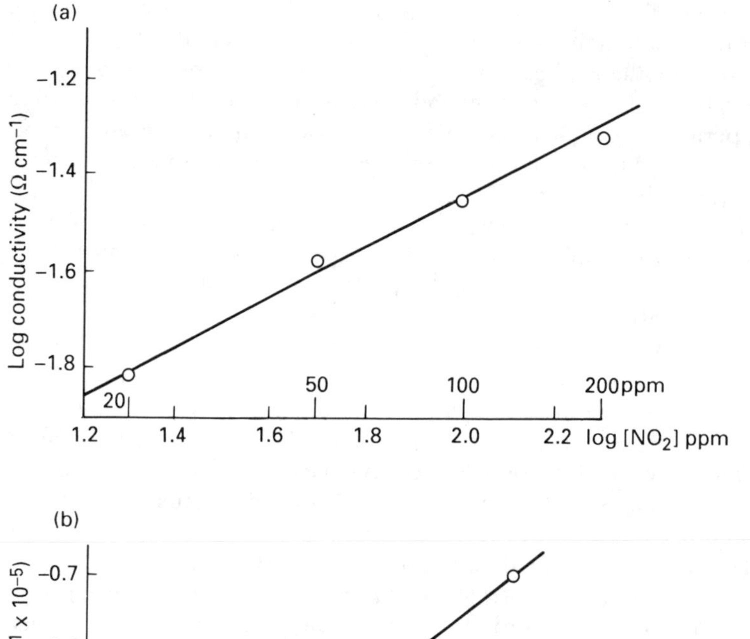

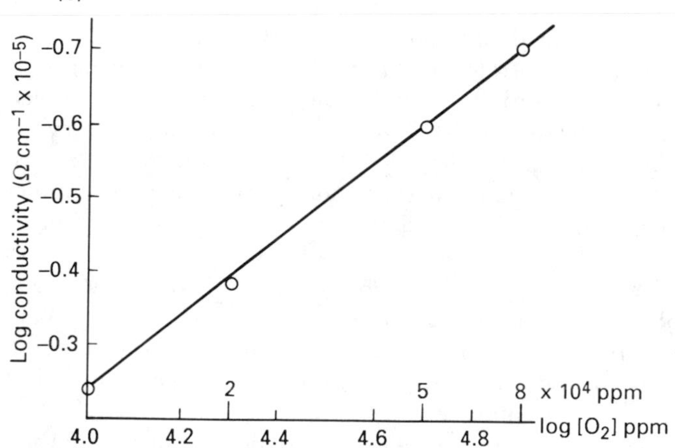

Fig. 11.3 Conductivity response by poly(fluoroaluminium)phthalocyanine, deposited over interdigitated electrodes to (a) NO_2 and (b) O_2.

with a reference sensor (where the film is passivated with a protective coating) was shown experimentally to be stable.

The charge-transfer salts involving TCNQ have also been encountered in earlier chapters. LB films of *N*-docosylpyridinium–TCNQ can be formed with conductive properties (Henrion *et al.*, 1988) and can be partially oxidized ($TCNQ^- \rightarrow TCNQ^0$) with increase in conductance, or reduced with decrease in conductance. Exposure to iodine causes an increase in conductance which is stable for several months; NO_2 has a similar effect. Both of these 'pre-exposed' films could be employed to detect low additional concentrations of I_2 or NO_2.

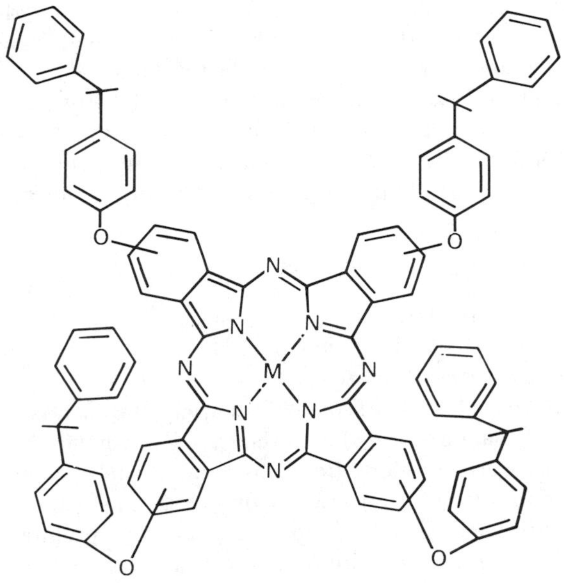

Fig. 11.4 Tetracumylphenoxyphthalocyanine.

Alternatively NH_3 causes the conversion $TCNQ^0 \rightarrow TCNQ^-$ and thus a decrease in conductance.

As may be deduced from this and earlier discussions, many of the conducting and semiconducting matrices that have been employed for other signal transducing contexts may be applicable to this type of measurement.

Piezoelectric-based Sensors

The use of frequency as the transduction parameter can be a particularly sensitive measurement in a number of applications. Piezoelectric materials, for example, are mass-to-frequency transducers, sensitive to changes in mass, density or viscosity. Piezoelectricity was first observed by the Curie brothers in 1880, who showed that an electric dipole could be developed in anisotropic crystals under stress. Quartz is probably the most frequently employed naturally occurring anisotropic piezoelectric crystal, but ceramics and more recently piezoelectric polymer materials have been applied. Since piezoelectric materials are anisotropic, the transduction of mechanical or electrical impulses will depend on the orientation of the impulse with respect to the plane of the crystal. AT quartz and BT quartz, for example, refer to the orientation of the plate with respect to the crystal plane. The AT-cut ($+35°15'$) and BT-cut ($-49°00'$) crystals have the best temperature coefficients, the former showing about 1 ppm/°C in the range 10–50°C.

One of the most straightforward applications of a piezoelectric device is in the microbalance. Crystals placed in an oscillator circuit will show a resonant frequency which depends to a large extent on the total mass of the crystal and any surface coverings. The measurement can be extremely sensitive. The relationship between surface mass change, Δm(g) and resonant frequency f(Hz) is given by the Sauerbrey equation, and calculated for quartz crystals (Guilbault *et al.*, 1980; Alder and McCallum, 1983) the change in resonant frequency on adsorption of an analyte is given by:

$$\Delta f = -2.3 \times 10^6 f^2 \Delta m / A$$

where Δf is in Hz, f is in MHz, and A is the area (cm^3). This suggests that a resolution of the order of 2500 Hz/μg for a 15 kHz crystal can be expected, and a detection limit of 10^{-12} g is estimated.

The majority of piezoelectric sensor devices operating in this intrinsic mode have been directed at vapour-phase measurements. They generally employ pairs of matched crystals so that a differential measurement can be made to eliminate spurious signals arising from electronic fluctuations, temperature changes, etc. and are primarily directed at the detection of atmospheric pollutants (Guilbault and Jordan, 1987). The gaseous pollutant is selectively adsorbed by a coating on the surface of the piezoelectric crystal, which is specific for the target analyte.

In general, the devices require a constant and low relative humidity, or else the principal signal is due to H_2O! Schuman and Fogelman (1976) have measured CO_2 at a crystal modified with didodecylamine or dioctadecylamine, and contained behind a hydrophobic Teflon membrane. This membrane barrier allowed the use of the device in an aqueous media, but would not be suitable for larger analyte molecules that do not penetrate 'water barrier' membranes.

Biological molecules can also be employed to induce specificity. In fact many biological molecules such as proteins and DNA are anisotropic and could themselves be piezoelectric materials. To date, however, these compounds are generally utilized in conjunction with piezoelectric crystals such as AT quartz.

FORMALDEHYDE DEHYDROGENASE-MODIFIED CRYSTALS

The oxidation of formaldehyde to formic acid is catalysed by the NAD-dependent enzyme, formaldehyde dehydrogenase. The enzyme, together with glutathione as cofactor, was immobilized on a 9 MHz AT-cut quartz crystal by glutaraldehyde crosslinking, to give a piezoelectric device with excellent specificity for formaldehyde (Guilbault, 1983).

CHOLINESTERASE-MODIFIED CRYSTALS

Organophosphorus pesticides can be detected via their reaction with the cholinesterase enzymes. For example, acetylcholinesterase, immobilized by glutaraldehyde crosslinking on a quartz piezoelectric crystal responded to malathion in the 'gas phase' within 1 min (Guilbault *et al.*, 1986). In order to obtain a good enzyme-linked response, at least 5 ppm water was found to be required and, since the piezoelectric signal is related to the relative humidity, the instrument must be calibrated at constant humidity.

IMMUNOSENSITIVE CRYSTALS

With the antibody anti-parathion immobilized on the piezoelectric crystal, an alternative approach to the detection of organophosphorus pesticides is possible. The resulting piezoelectric device would respond to parathion and similarly structured pesticides within 2–3 min (Ngeh-Ngwainbi *et al.*, 1986).

As would be anticipated for these 'gas-phase' bio-linked measurements, a few ppm H_2O were required for the bioassay to begin to function. However, the signal response was larger at higher humidities, so that as in the previous example the device needed to be calibrated and employed at constant humidity.

SOLUTION ASSAY

The application of piezoelectric crystals, operating in this intrinsic mode in solution phase assay is impaired by the device response to the solution components such as electrolyte concentration and specific conductivity, or the increased occurrence of non-specific adsorption.

While these factors have usually inhibited the use of piezoelectric crystals in solution for bio-linked assays, they do not prevent the development of the technique for ion concentration and ion-selective measurements. Nomura and Nagamune (1983) have related Δf to KCl concentration up to 2 mM. At concentrations >20 mM, the crystals are reported to short-circuit, and lead to apparent excessive changes in resonant frequency. In fact Shou-Zhuo and Zhi-Hong (1987) have shown that for a solution of given capacitance characteristics, the change in frequency is proportional to the total salt concentration and irrespective of the nature of the salt. Plots of Δf versus salt concentration are linear, but with a slope ranging from positive through to negative, with increasing capacitance.

The authors suggest that the technique would be applicable to the determination of total salt concentration in natural waters of known capacitance. Since the response was independent of the type of salt, a device calibrated with respect to NaCl was shown to give an estimated total salt concentration which was in good agreement with the conductance method.

The above utilization of intrinsic frequency measurements at piezoelectric crystals does, however, illustrate the potential problem involved with performing bio-linked assays in solution.

Surface Acoustic Waves (SAW)

Rather than employing the bulk wave piezoelectric microbalance model as the basic transducer, an alternative approach is that of surface acoustic waves.

Two sets of interdigitated electrodes are deposited on a piezoelectric crystal

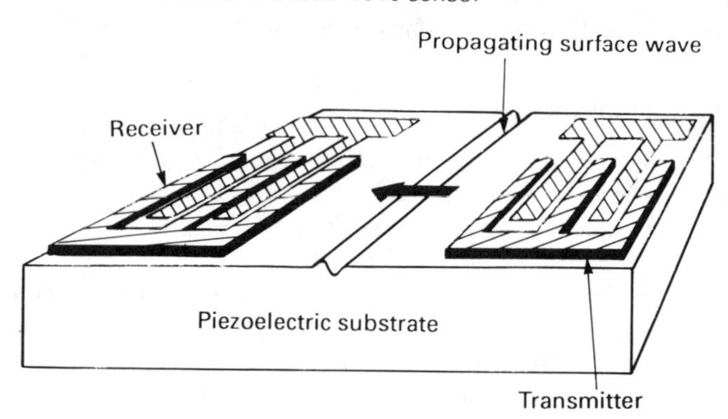

Fig. 11.5 Model of a surface acoustic wave (SAW) device.

surface, as transmitter and receiver. Radiowave frequencies applied to the transmitter electrodes produce a synchronous mechanical stress in the crystal, which produces an acoustic wave with both longitudinal and vertical shear components. This Rayleigh surface wave propagates along the surface of the piezoelectric crystal and is received at the second set of electrodes, where mechanical vibrations are translated to an electrode voltage (Fig. 11.5).

The penetration depth for the surface wave is of the order of a wavelength (compare with evanescent optical wave, see Chapter 6), so that species immobilized on the surface of the crystal will effect the transmission of the wave, and if the crystal is sufficiently thick, it will be independent of any species on the reverse side of the device.

In fact interdigitated electrode arrays are suitable for launching waves other than Rayleigh waves, which involve bulk 'plate modes'. Indeed, Calabrese *et al.* (1987) have observed that in thin piezoelectric crystals (<6 wavelengths) a signal attenuation was observed on loading the back surface with water. This is inconsistent with exclusive Rayleigh wave propagation. For thicker devices 'Rayleigh-type' behaviour was observed, with attenuation by water on the surface typically causing 99% of the signal to be lost. This contrasts with the thin devices where 'front-surface' loading had little effect.

Calabrese *et al.* proposed that propagation in the thin devices is not by the same Rayleigh mode mechanism, and in the thicker devices Rayleigh wave energy losses occur through viscous interactions with the surface film, or possibly more significantly for thicker surface films, due to the longitudinal component of the Rayleigh wave generating a compressional wave in the film. When the velocity of the wave in the film is greater than in the piezoelectric substrate, the wave will be directed into the film. This suggests that considerable limitations could arise in the use of SAW devices in aqueous media.

Excitation of the plate modes is achieved by the Rayleigh waves, but for substrate thicknesses that are very small compared with the wavelength (Fig.

11.6), the excitation frequency is much lower (*Lamb mode*). Zellers *et al.* (1988) have evaluated polymer-coated ZnO/Si SAW oscillators with ZnO/Si Lamb-mode oscillators and concluded that higher sensitivity and greater flexibility of sensor design could be achieved with a Lamb-mode device.

Nevertheless, in the gaseous phase numerous examples of SAW device operation exist and in both gaseous and aqueous phase devices, with a modified chemically sensitive or biochemically sensitive interface, the velocity of the wave in the sensitive surface layer may be more comparable with that in the piezoelectric material.

In fact, the general trend with SAW sensors is towards surface-modified devices. The phthalocyanines, for example, have already been identified as *p*-type semiconductors, which interact strongly with electron-donor compounds, and copper phthalocyanine (CuPc) has been employed in a conductimetric sensor for ammonia. Nieuwenhuizen and co-workers (Nieuwenhuizen and Nederlof, 1988; Nieuwenhuizen *et al.*, 1988) have investigated the use of vapour deposited metal phthalocyanines (MPc) in SAW devices sensitive to electronegative compounds, and have made similar observations concerning selectivity and sensitivity to those found for the equivalent MPc conductimetric devices. As with many of the other 'recognition surface–analyte' interactions which have been cited, the process may be linked with a number of different transducers, often without essentially altering the actual recognition event. In these SAW devices the phthalocyanines were deposited by vapour deposition, although the authors suggest that better stability would be obtained by functionalization of the phthalocyanine and covalent immobilization via a spacer to the piezoelectric support. In the conductimetric device previously described, the Pc was derivatized to make it compatible with the Langmuir–Blodgett film deposition method. In fact, all three immobilization methods could be interchanged between the two transduction techniques, and their evaluation would show many common features, whether employed in the SAW or the conductimetric mode.

Nieuwenhuizen *et al.* (1988) found, however, that although a change in conductivity was associated with the interaction between the various Pc's and the

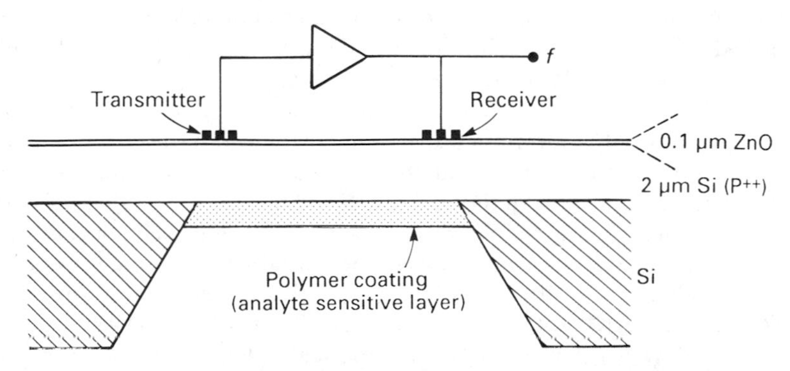

Fig. 11.6 Transverse section through a Lamb-mode device, showing the analyte sensitive layer on the underside (after Zellers *et al.*, 1988).

analyte vapours, no direct correlation could be made between the SAW response and conductivity data. Obviously, the choice of transducer itself can impart some degree of selectivity to the measurement that is performed, and even though changes in conductivity are associated with the interaction of the analyte with the PC-modified surface of the SAW device, this parameter does not account fully for the frequency change mechanism. Δf is a compound parameter, which will also include the influence of factors such as change in mass, dielectric, elastic and piezoelectric properties of the interface layer.

Measurements based on small changes in mass would be particularly suitable for immunoassay, where the primary event is the binding between antibody and antigen. In this instance the biochemically specific surface is provided by the immuno-reagent—for example, for a target antigen—by the corresponding antibody.

Roederer and Bastiaans (1983) have demonstrated that the antigen, human immunoglobulin G (IgG) can be detected at a silanized quartz SAW device modified with anti-IgG. The measurement was compared with a separate reference device, where the non-specific response of borate-buffered saline was recorded at a non-protein modified surface. The report identifies some of the possible limitations of the methods. A true reference is ideally required, which provides an accurate baseline in the sample solution and, the immobilized biorecognition layer must be of sufficient quality to prevent the non-specific adsorption of either target antigen or any other sample species which will cause a frequency response. Solutions to such deficiencies may provide the sought after improvement in the detection limit, which would be necessary to make this a viable alternative immunoassay method.

The damping of crystal resonance in fluid media by re-direction of the wave into the fluid, has been highlighted as a particularly serious sensitivity limiter. Ishimori *et al.* (1981) have, however, employed such a wave in a rather different piezoelectric cell geometry. Two piezoelectric membranes were mounted parallel to one another at a distance of 2.5 mm. A frequency of 40 kHz was applied to the system, so that the signal was launched at one membrane and received at the other. The model was developed for the monitoring of cell population in a fermenter, an important parameter in the regulation of fermentation processes, and showed a linear relationship between the output voltage and the cell population in the range 10^6–10^8 cells cm^{-3}.

Possible interferents such as pH, ionic strength and solvent density changes were shown to play only a minimum role in the response, but the system was found to be significantly effected by the adiabatic compressibility of the solution between the piezoelectric membranes, and it was concluded that, at least in part, the signal was due to the decrease in compressibility with increase in cell concentration.

With no immobilized biorecognition component, this sensor can be more easily heat sterilized than those with a heat-sensitive surface layer, and it is thus suitable for *in situ* (Zone 1, see Chapter 1, p. 11) fermenter use. The development of such non-specific cell-count devices would be applicable for fermentation processes where an estimate of biomass—the total microbial population—can be used to regulate the process. The use of this same cell geometry but with biorecognition molecule modified membranes for biospecific interface reactions requires some further ingenuity if the signal is to be unaffected by other solution sample species.

References

Alder, J.F. and McCallum, J.F. (1983). *The Analyst* **108**, p. 1169.

Blanc, J.P., Blasquez, G., Germain, J.P., Larbi, A., Maleysson, C. and Robert, H. (1988). *Sensors and Actuators* **14**, p. 143.

Bott, B. and Jones, T.A. (1984). *Sensors and Actuators* **5**, p. 43.

Bott, B. and Jones, T.A. (1985). *Transducers '85*, p. 414.

Calabrese, G.S., Wohltjen, H. and Roy, M.K. (1987). *Anal. Chem.* **59**, p. 833.

Chin, W.-T. and Kroontje, W. (1961). *Anal. Chem.* **33**(12), p. 1757.

Guilbault, G.G. (1983). *Anal. Chem.* **55**, p. 1682.

Guilbault, G.G., Ho, M. and Rietz, B. (1980). *Anal. Chem.* **52**, p. 1489.

Guilbault, G.G. and Jordan, J. (1987). *A Review of Piezoelectric Crystals in Analytical Chemistry*. Cleveland, CRC Press.

Guilbault, G.G., Ngeh-Ngwainbi, J., Foley, P. and Jordan, J. (1986). *Proc. Int. Conf. Electroanalysis na h'Eireann*, Eds Smyth, M.R. and Vos, J.G. Dublin, Elsevier.

Hanss, M. and Rey, A. (1971). *Biochim. Biophys. Acta* **227**, p. 630.

Henrion, L., Derost, G., Ruaudel-Teixier, A. and Barraud, A. (1988). *Sensors and Actuators* **14**, p. 251.

Ishimori, Y., Karube, I. and Suzuki, S. (1981). *Appl. Environ. Microbiol.* **42**(4), p. 632.

Lawrence, A.J. (1971). *Eur. J. Biochem.* **18**, p. 221.

Lawrence, A.J. and Moore, G.R. (1972). *Eur. J. Biochem.* **24**, p. 538.

Malmros, M.K., Guilbinski, J. and Gibbs, W.B. (1987/8). *Biosensors*, **3**, 71.

Miasik, J.J., Hooper, A. and Tofield, B.C. (1986). *J. Chem. Soc. Faraday Trans, 1* **82**, 1117.

Miike, H., Asashiba, Y., Hashimoto, H. and Ebina, Y. (1984). *Jap. J. Appl. Phys.* **23**, p. 386.

Ngeh-Ngwainbi, J., Foley, P.H., Kuan, S.S. and Guilbault, G.G. (1986). *J. Am. Chem. Soc.* **108**, p. 5444.

Nieuwenhuizen, M.S. and Nederlof, A.J. (1988). *Anal. Chem.* **60**, p. 236.

Nieuwenhuizen, M.S., Nederlof, A.J. and Barendsz, A.W. (1988). *Anal. Chem.* **60**, p. 230.

Nomura, T. and Nagamune, T., (1983). *Anal. Chim. Acta* **155**, p. 231.

Roederer, J.E. and Bastiaans, G.J. (1983). *Anal. Chem.* **55**, p. 2333.

Schuman, M.S. and Fogelman, W.W. (1976). *J. Water Pollut. Control Fed.* **49**(6), p. 901.

Shou-Zhuo, Y. and Zhi-Hong, M. (1987). *Anal. Chim. Acta* **93**, p. 97.

Watson, L.D., Maynard, P., Cullen, D.C., Sethi, R.S., Brettle, J. and Lowe, C.R. (1987/88). *Biosensors* **3**, p. 101.

Wohltjen, H., Barger, W.R., Snow, A.W. and Jarvis, N.L. (1985). *IEEE Trans. Electron. Devices* **ED32**(7), p. 1170.

Zellers, E.T., White, R.M. and Wenzel, S.W. (1988). *Sensors and Actuators* **14**, p. 35.

Index